Nutrient Deficiencies of Field Crops

Guide to Diagnosis and Management

MIX
Paper from
responsible sources
FSC® C022612
FSC
www.fsc.org

Dedicated to the late **Professor P.C. Gupta**
Honorable teacher of both authors and the father of Dr Prakash Kumar
We remember his learning

'Hard work is the scale that measures your height.'

Late Professor P.C. Gupta, a renowned botanist and great teacher
Former Head, Department of Agricultural Botany, Amar Singh Post Graduate College, Lakhaoti, Bulandshahr, Uttar Pradesh, India

Nutrient Deficiencies of Field Crops

Guide to Diagnosis and Management

Dr Prakash Kumar

MSc (Ag) Agronomy, PhD
Deputy Director, Agriculture (Agronomy),
Research Building Sirohi
Department of Agriculture
Government of Rajasthan
Rajasthan, India – 307001

Dr Manoj Kumar Sharma

MSc (Ag) Soil Science, PhD
Assistant Director, Soils. IMTI, Kota
Department of Agriculture
Government of Rajasthan
Rajasthan, India – 324009

www.cabi.org

CABI is a trading name of CAB International

CABI
Nosworthy Way
Wallingford
Oxfordshire OX10 8DE
UK

Tel: +44 (0)1491 832111
Fax: +44 (0)1491 833508
E-mail: info@cabi.org
Website: www.cabi.org

CABI
38 Chauncey Street
Suite 1002
Boston, MA 02111
USA

Tel: +1 800 552 3083 (toll free)
Tel: +1 617 395 4051
E-mail: cabi-nao@cabi.org

A catalogue record for this book is available from the British Library, London, UK.

Library of Congress Cataloging-in-Publication Data

Kumar, Prakash, 1969-
Nutrient deficiencies of field crops : guide to diagnosis and management / Prakash Kumar, Manoj Kumar Sharma.
p. cm.
ISBN 978-1-78064-278-9 (hbk : alk. paper) 1. Deficiency diseases in plants. 2. Field crops--Diseases and pests--Identification. 3. Field crops--Nutrition. I. Sharma, Manoj Kumar, 1966- II. Title.

SB742.K86 2013
633--dc23

2013020835

ISBN: 978 1 78064 278 9

Commissioning editor: Rachel Cutts
Editorial assistant: Emma McCann
Production editor: Tracy Head

Typeset by SPi, Pondicherry, India.
Printed and bound by Gutenberg Press Ltd, Tarxien, Malta.

A few photographs and some text in this publication are reproduced from an earlier work by the authors entitled *A Guide to Identifying and Managing Nutrient Deficiencies in Cereal Crops*, published by International Plant Nutrition Institute, Norcross, Georgia, USA. This material is reproduced here with the kind permission of the publisher.

Contents

Foreword

High-yielding varieties and intensive cultivation have resulted in depletion not only of macro- and secondary nutrients but also some of the micronutrients like zinc, iron, copper, boron, manganese and molybdenum, resulting in widespread deficiencies of such elements across the globe. Reduction in soil organic matter content and poor quality of groundwater have further caused imbalance in nutrient supply capacity. Nutrient deficiency symptoms appear though, but often go unmarked for want of adequate identification, causing low yield levels and profitability. The situation becomes compounded if deficiency disorders are misunderstood as prevalent diseases, resulting in wasteful expenditure on the use of chemicals. Therefore, scientists, plant protection advisors, extension functionaries and farmers need to be trained about a proper diagnostic system for a particular nutrient deficiency.

The present book entitled *Nutrient Deficiencies of Field Crops: Guide to Diagnosis and Management*, authored by Dr Prakash Kumar, Deputy Director, Research (Agronomy), Department of Agriculture, Government of Rajasthan, Sirohi and Dr Manoj Kumar Sharma, Assistant Director, Soils, Irrigation Management Training Institute, Government of Rajasthan, Kota, contains plant nutrient-specific descriptions of deficiency disorders supported by distinct illustrations pertaining to 27 important field crops, which will be of immense help to various stakeholders.

This book is in the series of previous books entitled *A Guide to Identifying and Managing Nutrient Deficiencies in Cereal Crops*, jointly published by the International Maize and Wheat Improvement Center (CIMMYT), Mexico and the International Plant Nutrition Institute (IPNI), Georgia, USA, covering six important cereal crops and is available on the CIMMYT web site.

I hope this book will become a popular handbook for different users including the students and will help in judicious management of chemical fertilizers.

Dr Arvind Kumar
Deputy Director General (Education)
Indian Council of Agriculture Research (ICAR)
Division of Education, Krishi Anusandhan Bhawan – II
New Delhi, India – 110012

Messages

Soil nutrient deficiencies are often manifested as visual symptoms on plant parts, but are rarely diagnosed correctly. The reason is often poor identification of nutrient deficiency symptoms by the farmers.

Correct identification of nutrient deficiency through visual symptoms in the plants can make farmers aware of nutrient imbalance and deficiency in their soils. Timely identification of plant nutrient deficiency symptoms and quick correction by appropriate nutrient management strategies may help prevent significant yield losses. Dr Prakash Kumar and Dr Manoj Kumar Sharma have done excellent work in this direction. Their latest work, *Nutrient Deficiencies of Field Crops: Guide to Diagnosis and Management*, being published by CABI, is a major contribution in this field. I wish them all success.

Dr Kaushik Majumdar
Director, IPNI
South Asia Programme
354, Sector 21, Huda
Gurgaon, 122016, Haryana, India

The International Plant Nutrition Institute (IPNI) is committed to 'the responsible management of plant nutrition for the benefit of the human family'. We are pleased to work with plant nutrition scientists around the world in furthering this important work. We especially appreciate Dr Prakash Kumar and Dr Manoj Kumar Sharma for their exemplary contribution to our mandate for nutrient stewardship by helping us publish *A Guide to Identifying and Managing Nutrient Deficiencies in Cereal Crops* in collaboration with the International Maize and Wheat Improvement Center (CIMMYT).

I am delighted to know that CABI is now publishing their recent book, *Nutrient Deficiencies of Field Crops: Guide to Diagnosis and Management*, covering 27 field crops. I extend my best wishes to the authors and the publisher for their significant contribution in the field of plant nutrition.

Dr Terry L. Robert
President, IPNI
IPNI Headquarters, 3500 Parkway Lane, Suite 550
Norcross, GA 30092-2806, USA

Preface

The identification of nutrient deficiencies of crops in field conditions through visual symptoms is difficult even today. The environmental conditions, growth stages, types of plant species or varieties and many other interactions may cause variations in visual deficiency symptoms. All renowned authors on the subject from Gove Hambidge (the editor of *Hunger Signs in Crops*), N.J. Grundon (the author of *Hungry Crops: A Guide to Nutrient Deficiencies in Field Crops*) to William F. Bennet (the editor of *Nutrient Deficiencies and Toxicities in Crop Plants*) have rightly emphasized that the visual symptoms should not be the sole base of any conclusion when identifying a nutrient deficiency problem in field conditions. The visual symptoms should be used with other identification tools such as plant analysis, soil analysis, etc. If treatments are given on the basis of visual diagnosis, the treatments must be applied first on a small strip of the crop to avoid any possible loss due to wrong diagnosis. Following the elders' advice, we also suggest the same. But our purpose in presenting this book is to create mass awareness among the farming community and extension functionaries about the fast emerging deficiency problems around the globe, especially in those areas where soil and plant testing facilities or other diagnostic tools are not promptly available. A segment of our work, entitled *A Guide to Identifying and Managing Nutrient Deficiencies in Cereal Crops*, published by the International Maize and Wheat Improvement Center (CIMMYT), Mexico and the International Plant Nutrition Institute (IPNI), Georgia, USA, was purposely released for use in mass awareness campaigns on food crops in developing countries.

The management recommendations given in this book are quite generalized to suggest only a line of management. The specific nutrient management recommendations should be drawn from the local district recommendations as the input availability and methods of application may vary from place to place. We have tried our best to simplify the nutrient deficiency identification by providing developmental stages with photographs. This large collection is the result of 15 years of our hard work in field with the sole aim of bringing out this kind of compilation. With this much hard work behind us we are extremely happy to present this book.

We are highly grateful to all seniors, plant nutrient experts and friends for technical guidance and constant encouragement. We are thankful to Dr Kaushik Majumdar, Dr T. Satyanarayana, Dr Raj Gupta, Dr M.L. Jat and Gavin D. Sulewski who edited our previous publication on nutrient deficiency.

We wish to especially thank our elders and family members, the late Professor P.C. Gupta, Smt. Chandrawati, Shri Makkhan Lal Varshney, Smt. Satyawati, Dr Pramod Kumar Gupta, Dr Brijesh Kumar, Shri Sanjeev Varshney, Shri Nagesh Gupta, the late Shri Sharad Mittal, Dr Seema and Master Prakhar Ganesh on the one hand and on the other, the late Shri Amichand Sharma, the late Smt. Kasturi Devi, Shri K.P. Sharma, Smt. Saroj Sharma, R.K. Sharma, A.K. Sharma, Smt. Madhulika Sharma, Master Mudit Sharma and Master Pulkit Sharma for their best wishes and ever increasing encouragement for this project.

Dr Prakash Kumar
Dr Manoj Kumar Sharma

1 An Introduction to Plant Nutrition

Manoj Kumar Sharma

The biggest challenge for agriculture over the coming decades will be to meet the world's increasing demand for food in a sustainable manner. Therefore, our goal will be to produce more to feed the growing population. In order to achieve this uphill task, there are two options before us. The first option is to bring more land under cultivation and the second option is to increase production per unit cultivated area by adopting intensive cultivation. Because of the continuous increasing demand on land for other developmental activities, the scope for increasing cultivated area is limited. Accordingly, greater attention will have to be paid to increase the production per unit area of cultivated land by adopting exhaustive use of agricultural inputs.

Deteriorating soil fertility and improper management of plant nutrients have further aggravated the problem. Large increases in productivity cannot be attained without ensuring that plants are supplied with adequate and balanced nutrition. Soils are the storehouse of most of the plant nutrients essential for plant growth and development and the way in which nutrients are managed will have a great impact on plant growth, soil fertility and agricultural sustainability. Plant growth is considered the result of a complex process by which the plant synthesizes food by using solar energy, carbon dioxide, water and nutrients from the soil.

Essential Plant Nutrients

Plants require a number of elements for their normal growth and development. Nearly 90 elements are taken up by plants but all are not essential. To differentiate between essential and non-essential elements, Arnon and Stout (1939) proposed the following criteria of essentiality:

1. The plant must be unable to grow normally or complete its life cycle in the absence of the element.
2. The element is specific and cannot be replaced by another.
3. The element plays a direct role in plant metabolism.

But it seems that an element would have to be considered essential even if it does not fulfil the second criterion of essentiality. For example, vitamin B_{12} is known to be essential for many bacteria, diatoms and other algae, but the essentiality of cobalt by itself has not been demonstrated. According to this criterion, molybdenum and chlorine cannot be considered as essential because molybdenum can be replaced by vanadium and chlorine can be replaced by halides, however, they are functional in plant metabolism.

In order to overcome the above difficulties, Nicholas (1961) presented a more accurate definition of essential elements and used the term 'functional or metabolic nutrient' to include any mineral element that functions in plant metabolism, whether its function is specific or not.

The chemical elements that are now considered to be essential for higher plants are presented in Table 1.

Classification of Essential Nutrients

The essential nutrients are the chemical compounds necessary for the growth and metabolic activities of an organism. The essential plant nutrients may be divided into macronutrients (primary and secondary nutrients) and micronutrients depending on their concentrations in plant tissues.

Table 1. The list of chemical elements essential for higher plants.

Carbon (C)	Potassium (K)	Zinc (Zn)
Hydrogen (H)	Calcium (Ca)	Molybdenum (Mo)
Oxygen (O)	Magnesium (Mg)	Boron (B)
Nitrogen (N)	Iron (Fe)	Chlorine (Cl)
Phosphorus (P)	Manganese (Mn)	Sodium (Na)
Sulphur (S)	Copper (Cu)	Silicon (Si)
Cobalt (Co)		

The elements sodium, silicon and cobalt have not been stated as essential elements for all higher plants. Vanadium has been established as an essential element for some microorganisms.

Macronutrients

Macronutrients or major nutrients are found and required in plants in relatively higher amounts than micronutrients. The content of a macronutrient in plant tissue can be a thousand times greater than the content of a micronutrient. These include carbon, hydrogen, oxygen, nitrogen, phosphorus, sulphur, potassium, calcium and magnesium. Carbon, hydrogen and oxygen constitute 90–95% of the plant dry matter weight and the remaining six macronutrients are further sub-divided into primary and secondary nutrients.

- *Primary nutrients*: Nitrogen, phosphorus and potassium are called primary nutrients because they are the major constituents of commercial fertilizers that are used in large amounts to correct their wide-spread deficiencies.
- *Secondary nutrients*: Calcium, magnesium and sulphur are called secondary nutrients because they are required in moderate quantities by plants, their deficiencies are localized and they can be supplemented through carriers of the primary nutrients. For example, single super phosphate is a phosphatic fertilizer and contains both calcium and sulphur. Likewise, ammonium sulphate is a nitrogenous fertilizer that also supplements sulphur.

Micronutrients

Micronutrients are required by plants in relatively small quantities but are as essential as the macronutrients. These elements are also known as trace elements. They are further classified into micronutrient cations (iron, manganese, zinc and copper) and micronutrient anions (boron, molybdenum and chlorine).

The classification of nutrients into macro- and micronutrients seems to be arbitrary because there are many cases in which the differences in contents of the nutrients are not clearly defined so as to categorize them into macro- or micronutrients. For example, in many cases, the iron or manganese content in plant tissues is sometimes as high as the content of sulphur or magnesium.

Nutrient Mobility in Soil

The movement of nutrients in soil varies greatly and largely influences their availability to the plants. Knowledge about the mobility of nutrients in soil is very important from the nutrient management point of view when deciding the methods, time and frequency of nutrient supplying sources. On the basis of their mobility in soil, nutrients can be broadly categorized as: mobile, less mobile and immobile.

- *Mobile nutrients*: These nutrients are highly soluble and their large fraction is found in soil solution. Because of their high mobility, they become readily available to plants and are very prone to leaching losses. Such nutrient ions are NO_3^-, SO_4^{2-}, Cl^- and $H_2BO_3^-$.
- *Less mobile nutrients*: These are also soluble but are found in lower quantities in soil solution as they are adsorbed on clay complexes and easily get released into soil solution. Their availability to plants is moderate. These include NH_4^+, K^+, Ca^{2+}, Mg^{2+} and MoO_4^{2-} ions.
- *Immobile nutrients*: These nutrients are very tightly held by soil particles and are not easily released into the soil solution. Therefore, the availability to plants is low. They are Fe^{2+}, Mn^{2+}, Zn^{2+}, Cu^{2+}, HPO_4^{2-} and $H_2PO_4^-$ ions.

Table 2. Nutrient elements and their uptake forms. (Developed on the basis of information given by Ramamoorthy and Velayutham, 1976; Tisdale *et al.*, 1985; Mengel and Kirkby, 2001; and many others.)

Nutrient element	Uptake form(s)	Nutrient element	Uptake form(s)
Carbon (C)	CO_2	Manganese (Mn)	Mn^{2+}
Hydrogen (H)	H_2O	Copper (Cu)	Cu^{2+}
Oxygen (O)	O_2	Zinc (Zn)	Zn^{2+}
Nitrogen (N)	NH_4^+ and NO_3^-	Molybdenum (Mo)	MoO_4^{2-}
Phosphorus (P)	HPO_4^{2-} and $H_2PO_4^-$	Boron (B)	$H_2BO_3^-$ and $B(OH)_3$
Sulphur (S)	SO_4^{2-}	Chlorine (Cl)	Cl^-
Potassium (K)	K^+	Sodium (Na)	Na^+
Calcium (Ca)	Ca^{2+}	Silicon (Si)	$Si(OH)_4$
Magnesium (Mg)	Mg^{2+}	Cobalt (Co)	Co^{2+}
Iron (Fe)	Fe^{2+}		

Nutrient Mobility in Plants

The appearance of deficiency symptoms in plants chiefly depends upon the extent and the rate of retranslocation of nutrients from older to younger tissues within plants. Nutrients vary greatly in their mobility within plants. On the basis of their mobility, the nutrients are classified into two categories: mobile nutrients and immobile nutrients.

- *Mobile nutrients*: Mobile nutrients are those that are capable of being translocated from older to younger tissues within the plant. When the plant becomes deficient in these nutrients, as a consequence the deficiency symptoms are observed on the older leaves. These include nitrogen, phosphorus, potassium and magnesium.
- *Immobile nutrients*: Immobile nutrients are those that are not capable of being translocated from older to younger tissues within the plant. When the plant becomes deficient in these nutrients, the deficiency symptoms are thus evident on the younger tissues. These include calcium, sulphur, iron, manganese, zinc, copper, boron and molybdenum.

Functions of Nutrients in Plants

Carbon, hydrogen and oxygen (C, H and O)

Carbon, hydrogen and oxygen constitute about 90–95% of the dry matter of the plant. They are the constituents of organic components in plants and are involved in many enzymatic processes. Carbon and oxygen mainly take part as components of the carboxylic group; hydrogen and oxygen are involved in oxidation–reduction processes. Carbon is taken up in the form of CO_2 from the atmosphere and possibly in the form of HCO_3^- from the soil solution. Carbon is assimilated by plants as CO_2 in the carboxylation process. Hydrogen is taken up in the form of water from the soil solution. During photosynthesis H_2O is reduced to H, which then passes through a series of steps and is transferred to an organic compound resulting in the reduction of $NADP^+$ to a reduced form (NADPH). This coenzyme is very important in oxidation–reduction processes in which H can be transferred to a large number of compounds.

Nitrogen (N)

The uptake of nitrogen by the plant occurs in the form of nitrate ions (NO_3^-) or ammonium ions (NH_4^+) from the soil solution or as gaseous ammonia or gaseous nitrogen (N_2) from the atmosphere. The fixation of atmospheric molecular N_2 is dependent on the presence of specific nitrogen-fixing microorganisms in symbiotic association with higher plants. The total content of nitrogen varies from 2 to 4% in plant dry matter. Nitrogen is an essential constituent of amino acids, nucleic acids, nucleotides and chlorophy l. It promotes rapid growth. It is also involved in enzymatic processes because all enzymes and coenzymes contain nitrogen. Both NH_4^+ and NO_3^- forms can be taken up and metabolized by plants. The NO_3-N is assimilated by the processes of reduction and amination. The NH_4-N is also assimilated by the amination process.

The uptake of NO_3^- and NH_4^+ forms is very sensitive to pH. The best uptake of NH_4^+-N takes place in neutral conditions while rapid uptake of NO_3^--N occurs at low pH values. The NO_3^- uptake at high pH values may be reduced due to the competitive effect of hydroxide (OH^-) ions.

Almost all of the NH_4^+-N absorbed is assimilated in the root tissues and redistributed as amino acids. The NO_3^--N can be translocated as such to the upper parts of the plant but this depends upon the nitrate reduction potential of the roots. Thus, nitrate and amino acids are the main forms in which nitrogen is translocated in the vascular system of higher plants. NO_3^--N may also contribute to maintaining cation–anion balance and osmoregulation.

Sulphur (S)

The total content of sulphur varies from 0.2 to 0.5% in plant dry matter. Plants absorb sulphur mainly in the form of SO_4^{2-} ions from the soil solution. Other plant nutrients hardly affect the absorption of SO_4^{2-}; however, selenium depresses SO_4^{2-} uptake substantially. Plants can also absorb SO_2 from the atmosphere through the stomata. Sulphur is a constituent of organic components and is also involved in enzymatic processes. The absorbed SO_4^{2-}-S gets reduced in the plant and then is readily incorporated into an organic molecule. Sulphur is a constituent of essential amino acids (cysteine, methionine and cystine) involved in chlorophyll production, thus is required for protein synthesis and plant function and structure. It is also a constituent of coenzymes required for protein synthesis. Sulphur is contained in the plant hormones thiamin and biotin, both of which are involved in carbohydrate metabolism. Sulphur is also involved in some oxidation–reduction reactions. Sulphur plays an important role in the formation of disulphide bonds between polypeptide chains and thus it stabilizes polypeptide structure. Ferredoxins are also an important group of sulphur-containing compounds that are involved in photosynthesis. Several plant species contain small amounts of volatile sulphur compounds mainly in the form of disulphides or polypeptides.

Phosphorus (P)

Plant roots absorb phosphate from soil solution even at very low phosphate concentrations. The uptake of phosphorus by the plant occurs in the form of HPO_4^{2-} and $H_2PO_4^-$ ions from the soil solution. The HPO_4^{2-} ion dominates in calcareous soils while the $H_2PO_4^-$ ion dominates in acid soils. The phosphate is thus absorbed by plant roots against a very steep concentration gradient. Phosphate absorption is regarded as mediated by H^+ co-transport. The absorbed phosphate is rapidly involved in metabolic processes. Phosphate is readily mobile in plants and can be translocated in an upward or downward direction.

Phosphorus is an essential constituent of adenosine triphosphate (ATP), nucleotides, nucleic acids and phospholipids. Its major functions are in energy storage and transfer and the maintenance of membrane integrity. ATP is the most important compound, in which phosphate groups are linked by pyrophosphate bonds. The nucleotide triphosphates are also involved in the synthesis of RNA and DNA. In nucleic acids, phosphate groups act to bridge the RNA or DNA with another RNA or DNA by two ester bonds. DNA is the carrier of genetic information from one generation to another whereas RNA is involved in protein synthesis. Nucleic phosphates dominate in the nucleus, phospholipids in the chloroplasts, sugar phosphates in the cytoplasm and inorganic phosphates in the vacuole. Phytin is also an organic phosphorus compound that occurs in seeds.

Potassium (K)

Potassium in the form of the K^+ ion can be taken up readily by plant roots from the soil solution. The K^+ concentration in soil solution controls the K^+ diffusion rate towards the plant roots and also the uptake of K^+. Potassium is the most abundant cation in plant tissues and also the most important ion with respect to its physiological and biological functions. The high permeability of plant membranes to K^+ results in its extreme mobility throughout the whole plant. This unique feature is important to various physiological processes such as meristematic growth, photosynthesis, water status and translocation of photosynthates that are influenced by the K^+ concentration in plants. Potassium provides strength to plant cell walls and is involved in the lignification of sclerenchyma tissues. On the whole-plant level, potassium increases leaf area and leaf chlorophyll content, delays leaf senescence, and therefore contributes to greater canopy photosynthesis and crop growth. Potassium controls water loss from plants as the K^+ ion plays a crucial role in opening and closing of stomata. Higher potassium uptake results in accumulation of K^+ ions in appreciable concentrations in guard cells. Potassium plays a crucial role in translocation of photosynthates and also promotes mobilization of stored material. The activation of various enzyme systems in the plant is also influenced by the K^+ ion.

Calcium (Ca)

The calcium content in higher plants generally ranges from 0.5 to 3.0% in plant dry matter. The uptake of calcium occurs in the form of Ca^{2+} ions from the soil solution. Calcium can be absorbed by young root tips only. The presence of other cations such as K^+ and NH_4^+ in the soil solution suppresses calcium absorption by plant roots. Calcium plays a very significant role in cell elongation and cell division. Calcium is a constituent of calcium pectates, which are important cell wall constituents that are also involved in biomembrane maintenance. Calcium is important for maintaining cell wall integrity, is an enzyme activator and is required for osmoregulation and the maintenance of cation–anion balance in cells.

Magnesium (Mg)

Magnesium in the form of the Mg^{2+} ion can be absorbed by plant roots from the soil solution. The Mg^{2+} concentration in soil solution is higher than that of K^+, but the Mg^{2+} absorption rate is much lower than the absorption rate of K^+. The lower uptake rate of Mg^{2+} may be attributed to the lack of a special uptake mechanism for transporting Mg^{2+} across the plasma membrane. The uptake of Mg^{2+} can be largely depressed by an excess of other cations, especially of K^+ and NH_4^+. The translocation of Mg^{2+} from roots to the upper plant parts can also be restricted by K^+ and Ca^{2+} ions. Magnesium activates several enzymes. It is a constituent of chlorophyll and acts as the coordinating metal ion in the chlorophyll molecule. Thus it is involved in CO_2 assimilation and protein synthesis. Magnesium also regulates cellular pH and cation–anion balance.

Iron (Fe)

Iron in the soil must be reduced before it can be taken up into the root cell. Iron may be found in ionic form or as chelates in the free space. The reduction of ferric (Fe^{3+}) to ferrous (Fe^{2+}) results in breakup of the chelate complex and subsequently Fe^{2+} ions can be taken up by plant roots. The uptake of iron is significantly influenced by the presence of Mn^{2+}, Cu^{2+}, Ca^{2+}, Mg^{2+}, K^+ and Zn^{2+}. Iron is not readily translocated to younger plant parts and ferric citrate appears to be the major form in which iron translocates in xylem. HCO_3^- is the most important factor which induces iron chlorosis by depressing iron uptake and translocation in the plant.

The two important characteristics of iron are its ability to undergo a valency change and its tendency to form chelate complexes. They provide opportunity to involve iron in numerous physiological processes. The most well-known function of iron is in enzyme systems in which haem or haemin acts as a prosthetic group, where iron plays a somewhat similar role to magnesium in the porphyrin structure of chlorophyll. The haem enzyme systems include catalase, peroxidase, cytochrome oxidase and various cytochromes. Iron–sulphur proteins play a major role in redox reactions. Ferredoxin functions as a redox system in photosynthesis, in nitrite reduction, sulphate reduction and nitrogen assimilation.

Manganese (Mn)

The most important fraction of manganese in the soil solution occurs as Mn^{2+} ions and is important in plant nutrition. In addition to the Mn^{2+} form, the other easily reducible forms of manganese also contribute appreciably in plant nutrition. The availability of manganese in soil solution is influenced by oxidation–reduction reactions in soil. Plants absorb manganese in the form of Mn^{2+} ions and the uptake is metabolically mediated. The presence of other cation species in soil solution influences manganese availability, particularly magnesium depresses manganese uptake by the plants. Manganese is an immobile element which is not retranslocated in plants. Manganese and magnesium resemble one another in their biochemical functions as both ions are involved in bridging ATP with enzyme complexes such as phosphokinases and phosphotransferases. Manganese also activates decarboxylase and dehydrogenase enzymes in the tricarboxylic acid cycle but in most cases it is not specific for these enzymes and can be substituted by magnesium. Manganese is involved in redox reactions within the photosynthetic electron transport system in plants. In photosynthesis, manganese deficiency seriously disturbs the electron transport chain in the light-activated reaction.

Zinc (Zn)

The zinc content found in most plants is low, generally up to 100 ppm in plant dry matter. The uptake of zinc from soil solution occurs in the form of Zn^{2+} ions. Zinc is essential for several biochemical processes in the plant, such as cytochrome and nucleotide synthesis, auxin metabolism, chlorophyll production, enzyme activation and the maintenance of membranes. The carbonic anhydrase enzyme is specifically activated by the Zn^{2+} ion. Other enzymes such as alcohol dehydrogenase, superoxide dismutase and RNA polymerase also contain bound zinc.

Copper (Cu)

The copper content in most plants is very low and generally ranges from 2 to 20 ppm in plant dry matter. The uptake of copper from soil solution occurs in the form of Cu^{2+} ions. Copper uptake is regarded as a metabolically mediated process. It is very evident that copper and zinc strongly inhibit the uptake of each other. Copper is not readily mobile in the plant although it can be translocated from lower to upper leaves. It plays a key role in the following processes: nitrogen, protein and hormone metabolism; photosynthesis and respiration; and pollen formation and fertilization. By virtue of its ability to undergo a valency change, enzymatically bound copper participates in oxidation–reduction reactions. Copper is involved in numerous enzymes, the most important copper-containing enzymes being plastocyanin, superoxide dismutase and amine oxidases. Other copper-containing proteins are phenolase and laccase enzymes, which are involved in lignin synthesis. Amine oxidases are also copper proteins that catalyse oxidative deaminations.

Boron (B)

Boron is present mainly in an undissociated form in soil solution. The uptake of boron occurs in the forms of $H_2BO_3^-$ and $B(OH)_3$. Boron is considered relatively immobile in plants. It is not a component of enzymes unlike many other essential nutrients. Boron more likely plays an important role in nucleic acid metabolism, protein metabolism, photosynthesis, carbohydrate biosynthesis and in cell membrane stability. Uracil is an essential component of RNA and in its absence ribosomes cannot be formed, which affects protein synthesis. Boron is very closely involved in uracil synthesis. In meristematic tissues, protein synthesis, RNA synthesis and ribose formation are very important processes which are influenced by boron and thus the boron content affects meristematic growth. Boron is involved in RNA metabolism.

Molybdenum (Mo)

Molybdenum is absorbed by plants in the form of MoO_4^{2-} ions from soil solution. The uptake of molybdenum can be reduced by the presence of sulphate ions. Molybdenum possibly translocates in the form of MoO_4^{2-}, molybdenum–sulphur amino acid complexes or molybdate complexes with sugars or other polyhydroxy compounds. The mobility of molybdenum in plants is considered only moderate. The molybdenum content in most plants is usually low and is less than 1 ppm in plant dry matter. Several molybdoenzymes including nitrogenase, nitrate reductase, xanthine dehydrogenase, aldehyde oxidase and possibly sulphite oxidase are of significance to plants. Because of its involvement in the processes of N_2 fixation, nitrate reduction and the transport of nitrogen compounds in plants, molybdenum plays a crucial role in the nitrogen metabolism of plants.

Chlorine (Cl)

Chlorine is absorbed by plants in appreciable amounts in the form of Cl^- ions from soil solution. The uptake is metabolically controlled and is sensitive to both variations in temperature and metabolic inhibitors. In photosynthesis, chloride is an essential cofactor for the activation of the oxygen-evolving enzyme associated with photosystem II. In some plant species the Cl^- ion has an effect on stomatal regulation of the guard cells which in turn indirectly influences photosynthesis. The stimulation of ATPase located in tonoplasts is specifically brought about by Cl^- ions. The ATPase probably functions as an electrogenic H^+ pump transporting H^+ and Cl^- ions from cytoplasm into the vacuole and maintains cytoplasm at a higher pH than the vacuole. By virtue of its biochemically inertness Cl^- has the ability to fulfil osmotic and cation neutralization roles.

Cobalt (Co)

Cobalt is required by plants in very small quantities varying from 0.02 to 0.5 ppm in plant dry matter. The mechanism of cobalt uptake by plants is not clearly known. It is clearly evident that cobalt is essential for symbiotic N_2 fixation and for rhizobial growth. The effect of cobalt on N_2 fixation appears to be mediated by rhizobial growth and vitamin B_{12} and its coenzyme forms. Cobalt is a constituent of vitamin B_{12}. The coenzyme cobalamin forms complexes with cobalt. The three important enzyme systems in rhizobium are cobalamin dependent. This signifies the role of cobalt in nodulation and N_2 fixation in legumes. The methylmalonyl-coenzyme mutase is involved in the synthesis of haem in the bacteriods needed for leghaemoglobin production, ribonucleotide reductase is involved in the reduction of ribonucleotides to deoxyribonucleotides (DNA synthesis) and methionine synthetase is involved in protein synthesis.

Silicon (Si)

Silicon is regarded as the second most abundant element in the lithosphere and occurs in almost all minerals. Silicon is absorbed by plants in the form of monosilicic acid ($Si(OH)_4$). The uptake mechanism of silicon is not well understood. Silicon has many beneficial effects on plant growth, the most important of which is the epidermal accumulation of silicon which reduces cuticular water loss. In cereals, particularly in rice, silicon plays an important role in maintaining the leaves erect and in increasing tolerance to lodging. It also increases plants' tolerance to high levels of manganese. Rice is considered a highly responsive crop to silicon. The reproductive organs of rice are specifically promoted by silicon nutrition.

Further Reading

Ahmed, D.S. and Evans, H.J. (1960) Cobalt: a micronutrient element for the growth of soybean plants under symbiotic conditions. *Soil Science* 90, 205–210.

Amesz, J. (1993) The role of manganese in photosynthetic oxygen evolution. *Biochimica et Biophysica Acta* 726, 1–12.

Arnon, D.I. and Stout, P.R. (1939) The essentiality of certain elements in minute quantity for plants with special reference to copper. *Plant Physiology* 14, 371–375.

Bieleski, R.L. and Ferguson, I.B. (1983) Physiology and metabolism of phosphate and its compounds. In: Lauchli, A. and Bieleski, R.L. (eds) *Inorganic Plant Nutrition. Encyclopedia of Plant Physiology, New Series*. Vol. 15A. Springer Verlag, Berlin, pp. 422–449.

Bowen, J.E. (1969) Absorption of copper, zinc, and manganese by sugarcane leaf tissue. *Plant Physiology* 44, 255–261.

Burstrom, H.G. (1968) Calcium and plant growth. *Biological Reviews* 43, 287–316.

Chaney, R.L., Brown, J.C. and Tiffin, L.O. (1972) Obligatory reduction of ferric chelates in iron uptake by soybeans. *Plant Physiology* 50, 208–213.

Cowling, D.W. and Lockyer, D.R. (1976) Growth of perennial ryegrass (*Lolium perenne* L.) exposed to a low concentration of sulphur dioxide. *Journal of Experimental Botany* 27, 411–417.

Izawa, S., Heath, R.L. and Hind, G. (1969) The role of chloride ion in photosynthesis. III. The effect of artificial electron donors upon electron transport. *Biochimica et Biophysica Acta* 180, 388–398.

Koch, K. and Mengel, K. (1977) Effect of K on N utilization by spring wheat during grain formation. *Agronomy Journal* 69, 477–480.

Leggett, J.E. and Epstein, E. (1956) Kinetics of sulfate absorption by barley roots. *Plant Physiology* 31, 222–226.

Marschner, H. (1995) *Mineral Nutrition in Higher Plants*. Academic Press, San Diego, California.

Mengel, K. and Kirkby, E.A. (2001) *Principles of Plant Nutrition*, 5th edn. Kluwer Academic Publishers, Dordrecht, the Netherlands.

Mengel, K. and Von Braunschweig, L.C. (1972) The effect of soil moisture upon the availability of potassium and its influence on the growth of young maize plants (*Zea mays* L.). *Soil Science* 134, 142–148.

Nable, R.O. and Loneragan, J.F. (1984) Translocation of manganese in subterranean clover. I. Redistribution during vegetative growth. *Australian Journal of Plant Physiology* 11, 101–111.

Nicholas, D.J.D. (1961) Minor mineral elements. *Annual Review of Plant Physiology* 12, 63–90.

Outlaw, W.H. Jr (1983) Current concepts on the role of potassium in stomatal movements. *Physiologia Plantarum* 59, 302–311.

Page, E.R., Schofield-Palmer, E.K. and McGregor, A.J. (1962) Studies in soil and plant manganese. I. Manganese in soil and its uptake by oats. *Plant and Soil* 16, 238–246.

Pilbeam, D.J. and Kirkby, E.A. (1983) The physiological role of boron in plants. *Journal of Plant Nutrition* 5, 563–582.

Ramadan, A., Volker Romheld, A., Kirkby, E.A. and Marschner, H. (1997) Influence of increasing bicarbonate concentrations on plant growth, organic acid accumulation in roots and iron uptake by barley, sorghum, and maize. *Journal of Plant Nutrition* 20, 1731–1753.

Ramamoorthy, B. and Velayutham, M. (1976) Nitrogen, phosphorus and potassium in soil – chemistry, forms and availability. In: Kanwar, J.S. (ed.) *Soil Fertility – Theory and Practice*. Indian Council of Agricultural Research, New Delhi, pp. 156–201.

Rao, K.P. and Rains, D.W. (1976) Nitrate absorption by barley II. Influence of nitrate reductase activity. *Plant Physiology* 57, 59–62.

Sandmann, G. and Boger, P. (1983) The enzymological function of heavy metals and their role in electron transfer processes of plants. In: Lauchli, A. and Bieleski, R.L. (eds) *Inorganic Plant Nutrition. Encyclopedia of Plant Physiology, New Series*. Vol. 15B. Springer Verlag, Berlin, pp. 563–596.

Srivastava, P.C. (1997) Biochemical significance of molybdenum in crop plants. In: Gupta, U.C. (ed.) *Molybdenum in Agriculture*. Cambridge University Press, New York, pp. 47–70.

Stout, P.R., Meagher, W.R., Pearson, G.A. and Johnson, C.M. (1951) Molybdenum nutrition of crop plants. *Plant and Soil* 3, 51–81.

Tisdale, S.L., Nelson, W.L. and Beaton, J.D. (1985) *Soil Fertility and Fertilizers*. Macmillan, New York.

Ullrich-Eberius, C.I., Novacky, A., Fischer, E. and Luttge, U. (1981). Relationship between energy dependent phosphate uptake and the electrical membrane potential in *Lemna gibba* G1. *Plant Physiology* 67, 797– 801.

Watanabe, F. and Nakano, Y. (1999) Vitamin B_{12}. *Nippon Rinsho* 57, 2205–2210.

2 How to Identify Plant Nutrient Deficiencies in Field Conditions

Prakash Kumar

Visual Diagnosis and Difficulties

The term 'clinical diagnosis' is used in medical sciences to describe the way of diagnosis based on the appearance of the clinical symptoms, without any laboratory tests or X-ray films. In the process of clinical diagnosis, the practising doctor matches the appearing clinical symptoms of the patient with the symptoms of the diseases known to him and makes a preliminary idea of the probable disease. Then, the doctor suggests some laboratory tests to verify the disease. After confirmation, the treatment is prescribed. Though the clinical diagnosis is a preliminary assumption by the doctor, it is a very important observation as only this provides the right direction to the tests and treatments. Therefore the clinical diagnosis is the most essential skill of a doctor, which is based on his/her 'knowledge' and 'experience'. At times, experienced doctors are so sure and confident that they prescribe the treatment without any laboratory test.

This is also applicable to the diagnosis of crop abnormalities and disorders. A crop expert needs to acquire vast knowledge and experience to develop the skill of visual diagnosis. The crop expert should know the commonly occurring crop problems and their related visual symptoms.

A crop plant may become abnormal because of many reasons, such as the following:

1. *Unsatisfactory environment*: Too cold, too hot, too wet, too dry, excess rains, flood, drought, frost, low light, lack of oxygen, lack of carbon dioxide, too much sulphur dioxide, too much winds, improper soil pH, saline or sodic soils, adverse soil physical conditions, shallow soil depth, presence of hard pan and root penetration problems, etc.
2. *Nutritional starvation*: Deficiency of nitrogen, phosphorus, potassium, calcium, magnesium, sulphur, iron, zinc, boron, molybdenum, manganese, cobalt, copper, etc.
3. *Nutritional and other toxicities*: Toxicity of iron, manganese, copper, boron, sodium and chlorine, etc.
4. *Wrong or excess application of manures, fertilizers and other soil amendments*: Application of non-decomposed organic manure, excess application of fertilizers, over-liming of acid soils, etc.
5. *Wrong or excess application of herbicides or pesticides*: Toxic doses, drift effects, wrong selection, residual effects, etc.
6. *Biological hazards*: Fungal diseases, bacterial diseases, viral diseases, mycoplasma diseases, nematode infestations, insect infestations, mite attack, weed competition, bird and animal damage, etc.
7. *Genetic deformity*: Inbreeding depression, chlorophyll-less mutants, variegated mutants, etc.

Two different types of problem may produce similar types of visual symptoms. Moreover, many problems may occur at a time on one plant, producing a confusing complex of different symptoms.

When we discuss the identification of a nutrient deficiency through visual symptoms, we generally explain its common characteristics such as the following.

1. *Nutrient deficiency symptoms appear in patches*: It is generally explained by plant nutrition experts that nutritional disorders appear in patches in field conditions. But similar patches may appear due to many other reasons. For example:

- **a.** root-infesting nematodes always appear in patches and produce symptoms resembling nutritional disorders; and
- **b.** pathogens causing root rot and wilt cover patches in the field, and at the initial stage of infection they are often confused with nutritional disorders.

2. *Nutrient deficiency symptoms appear in a certain pattern*: Nutritional disorders appear on a plant in a certain pattern depending upon the mobility of the nutrient in the plant body. For example, the deficiency symptoms of nitrogen, phosphorus, potassium and magnesium always appear first on the lower leaves and proceed upwards and the deficiency symptoms of calcium, sulphur, iron, manganese, copper and boron appear first on the upper leaves and proceed downwards. However there are numerous examples of non-nutritional causes showing the same type of patterns. For examples, the fungal disease 'banded leaf and sheath blight' in maize starts from the lower leaves to proceed upwards and the symptoms of the viral disease 'maize rayado fino marafivirus (mrfv)' appear first on the upper leaves to proceed downwards.

3. *Nutrient deficiency symptoms are specific and unique*: It is generally told that 'on one type of plant, the nutrient deficiency symptoms of one nutrient are specific and unique', but this is not always true. For example, zinc deficiency symptoms of wheat differ from variety to variety and climate to climate. Moreover, some entirely different factors may produce exactly the same type of symptoms. For example, the symptoms of 'chlorine toxicity' are almost similar to the symptoms of 'potassium deficiency' in many crops, vegetables and fruits.

Now it is clear that just one parameter of the 'nutrient deficiency identification' is not sufficient to draw a conclusion. So, in this scenario of vast confusion, we should draw a line of diagnosis that can lead us near to the fact.

Identification of Nutrient Deficiencies: a Line of Diagnosis

The nutritional disorders have the following four important characters:

1. They occur in certain conditions.
2. They appear on plants in a certain pattern.
3. Their symptoms are specific.
4. The deficiency symptoms pass through well-established 'developmental stages' and all stages are often found in the field at a time.

A plant nutrition expert, in identifying nutritional disorders in field conditions, should consider four basic parameters for visual diagnosis in the following sequence:

1st Conditions of occurrence.
2nd Pattern of appearance.
3rd Specificity of the symptom.
4th Stages of development.

The author describes this sequence as a *'line of diagnosis'* for the identification of nutrient deficiencies in field conditions. All four parameters should positively match with the 'nutrient deficiency in question' to conclude a diagnosis under observation.

Conditions of occurrence

A nutrient may be deficient only because of two reasons: (i) the nutrient is totally low in the soil or in its parent material; and (ii) the nutrient is present in sufficient quantity but the plant is unable to use it because of certain reasons. One of the most common reasons for the latter is that the nutrient is not present in 'plant available form'. The plant can use nutrients only when they are present in the required plant available chemical form. The transformation of these forms from 'available' to 'unavailable' and vice versa is governed by many factors, such as soil chemical reactions, soil microbial activities, changing soil physical conditions and changing weather conditions. So, whenever a deficiency occurs, there are certain conditions that are responsible for such occurrence.

Look at the following statements from some experts to understand the above conditions.

1. If there are regular heavy rains in light-textured, low-organic-matter, well-drained soils then nitrogen deficiency is certain to occur (Plate 1).
2. In recently levelled lands, the deficiency of zinc is most prominent.
3. Heavy and excess application of phosphatic fertilizers may cause zinc deficiency.
4. Heavy and excess application of potassium fertilizers may cause magnesium deficiency.
5. Excess application of urea may cause calcium deficiency.
6. Over-liming of acid soils may cause iron deficiency.
7. Molybdenum deficiency is mostly found in acid soils where soil pH is below 6.5.
8. Iron deficiency is mostly found in alkaline soils where soil pH is above 7.5.

Plate 1. Nitrogen deficiency in maize (*Zea mays* Linn.) after regular heavy rains. (Photo by Dr Prakash Kumar.)

9. Deficiencies of phosphorus, potassium, sulphur, calcium and magnesium are more common in acid soils where soil pH is below 6.0.

10. Waterlogged conditions may cause phosphorus deficiency.

All of these statements indicate that, if the specified situation prevails, there is a huge chance that the related nutrient may become deficient. A plant expert, while suspecting and examining nutrient deficiency symptoms in the field conditions, should first look at the probable conditions that may lead to a deficiency. The crop expert generally lists down such conditions carefully in his/her notebook for ready use. In this book, with every nutrient deficiency, we list such conditions of occurrence with the heading 'Likely to occur in'.

Pattern of appearance

Nutrient deficiency symptoms always appear in a definite pattern. The patterns are generally nutrient specific (zinc is an exception where patterns are crop specific).

The deficiency symptoms of nitrogen, phosphorus, potassium and magnesium always appear first on the lower leaves and proceed upwards and deficiency symptoms of calcium, sulphur, iron, manganese, copper, boron and molybdenum appear first on the upper leaves and proceed downwards. These patterns are very important for visual diagnosis of nutrient deficiencies as they are rigidly definite, hardly with any exception. For example, calcium deficiency symptoms on maize plants appear first on the upper leaves (Plate 2). They will never appear first on the lower leaves in any condition. If deficiency is severe, the symptoms may eventually cover the entire plant including the lower leaves, but the pattern will be the same. We cannot find any plant in the field suffering with calcium deficiency and having the upper leaves normal with the lower leaves showing deficiency symptoms.

So, the pattern of appearance is definite in all plants. Even in perennial bushes and large trees, the pattern of appearance is the same. The only difference in the perennial bush and tree plants is that the pattern appears on every branch and not on the entire plant as a whole. Suppose a perennial tree is suffering from iron deficiency, then every branch will show the pattern of appearance. The upper emerging leaves of that branch will suffer first in the case of iron deficiency (Plate 3).

The pattern of appearance is one of the strongest tools in the visual diagnosis of deficiency symptoms. In this book, the large poster pictures given on the left page with every entry are purposefully designed to show the patterns of appearance clearly.

Plate 2. Calcium deficiency in maize (*Zea mays* Linn.) showing the pattern of appearance. (Photo by Dr Prakash Kumar and Dr Manoj Kumar Sharma.)

Plate 3. A branch of tree plant *Dalbergia sissoo* Roxb. showing the iron deficiency pattern: lower leaves normal and upper leaves with interveinal chlorosis. (Photo by Dr Prakash Kumar.)

Specificity of the symptom

Every plant, while suffering with any nutrient deficiency, produces one or a few characteristic symptoms that are 'specific' in nature. For example:

1. *'White specks' of maize*: 'White specks' that develop in interveinal areas of the leaf is the characteristic symptom of manganese deficiency in maize (Plate 4).

2. *'White bands' of maize*: White bands developed on both sides, between the midrib and margin of the leaf, are the characteristic symptom of zinc deficiency in maize (Plate 5).

3. *'Interveinal chlorosis' of maize*: A pale yellow chlorosis develops in the interveinal tissues (tissues between the veins), leaving the veins green and prominent. The symptom is described as interveinal chlorosis – the characteristic symptom of the iron deficiency in maize (Plate 6).

The close observation of these specific symptoms is the most important part of visual diagnosis. In this book, we provide 'close-up' pictures and description of these 'specific' symptoms for every nutrient deficiency.

Plate 4. White specks in maize (*Zea mays* Linn.) due to manganese deficiency. (Photo by Dr Prakash Kumar and Dr Manoj Kumar Sharma.)

Plate 5. White bands in maize (*Zea mays* Linn.) due to zinc deficiency. (Photo by Dr Prakash Kumar and Dr Manoj Kumar Sharma.)

Fourth stages of development

Based on the severity of the deficiency, all nutrient deficiency symptoms go through well-established developmental stages. In mild deficiency conditions, only one or two stages are visible but in the severe deficiency condition all developmental stages are often found at a time in the deficient field.

For example, we can distinguish the following four developmental stages of iron deficiency.

In maize (*Zea mays* Linn.):

Stage I: In the mild iron deficiency condition, the topmost leaves of the plants develop temporary fading of interveinal tissues with prominent green veins (Plate 7). The plant may recover or regain normal condition after some time.

Stage II: When deficiency becomes more severe, a pale yellow chlorosis develops in interveinal tissues (tissues between the veins) leaving the veins green and prominent (Plate 8).

Stage III: In the later stages of deficiency, the prominent green veins also fade and become light green to pale yellow (Plate 9).

Plate 6. Interveinal chlorosis in maize (*Zea mays* Linn.) due to iron deficiency. (Photo by Dr Prakash Kumar and Dr Manoj Kumar Sharma.)

Plate 7. Stage I – temporary fading of interveinal tissues. (Photo by Dr Prakash Kumar.)

Plate 8. Stage II – interveinal chlorosis with green and prominent veins. (Photo by Dr Prakash Kumar and Dr Manoj Kumar Sharma.)

Stage IV: In acute deficiency conditions the entire leaf bleaches to papery white (Plate 10).

Plate 9. Stage III – veins fade and become light green to pale yellow. (Photo by Dr Prakash Kumar and Dr Manoj Kumar Sharma.)

Plate 10. Stage IV – entire leaf bleaches to papery white. (Photo by Dr Prakash Kumar and Dr Manoj Kumar Sharma.)

In wheat (*Triticum aestivum* Linn.) – see Plate 11.

Stage I

Stage II

Stage III

Stage IV

Plate 11. Wheat leaves showing stages of iron deficiency. (Photos by Dr Prakash Kumar and Dr Manoj Kumar Sharma.)

In rice (*Oryza sativa* Linn.) – see Plate 12.

Stage I

Stage II

Stage III

Stage IV

Plate 12. Rice leaves showing the stages of iron deficiency. (Photos by Dr Prakash Kumar.)

In cauliflower (*Brassica oleracea* Linn.) – see Plate 13.

Affected top leaf

Stage I

Stage II

Stage III

Plate 13. Cauliflower leaves showing the stages of iron deficiency (Photos by Dr Prakash Kumar and Dr Manoj Kumar Sharma.)

In bottle gourd (*Lagenaria siceraria* (Molina) Standley) – see Plate 14.

Affected top leaf

Stage I

Stage II

Stage III

Plate 14. Bottle gourd plant showing the stages of iron deficiency. (Photos by Dr Prakash Kumar and Dr Manoj Kumar Sharma.)

In lime (*Citrus aurantifolia* (Christm.) Swingle) – see Plate 15.

Plate 15. Lime leaves showing the stages of iron deficiency starting from the normal leaf. (Photos by Dr Prakash Kumar and Dr Manoj Kumar Sharma.)

In sissoo (*Dalbergia sissoo* Roxb.) – see Plate 16.

Normal leaf

Stage I

Stage II

Stage III

Plate 16. Sissoo leaves showing stages of iron deficiency. (Photos by Dr Prakash Kumar.)

In bamboo (*Bambusa spinosa* Roxb.) – see Plate 17.

Plate 17. A branch of bamboo showing all the stages of iron deficiency. (Photo by Dr Prakash Kumar.)

The developmental stages are important in diagnosis because they are also specific in nature and their presence further confirms the nutrient deficiency. In this book, with every entry of nutrient deficiency, the developmental stages are described with photographs.

Other Supportive Tools of Nutrient Deficiency Diagnosis

A plant expert, extension functionary or farmer may use the following supportive tools, which are very helpful in the diagnosis of nutrient deficiencies:

1. Digital soil pH meter.
2. Soil fertility survey reports and maps.
3. Indicator weeds.

Digital pH meter

Soil pH is one of the most important properties of the soil that affects the availability of nutrients to the plant (Table 3). A plant nutrient expert must consider this fact while assessing the nutrient deficiency symptoms. Digital pH meters are available today to measure soil pH *in situ*, on the spot. The correlation of soil pH values with nutrient availability to the plant is very helpful to make an idea of the deficiency.

Table 3. Soil pH and nutrient availability. (Developed on the basis of information given by Truog, 1946; Lucas and Davis, 1961; Sims, 1986; Brady, 1990; and many others.)

Name of nutrient	Sufficiently available (pH)	Moderately low availability (pH)	Severely low availability (pH)
Nitrogen	6.0–8.0	5.5–6.0 and 8.0–8.5	Below 5.5 and above 8.5
Phosphorus	6.0–7.5 and above 8.5	5.0–6.0 and 7.5–8.5	Below 5.0
Potassium	Above 6.0	5.5–6.0	Below 5.5
Calcium	6.5–8.5	6.0–6.5 and 8.5–9.0	Below 6.0 and above 9.0
Magnesium	6.5–8.5	6.0–6.5 and 8.5–9.0	Below 6.0 and above 9.0
Sulphur	Above 6.0	5.5–6.0	Below 5.5
Iron	Below 6.5	6.5–7.5	Above 7.5
Manganese	5.0–6.5	4.5–5.0 and 6.5–7.5	Below 4.5 and above 7.5
Boron	5.0–7.0 and above 9.0	4.5–5.0 and 7.5–8.0	8.0–9.0 and below 4.5
Copper	5.0–7.5	4.5–5.0 and 7.5–8.5	Below 4.5 and above 8.5
Zinc	5.0–7.5	4.5–5.0 and 7.5–8.5	Below 4.5 and above 8.5
Molybdenum	Above 7.0	5.5–7.0	Below 5.5

Soil fertility survey reports and maps

A plant nutrient expert should collect, if available, the soil fertility survey reports and related fertility maps of the area. These reports and maps are very important supporting tools for the nutrient deficiency diagnosis as they give the primary idea and knowledge of the probable deficiency problems of an area.

Indicator weeds

While collecting soil and plant samples for analysis from deficient fields, the author noticed that the common weeds in the field and the surrounding area are very good indicators of the nutrient deficiency and can be used as a reliable supporting tool.

Indicator weeds are especially helpful when 'hidden hunger' is prominent in the crop. Hidden hunger is a phenomenon in which the plant does not show deficiency symptoms even in the deficiency conditions. Hidden hunger is found in all plants for every deficiency, but the tolerance limits of plants to hide hunger are different. Those plants having weak hidden hunger quickly show deficiency symptoms even in mild deficiency conditions, but those having strong hidden hunger show deficiency symptoms only when the deficiency becomes severe. The plants having weak hidden hunger are described as 'indicator plants'.

Many common weeds are indicator plants and clearly indicate the nutrient deficiency conditions of the field. These weed indicators are very useful in those crops having strong hidden hunger. For example, greengram (*Vigna radiata* Linn.), a pulse crop, has very strong hidden hunger to phosphorus deficiency but many common rainy-season weeds associated with the crop indicate phosphorus deficiency in the field.

A lot of work is required to be done to mark the good indicator weeds. The author has conducted tests and come to the conclusion that the following weeds can be classified as indicator weeds.

1. Indicator weeds of nitrogen deficiency – see Plate 18.
2. Indicator weeds of phosphorus deficiency – see Plate 19.
3. Indicator weeds of sulphur deficiency – see Plate 20.
4. Indicator weed of manganese deficiency – see Plate 21.
5. Indicator weeds of iron deficiency – see Plate 22.

Plate 18. *Cleome viscosa* Linn. (Asian spider flower, on the left) and *Celosia argentea* Linn. (silver cockscomb, on the right) showing yellow and pink lower leaves, respectively, in nitrogen deficiency conditions. (Photos by Dr Prakash Kumar.)

Plate 19. Most grass species are good indicators of phosphorus deficiency (left) and *Euphorbia hirta* Linn. showing phosphorus deficiency (right). (Photos by Dr Prakash Kumar.)

Plate 20. *Amaranthus* sp. (left) and *Cannabis sativa* Linn. (right) are good indicators of sulphur deficiency. (Photos by Dr Prakash Kumar.)

Plate 21. *Solanum nigrum* Linn. is a good indicator of manganese deficiency. (Photo by Dr Prakash Kumar and Dr Manoj Kumar Sharma.)

Plate 22. *Commelina benghalensis* Linn. (left) and *Phyllanthus niruri* Linn. (right) are good indicators of iron deficiency. (Photos by Dr Prakash Kumar and Dr Manoj Kumar Sharma.)

Confirmation of Nutrient Deficiency

It is advised that soil analysis and plant analysis should be done to confirm the type of nutrient deficiency. No correction treatment should be given in the entire field without proper confirmation. As visual diagnosis does not provide the complete idea of the required nutrition, the agronomist always suggests soil analysis of the plot before sowing of the crop to calculate balanced doses of the required nutrients under integrated nutrient management (INM) practice. Therefore, farmers should be advised to adopt this practice to avoid deficiency or toxicity situations in their crops.

Further Reading

Abadia, J. and Barton, L.L. (2007) *Iron Nutrition in Plants and Rhizospheric Microorganisms*. Springer, Dordrecht, the Netherlands.

Allen, D.J., Ampofo, J.K.O. and Wortmann, C.S. (1996) *Pests, Diseases and Nutritional Disorders of the Common Bean in Africa: A Field Guide*. International Center for Tropical Agriculture (CIAT), Cali, Colombia.

Barker, A.V. and Pilbeam, D.J. (2006) *Handbook of Plant Nutrition*. CRC Press, Boca Raton, Florida.

Bennett, W.F. (1993) *Nutrient Deficiencies and Toxicities in Crop Plants*. American Phytopathological Society, St Paul, Minnesota.

Bould, C., Hewitt, E.J. and Needham, P. (1984) *Diagnosis of Mineral Disorders in Plants*. Vol. I, *Principles*. Chemical Publishing, New York.

Brady, N.C. (1990) *The Nature and Properties of Soils*, 10th edn. Macmillan Publishing Company, New York.

Dobermann, A. and Fairhurst, T. (2000) *Rice: Nutrient Disorders and Nutrient Management*. Potash & Phosphate Institute (PPI), Potash & Phosphate Institute of Canada (PPIC) and International Rice Research Institute (IRRI), Los Baños, Philippines.

Grundon, N.J. (1987) *Hungry Crops: A Guide to Nutrient Deficiencies in Field Crops*. Queensland Department of Primary Industries, Brisbane, Australia.

Grundon, N.J., Edwards, D.G., Takkar, P.N., Asher, C.J. and Clark, R.B. (1987) *Nutritional Disorders of Grain Sorghum*. Australian Centre for International Agriculture Research, Canberra.

Lucas, R.E. and Davis, J.F. (1961) Relationship between pH values of organic soils and availabilities of 12 plant nutrients. *Soil Science* 92, 177–182.

Marschner, H. (1995) *Mineral Nutrition of Higher Plants*. Academic Press, Orlando, Florida.

Nicholas, R. (1964) Studies on the major element deficiencies of the pigeonpea (*Cajanus cajan* L.) in sand culture – 1. Foliar symptoms of the major element deficiencies. *Plant and Soil* 21, 377–387.

Sharma, M.K. and Kumar, P. (2011) *A Guide to Identifying and Managing Nutrient Deficiencies in Cereal Crops*. International Maize and Wheat Improvement Center (CIMMYT), Mexico DF and International Plant Nutrition Institute (IPNI), Norcross, Georgia.

Sims, J.T. (1986) Soil pH effects on the distribution and plant availability of manganese, copper and zinc. *Soil Science of America Journal* 50, 367–373.

Snowball, K. and Robson, A.D. (1991) *Nutrient Deficiencies and Toxicities in Wheat: A Guide for Field Identification*. International Maize and Wheat Improvement Center (CIMMYT), Mexico DF.

Sprague, H.B. (1964) *Hunger Signs of Crops – A Symposium*, 3rd edn. David McKay Company, New York.

Tisdale, S.L., Nelson, W.L., Beaton, J.D. and Havlin, J.L. (2002) *Soil Fertility and Fertilizers*, 5th edn. Prentice-Hall of India Private Limited, New Delhi.

Truog, E. (1946) Soil reaction influence on availability of plant nutrients. *Soil Science Society of America Proceedings* 11, 305–308.

Truog, E. (1961) *Mineral Nutrition of Plants*. University of Wisconsin Press, Madison, Wisconsin.

Wallace, T. (1943) *The Diagnosis of Mineral Deficiencies in Plants by Visual Symptoms*. His Majesty's Stationery Office, London.

Weir, R.G. and Cresswell, G.C. (1995) *Plant Nutrient Disorders, Vols 1 to 5*. Inkata Press, Melbourne, Australia.

PART I

Nutrient Deficiencies in Cereal Crops

Plate 23. Nitrogen-deficient maize plant. (Photo by Dr Prakash Kumar.)

Plate 24. Young stage nitrogen-deficient crop. (Photo by Dr Prakash Kumar.)

Plate 25. Pale yellow chlorosis in a V-shaped pattern. (Photo by Dr Prakash Kumar.)

Plate 26. Brown necrosis starts from the leaf tip. (Photo by Dr Prakash Kumar.)

MAIZE (*Zea mays* Linn.) NITROGEN (N) DEFICIENCY

Symptoms

1. Maize is highly sensitive to nitrogen deficiency. Deficiency symptoms appear even in mild deficiency conditions. Nitrogen-deficient plants are stunted with thin, spindly stems and pale green to yellow leaves. Deficient plants produce hardly one small ear per plant and the ears have hardly any grains with reduced kernel size, resulting in a drastic reduction in crop yield.
2. Nitrogen is mobile in plants and under short supply conditions it is easily mobilized from older to younger leaves. The deficiency symptoms appear first and become more severe on older leaves (Plate 23).
3. If deficiency occurs during the young stage of the crop, the whole plant appears uniformly pale green to yellow (Plate 24). In later stages of the crop, older leaves become pale yellow while young leaves remain green.
4. If deficiency persists or occurs in a more mature crop stage, a pale yellow chlorosis develops at the tip of old leaves and proceeds towards the leaf base along the midrib in a V-shaped pattern (the specific symptom of nitrogen deficiency in maize; see Plate 25).
5. In mature deficient plants, pale green young leaves, pale yellow to pale brown middle leaves and dry old leaves can appear simultaneously (Plate 25).

Developmental stages

Stage I: In mild deficiency, the entire plant appears uniformly light green in colour (Plate 24).
Stage II: If the deficiency becomes severe, a pale yellow chlorosis begins at the tip of old leaves and proceeds towards the base along the midrib in a V-shaped pattern (Plate 25).
Stage III: Brown necrosis starts from the tip of the leaf and proceeds towards the leaf base (Plate 26).
Stage IV: The older affected leaves become pale brown and die (Plate 23).

Likely to occur in

1. Soils having low organic matter.
2. Light-textured sandy soils that have been leached by heavy rainfall or excessive irrigation.
3. Soils exhausted by intensive cropping.
4. Waterlogged conditions.
5. Acid soils having pH below 6.0.
6. Alkaline soils having pH more than 8.0.

Integrated nutrient management

1. Get the soil analysed before sowing to measure the amount of 'available' nitrogen in the soil.
2. Apply analysis-based recommended quantity of nitrogen as basal by using:
 a. organic manures;
 b. bio-fertilizers;
 c. nitrogenous fertilizers.
3. Top-dress soluble nitrogenous fertilizers such as urea in two split doses.
4. For quick recovery, apply urea (2% w/v solution) as a foliar spray in the standing crop. Foliar sprays are required to be repeated every 10–15 days.
5. Incorporate legume crops in rotation.

Further reading

Grundon, N.J. (1987) *Hungry Crops: A Guide to Nutrient Deficiencies in Field Crops*. Queensland Department of Primary Industries, Brisbane, Australia.

Uhart, S.A. and Andrade, F.H. (1995) Nitrogen deficiency in maize: I. Effects on crop growth, development, dry matter partitioning and kernel set. *Crop Science* 36, 1376–1383.

Uhart, S.A. and Andrade, F.H. (1995) Nitrogen deficiency in maize: II. Carbon–nitrogen interaction effects on kernel number and grain yield. *Crop Science* 35, 1384–1389.

Zhao, D., Reddy, K.R., Kakani, V.G., Read, J.J. and Carter, G.A. (2003) Corn (*Zea mays* L.) growth, leaf pigment concentration, photosynthesis and leaf hyperspectral reflectance properties as affected by nitrogen. *Plant and Soil* 257, 205–217.

Plate 27. Phosphorus-deficient maize plant. (Photo by Dr Prakash Kumar and Dr Manoj Kumar Sharma.)

Plate 28. Plant showing purpling on older leaves. (Photo by Dr Prakash Kumar.)

Plate 29. Plant showing purpling on margins of leaves. (Photo by Dr Prakash Kumar.)

Plate 30. Leaf showing purpling and brown necrosis. (Photo by Dr Prakash Kumar and Dr Manoj Kumar Sharma.)

MAIZE (*Zea mays* Linn.) PHOSPHORUS (P) DEFICIENCY

Symptoms

1. Deficient plants appear stunted, thin and spindly with dark green leaves. The number and size of stomata in leaves are decreased. Root growth is drastically reduced. Phosphorus-deficient plants bear hardly one small ear with few grains, resulting in very poor crop yields.
2. Phosphorus is mobile in plants and under short supply conditions it is easily mobilized from older to younger leaves.
3. The deficiency symptoms appear first and become more severe on older leaves while young leaves usually remain normal (Plate 28).
4. Old leaves develop a characteristic dark green to bluish green coloration.
5. Under severe deficient conditions, purple or purple–red colours develop on older dark green leaves. Purpling usually starts from the margins of older leaves (Plate 29).
6. In acute deficiency conditions or in favouring winter season, the purpling may cover the entire plant (Plate 27).
7. In the most advanced stage, affected leaves burn and die.

Developmental stages

Stage I: In mild deficiency conditions, a dark green to bluish green colour develops on older leaves.
Stage II: If deficiency persists and becomes more severe, a purple or purple–red colour develops on the margins of old leaves, usually beginning from the tip of the leaf and proceeding towards the base (Plate 29).
Stage III: As the symptoms advance the entire leaf becomes reddish brown or purple (Plate 27).
Stage IV: In acute deficiency conditions, a brown necrosis develops on the tip of the leaves and proceeds along the margins towards the base (Plate 30).

Likely to occur in

1. Soils having low organic matter.
2. Alkaline and calcareous soils.
3. Soils exhausted by intensive cropping.
4. Acid soils and highly weathered soils.
5. Soils where topsoil has been removed by erosion.
6. Acid soils having pH below 6.0.
7. Alkaline soils having pH between 7.5 and 8.5.

Integrated nutrient management

1. Get the soil analysed before sowing to measure the amount of 'available' phosphate in the soil.
2. Apply analysis-based recommended quantity of phosphorus as basal by using:
  a. organic manures;
  b. phosphate-solubilizing microbial cultures;
  c. phosphatic fertilizers.
3. In deficient standing crops apply soluble phosphatic fertilizers such as ammonium phosphate with irrigation water.

Further reading

Grundon, N.J. (1987) *Hungry Crops: A Guide to Nutrient Deficiencies in Field Crops*. Queensland Department of Primary Industries, Brisbane, Australia.

Mollier, A. and Pellerin, S. (1999) Maize root system growth and development as influenced by phosphorus deficiency. *Journal of Experimental Botany* 50, 487–497.

Plenet, D., Etchebest, S., Mollier, A. and Pellerin, S. (2000) Growth analysis of maize field crops under phosphorus deficiency. *Plant and Soil* 223, 117–130.

Sarkar, B.C., Karmoker, J.L. and Rashid, P. (2010) Effects of phosphorus deficiency on anatomical structures in maize (*Zea mays* L.). *Bangladesh Journal of Botany* 39, 57–60.

Plate 31. Potassium-deficient maize plant. (Photo by Dr Prakash Kumar.)

MAIZE (*Zea mays* Linn.) POTASSIUM (K) DEFICIENCY

Plate 32. Leaf showing marginal chlorosis. (Photo by Dr Prakash Kumar.)

Plate 33. Chlorosis on lower leaf with pale green foliage. (Photo by Dr Prakash Kumar and Dr Manoj Kumar Sharma.)

Plate 34. Chlorosis and necrosis on affected leaf. (Photo by Dr Prakash Kumar and Dr Manoj Kumar Sharma.)

Symptoms

1. Potassium deficiency causes shortening of the internodes and dwarfing of plants with a general loss of the dark green colour of foliage.
2. Affected plants produce small ears that are often very pointed and underdeveloped at the tip.
3. Potassium moves readily from old to young leaves, therefore deficiency symptoms appear first on old leaves (Plate 31).
4. Younger actively growing leaves draw potassium from old parts of the plant, therefore young leaves usually remain green and apparently healthy.
5. Symptoms begin as a pale yellow chlorosis on the tip of old leaves covering marginal tissue (Plate 32).
6. The chlorosis is followed by pale brown necrosis and both chlorosis and necrosis advance down the margins towards the base. Marginal chlorosis and necrosis of older leaves are the specific symptom of potassium deficiency (Plates 33 and 34).
7. In severe deficiency conditions prominent red strips develop on the lower stem and leaf sheaths.

Developmental stages

Stage I: Mild deficiency of potassium causes stunted growth, thin stems and pale green foliage.
Stage II: When deficiency persists and becomes more severe, marginal chlorosis develops on older leaves starting from leaf tips (Plate 32).
Stage III: Chlorosis is followed by pale brown necrosis. Both chlorosis and necrosis advance down the margins towards the base, leaving the mid-vein and surrounding tissue pale green (Plate 34).
Stage IV: In acute deficiency conditions, the affected lower leaves burn and die.

Likely to occur in

1. Soils formed from parent material low in potassium.
2. Light textured soils where potassium has been leached by heavy rainfall or excessive irrigation.
3. Soils low in organic matter.
4. Soils in which the Na:K, Mg:K or Ca:K ratio is wide.
5. Large bicarbonate concentration in irrigation water.
6. Acid soils having pH below 6.0.

Integrated nutrient management

1. Get the soil analysed before sowing to measure the amount of 'available' potassium in the soil.
2. Problematic acid/alkaline/saline soils should be reclaimed.
3. Add organic manures well before sowing.
4. Apply potassium nitrate, potassium sulphate or potassium chloride to the soil at or before sowing as per soil testing recommendations.
5. If potassium deficiency appears on a standing crop, apply soluble potassium salts such as potassium nitrate, potassium sulphate or potassium chloride with irrigation water. Foliar sprays of these salts are usually not recommended because a number of such sprays are needed to fulfil crop requirements.

Further reading

Grundon, N.J. (1987) *Hungry Crops: A Guide to Nutrient Deficiencies in Field Crops*. Queensland Department of Primary Industries, Brisbane, Australia.

Hsiao, T.C., Hageman R.H. and Tyner, E.H. (1970) Effect of potassium nutrition on protein and total free amino acids in *Zea mays*. *Crop Science* 10, 78–82.

Jordan-Meille, L. and Pellerin, S. (2008) Shoot and root growth of hydroponic maize (*Zea mays* L.) as influenced by K deficiency. *Plant and Soil* 304, 157–168.

Koch, D.W. and Estes, G.O. (1975) Influence of potassium stress on growth, stomatal behavior, and CO_2 assimilation in corn. *Crop Science* 15, 697–699.

Plate 35. Calcium-deficient maize plant with a 'bull-whip' on top. (Photo by Dr Prakash Kumar and Dr Manoj Kumar Sharma.)

MAIZE (*Zea mays* Linn.) CALCIUM (Ca) DEFICIENCY

Plate 36. Jointed younger leaves with normal older leaves. (Photo by Dr Prakash Kumar.)

Plate 37. Tip ends of leaves glued together. (Photo by Dr Prakash Kumar.)

Plate 38. A 'ladder-like' appearance of the plant. (Photo by Dr Prakash Kumar.)

Symptoms

1. Calcium deficiency in maize can destroy the entire crop. Calcium-deficient maize plants are very stunted with distorted, torn and ragged foliage. Mildly calcium-deficient plants develop small ears and distorted tassels, but if deficiency is severe, maize plants fail to grow and die before maturity.
2. Calcium is immobile in plants and under short supply conditions it is not easily mobilized from older to younger leaves.
3. The deficiency symptoms appear first and more severely on younger leaves.
4. Calcium deficiency symptoms begin with yellow to white interveinal lesions on young leaves.
5. If the deficiency persists and becomes more severe, the new emerging leaves fail to unroll and make a 'bull-whip-like' structure.
6. As symptoms advance, the new leaves develop holes in the lamina. The torn and malformed leaves give the plant a ragged appearance.
7. In acute deficiency conditions, the tips of the youngest leaves are glued together and do not separate from the whorl. This gives the plant a 'ladder-like' appearance.

Developmental stages

Stage I: In the early stage of deficiency, the young leaves become pale green and then develop yellow to white lesions in interveinal tissues. The yellow to white lesions become enlarged and the lamina tears easily from these areas.
Stage II: If the deficiency persists and becomes more severe, the youngest leaves remain rolled and make a 'bull-whip-like' structure (Plate 35).
Stage III: As the symptoms advance, the new leaves develop holes in the lamina. The torn and malformed leaves give the plant a ragged appearance (Plates 36 and 37).
Stage IV: In acute deficiency conditions, the tip ends of the leaves are glued together, causing a 'ladder-like' appearance (Plate 38).

Likely to occur in

1. Acid sandy soils that have been leached by heavy rainfall.
2. Sodic soils that are rich in exchangeable sodium.
3. Soils having high soluble aluminium and low exchangeable calcium.
4. Strongly acid peat and muck soils that are low in calcium.
5. Acid soils having pH below 6.5.
6. Alkaline soils having pH more than 8.5.

Integrated nutrient management

1. Get the soil analysed before sowing to measure the lime or calcium requirement of the soil.
2. Apply analysis-based recommended quantity of calcium-containing fertilizers well before sowing. Suitable calcium containing fertilizers may be gypsum (calcium sulphate), calcium nitrate or calcium chloride.
3. In acid soils, lime or limestone (calcium carbonate) and dolomite (a mixture of calcium carbonate and magnesium carbonate) are more suitable calcium supplements.
4. The foliar application of 2% w/v calcium sulphate (twice) is recommended in standing crops.

Further reading

Grundon, N.J. (1987) *Hungry Crops: A Guide to Nutrient Deficiencies in Field Crops*. Queensland Department of Primary Industries, Brisbane, Australia.

Kawaski, T. and Moritsugu, M. (1979) A characteristic symptom of calcium deficiency in maize and sorghum. *Communications in Soil Science and Plant Analysis* 10, 42–56.

Krantz, B.A. and Melsted, S.W. (1964) Nutritional deficiencies in corn, sorghum and small grains. In: Sprangue, H.B. (ed.) *Hunger Signs of Crops – A Symposium*, 3rd edn. David McKay Company, New York, pp. 25–57.

Maas, E.V. and Grieve, C.M. (1987) Sodium-induced calcium deficiency in salt-stressed corn. *Plant, Cell & Environment* 10, 559–564.

Plate 39. Magnesium-deficient maize plant. (Photo by Dr Prakash Kumar.)

Plate 40. Interveinal chlorosis on mid-section of the leaf. (Photo by Dr Prakash Kumar and Dr Manoj Kumar Sharma.)

Plate 41. Interveinal chlorosis covers the entire leaf. (Photo by Dr Prakash Kumar and Dr Manoj Kumar Sharma.)

Plate 42. Necrosis in interveinal areas. (Photo by Dr Prakash Kumar.)

MAIZE (*Zea mays* Linn.) MAGNESIUM (Mg) DEFICIENCY

Symptoms

1. Magnesium-deficient maize plants are stunted with thin, spindly stems and pale green foliage with rust brown lower leaves. Magnesium deficiency may cause severe losses to crop yields as deficient plants bear very small ears with reduced size of kernel.
2. Magnesium is mobile in plants and under short supply conditions it is easily mobilized from older to younger leaves. The deficiency symptoms appear first and more severely on older leaves.
3. The deficiency symptoms begin with a pale yellow interveinal chlorosis in the mid-section of the older leaves (Plate 40).
4. As the symptoms advance, interveinal chlorosis proceeds towards the tip and base of the leaf, ultimately covering the entire leaf (Plate 41). The youngest leaves usually remain unaffected and apparently healthy.
5. If the deficiency persists and becomes more severe, the symptoms advance towards the upper leaves and rust brown strips appear on affected older leaves.
6. In advanced stages, the interveinal tissue dies and appears as whitish to brown necrotic strips in interveinal areas of the leaf. The affected older leaves eventually burn and die.

Developmental stages

Stage I: In early stage of deficiency, the leaves become pale green and older leaves develop pale yellow interveinal chlorosis (Plate 39).
Stage II: If the deficiency persists and becomes more severe, rust brown strips develop on the older leaves.
Stage III: As the symptoms advance, the interveinal tissue dies and appears as whitish to brown necrotic strips on the affected leaves (Plate 42).
Stage IV: In acute deficiency conditions, the old leaves die and hang down.

Likely to occur in

1. Acid sandy soils that have been leached by heavy rainfall.
2. Peat and muck soils that are low in total magnesium.
3. Soils having a higher quantity of calcium or potassium.
4. Soils derived from parent material that is inherently low in magnesium.
5. Acid soils having pH below 6.5.
6. Alkaline soils having pH above 8.5.

Integrated nutrient management

1. Get the soil analysed before sowing to estimate the amount of soluble and exchangeable magnesium in the soil.
2. Apply analysis-based recommended quantity of magnesium before sowing using soluble salts such as magnesium sulphate or magnesium chloride.
3. In deficient standing crops, apply soluble salts such as magnesium sulphate, chloride or nitrate with irrigation water. Foliar sprays of these salts are usually not advised as many sprays at frequent intervals are required to fulfil the crop need.
4. Magnesium deficiency in acid soils may be corrected by applying dolomite (a mixture of calcium carbonate and magnesium carbonate, $CaCO_3 \cdot MgCO_3$) through broadcasting and mixing in the soil a few months before sowing.
5. Reclamation of problematic acid soils or alkaline soils should be done to regulate a proper supply of magnesium.

Further reading

Fox, R.H. and Piekielek, W.P. (1984) Soil magnesium level, corn (*Zea mays* L.) yield and magnesium uptake. *Communications in Soil Science and Plant Analysis* 15, 109–123.
Grundon, N.J. (1987) *Hungry Crops: A Guide to Nutrient Deficiencies in Field Crops*. Queensland Department of Primary Industries, Brisbane, Australia.
Jacob, A. (1958) *Magnesium: The Fifth Major Plant Nutrient* (translated from German to English by Walker, N.). Staples Press Limited, London.
Sigel, H. and Sigel, A. (1990) *Metal Ions in Biological Systems. Vol. 26. Compendium on Magnesium and Its Role in Biology, Nutrition and Physiology*. Marcel Dekker, Inc., New York.

Plate 43. Sulphur-deficient maize plant. (Photo by Dr Prakash Kumar.)

MAIZE (*Zea mays* Linn.)
SULPHUR (S) DEFICIENCY

Plate 44. Uniform paleness of leaves. (Photo by Dr Prakash Kumar and Dr Manoj Kumar Sharma.)

Plate 45. Sulphur-deficient pale green leaf. (Photo by Dr Prakash Kumar.)

Plate 46. Sulphur-deficient pale yellow leaf. (Photo by Dr Prakash Kumar.)

Symptoms

1. Sulphur-deficient maize plants are stunted with pale green to yellow foliage. Deficient plants lack vigour and produce small ears with few kernels, resulting in poor crop yields. Nitrogen application does not work if sulphur is deficient in a maize crop. One part of sulphur is required for about every 15 parts of nitrogen.
2. In mild sulphur deficiencies during the young stage of the crop, the whole plant becomes pale green to yellow-green. At this initial stage, the sulphur deficiency symptoms are often confused with those caused by nitrogen deficiency.
3. A close observation is required to see whether the older leaves are more dark green and the younger ones more pale (case of sulphur deficiency) or the younger leaves are more dark green and the older ones more pale (case of nitrogen deficiency).
4. Sulphur deficiency, even if severe, does not produce any type of necrosis or burning of leaves. The change in leaf colour is the only indicative symptom of sulphur deficiency.
5. Sulphur is immobile in plants and under short supply conditions it is not easily mobilized from older to younger leaves. So, the deficiency symptoms appear first and become more severe on the younger leaves.
6. Deficiency symptoms appear as an even and uniform, pale green to pale yellow chlorosis across the lamina of young leaves. The youngest leaves are the palest (Plates 45 and 46).
7. In acute deficiency conditions, a red–purple colour sometimes develops on affected younger leaves as suffusion over the yellow colour.

Developmental stages

Stage I: In mild deficiency, in the young plant stage, all leaves on the plant become pale green to yellow although older leaves remain comparatively darker (Plate 44).
Stage II: As symptoms advance within the growing crop, the younger leaves turn pale yellow with the older leaves being green and normal (Plate 43).
Stage III: If deficiency persists and becomes more severe, the deficiency symptoms move downwards covering more leaves.
Stage IV: In acute deficiency conditions, a red–purple colour sometimes develops on affected younger leaves as suffusion over the yellow colour.

Likely to occur in

1. Soils having low organic matter.
2. Light-textured sandy soils that have been leached by heavy rainfall or excessive irrigation.
3. Soils exhausted by intensive cropping.
4. Soils derived from parent material that is inherently low in sulphur.
5. Acid soils having pH below 6.0.

Integrated nutrient management

1. Get the soil analysed before sowing to measure the amount of 'available' sulphate in the soil and to predict the amount of fertilizer required.
2. Apply analysis-based recommended quantity of sulphur by mixing into the soil, well before sowing, either:
  a. elemental sulphur (flowers of sulphur); or
  b. gypsum (calcium sulphate).
3. In deficient standing crops apply ammonium sulphate, magnesium sulphate or potassium sulphate with the irrigation water.
4. Problematic acid soils should be reclaimed.

Further reading

Bennett, W.F. (1993) *Nutrient Deficiencies and Toxicities in Crop Plants*. American Phytopathological Society, St Paul, Minnesota.
Grundon, N.J. (1987) *Hungry Crops: A Guide to Nutrient Deficiencies in Field Crops*. Queensland Department of Primary Industries, Brisbane, Australia.
Kang, B.T. and Osiname, O.A. (1976) Sulphur response of maize in Western Nigeria. *Agronomy Journal* 68, 333–336.
Stewart, B.A. and Porter, L.K. (1969) Nitrogen–sulphur relationships in wheat (*Triticum aestivum* L.), corn (*Zea mays*), and beans (*Phaseolus vulgaris*). *Agronomy Journal* 61, 267–271.

Plate 47. Iron-deficient maize plant. (Photo by Dr Prakash Kumar and Dr Manoj Kumar Sharma.)

Plate 48. Iron-deficient leaf showing interveinal chlorosis. (Photo by Dr Prakash Kumar and Dr Manoj Kumar Sharma.)

Plate 49. Fading of prominent green veins to light colour. (Photo by Dr Prakash Kumar and Dr Manoj Kumar Sharma.)

Plate 50. Bleached leaf with papery white appearance. (Photo by Dr Prakash Kumar and Dr Manoj Kumar Sharma.)

MAIZE (*Zea mays* Linn.) IRON (Fe) DEFICIENCY

Symptoms

1. Iron-deficient maize plants are stunted with yellow foliage. Severely deficient plants produce hardly any ears.

2. Iron is immobile in plants. So, deficiency symptoms appear first and more severely on younger leaves. The older leaves remain normal and apparently healthy (Plate 47).

3. In mild deficiency conditions or at the initial stage of deficiency, the topmost younger leaves develop temporary fading of interveinal tissues into pale green to pale yellow colour.

4. If the deficiency persists and becomes more severe, a bright pale yellow chlorosis develops in interveinal tissues (tissues between the veins), leaving the veins green and prominent. Interveinal chlorosis of top leaves is the specific symptom of iron deficiency (Plates 47 and 48).

5. As the symptoms advance, the prominent green veins also fade and become light green to pale yellow (Plate 49). In acute deficiency conditions, the entire leaf bleaches to papery white (Plate 50).

Developmental stages

Stage I: The topmost leaves of the plants develop temporary fading of interveinal tissues with prominent green veins.

Stage II: Interveinal tissues of the affected leaf turn bright pale yellow with prominent green veins – described as 'interveinal chlorosis' (Plates 47 and 48). Interveinal chlorosis uniformly extends along the full length of the leaf.

Stage III: The prominent green veins also fade and become light green to pale yellow (Plate 49).

Stage IV: The entire leaf bleaches to papery white (Plate 50).

Likely to occur in

1. Sandy soils having low total iron.

2. Calcareous soils where solubility of iron is very low.

3. Peat and muck soils where organic matter ties up iron and reduces its availability in soil solution.

4. Acid soils where, despite high availability of iron in soil solution, excessively high levels of soluble zinc, manganese, copper or nickel hamper the uptake of iron by the plant.

5. Waterlogged soils.

6. Alkaline soils having pH above 7.5.

Integrated nutrient management

1. Get the soil analysed before sowing to measure the amount of 'available' iron in the soil.

2. Problematic alkaline soils should be reclaimed.

3. Inorganic soluble salts of iron such as iron sulphates or chlorides may be applied, but these forms of iron quickly become insoluble. Soil dressing with organic forms of iron such as iron chelates (10 kg/ha) is more effective than with inorganic forms of iron. These organic iron forms should be used as per soil pH. If the soil is neutral (around pH 7.0), the most suitable chelates are FeHEDTA and FeDTPA. For acid soils FeEDTA is the most effective chelate, while on alkaline calcareous soils FeEDDHA is recommended.

4. If deficiency appears on standing crop, apply organic iron chelates or inorganic soluble salts such as iron sulphates or chlorides (0.5 to 1.0% w/v solution) as a foliar spray. Apply two or three sprays at intervals of 10–15 days.

Further reading

Grundon, N.J. (1987) *Hungry Crops: A Guide to Nutrient Deficiencies in Field Crops*. Queensland Department of Primary Industries, Brisbane, Australia.

Mengel, K. and Geurtzen, G. (1988) Relationship between iron chlorosis and alkalinity in *Zea mays*. *Physiologia Plantarum* 72, 460–465.

Sprague, H.B. (1964) *Hunger Signs in Crops – A Symposium*. David McKay Company, New York.

Plate 51. Manganese-deficient maize plant. (Photo by Dr Prakash Kumar and Dr Manoj Kumar Sharma.)

MAIZE (*Zea mays* Linn.)
MANGANESE (Mn) DEFICIENCY

Plate 52. Manganese-deficient leaf with interveinal chlorosis. (Photo by Dr Prakash Kumar and Dr Manoj Kumar Sharma.)

Plate 53. White interveinal flecks on middle leaves. (Photo by Dr Prakash Kumar and Dr Manoj Kumar Sharma.)

Plate 54. A close-up of 'white flecks'. (Photo by Dr Prakash Kumar and Dr Manoj Kumar Sharma.)

Symptoms

1. Deficient plants appear stunted with short stems and pale green to yellow leaves. The deficient plant produces small ears with reduced size of kernels, resulting in low crop yields.
2. Manganese-deficient plants develop poor tassels. Development of anthers is delayed. Manganese deficiency also affects pollen viability in maize.
3. Manganese is partly mobile in maize and it is not easily mobilized from older to younger leaves. But some amount is transferred from old to young leaves, when the deficiency occurs.
4. The deficiency symptoms of manganese in maize appear first on middle leaves and spread to younger and older leaves. At initial stages of deficiency, the youngest and the oldest leaves remain normal.
5. In mild deficiency, a temporary fading of interveinal tissues occurs on the middle leaves. The plant recovers soon.
6. If deficiency persists and becomes more severe, the middle leaves develop interveinal chlorosis that extends the full length of the leaf (Plate 52).
7. As the symptoms advance, 'white flecks' develop within the interveinal chlorosis areas of the affected leaves. The appearance of white flecks is the specific symptom of manganese deficiency in maize.

Developmental stages

Stage I: In mild deficiency, the middle leaves develop temporary fading of interveinal tissues. The plant recovers soon.
Stage II: If the deficiency persists and becomes more severe, the middle leaves develop interveinal chlorosis which extends the full length of the leaf (Plates 51 and 53).
Stage III: As the symptoms advance, all chlorotic tissue dies and turns white resulting in white flecks on the interveinal area of affected leaves. The veins remain green and prominent (Plate 54).
Stage IV: Finally, the white flecks turn brown and affected leaves burn and die.

Likely to occur in

1. Light-textured sandy soils that have been leached by heavy rainfall or excessive irrigation.
2. Calcareous and alkaline soils where the solubility of manganese is very low.
3. Waterlogged peat soils where organic matter ties up manganese and reduces its availability in solution.
4. Soils derived from parent material that is inherently low in manganese.
5. Acid soils having pH below 5.0.
6. Alkaline soils having pH above 7.5.

Integrated nutrient management

1. Get the soil analysed before sowing to measure the amount of 'available' manganese in the soil.
2. Problematic alkaline soils should be reclaimed.
3. Add organic manures well before sowing.
4. Apply soluble salts of manganese such as manganese sulphate as basal.
5. In deficient standing crops apply manganese sulphate (0.2 to 0.3% w/v solution) as a foliar spray.

Further reading

Farshid, A. (2012) Manganese, iron and copper contents in leaves of maize plants (*Zea mays* L.) grown with different boron and zinc micronutrients. *African Journal of Biotechnology* 11, 896–903.

Grundon, N.J. (1987) *Hungry Crops: A Guide to Nutrient Deficiencies in Field Crops*. Queensland Department of Primary Industries, Brisbane, Australia.

Sharma, C.P., Sharma, P.N., Chatterjee, C. and Agrawala, S.C. (1991) Manganese deficiency in maize affects pollen viability. *Plant and Soil* 138, 139–142.

Sprague, H.B. (1964) *Hunger Signs in Crops – A Symposium*. David McKay Company, New York.

Plate 55. Zinc-deficient maize plant. (Photo by Dr Prakash Kumar and Dr Manoj Kumar Sharma.)

MAIZE (*Zea mays* Linn.)
ZINC (Zn) DEFICIENCY

Plate 56. Deficient plant with bands or streaks. (Photo by Dr Prakash Kumar.)

Plate 57. Initial stage of band formation. (Photo by Dr Prakash Kumar.)

Plate 58. Advanced stage of band formation. (Photo by Dr Prakash Kumar and Dr Manoj Kumar Sharma.)

Symptoms

1. Maize is one of the best indicator plants of zinc deficiency. The crop suffers badly if zinc is deficient.
2. Zinc deficiency may cause delayed and irregular silk emergence with distorted tassels devoid of anthers.
3. Deficiency prevents elongation of internodes, resulting in leaves being crowded together at the top of the plant and making a fan-shaped appearance.
4. Zinc deficiency symptoms in maize occur within 2 weeks after seedling emergence.
5. Because zinc is not readily transferred from old to young parts in the plant, symptoms develop first and more severely on young leaves. Younger leaves are most affected while older leaves remain green and apparently healthy (Plate 55).
6. White to yellow bands or streaks of bleached tissues appear on each side of the midrib beginning at the base of the leaf. The midrib and the leaf margins remain green. This is the specific symptom of zinc deficiency in maize (Plates 56, 57 and 58).
7. In acute deficiency conditions, the affected tissue eventually dies and turns pale grey, leaving the midrib and margins green.

Developmental stages

Stage I: Mild deficiency of zinc may show an interveinal striping on young leaves somewhat similar to iron or manganese deficiency; but in case of manganese and iron, the interveinal striping runs the full length of the leaf while in zinc deficiency it occurs mainly on the basal half of the leaf.
Stage II: When deficiency persists and becomes more severe, the youngest leaves turn pale green, and white to yellow bands or streaks appear between the midrib and margin in the basal half of the leaf (Plates 57 and 58).
Stage III: In acute deficiency conditions, the affected tissue eventually dies and turns pale grey.

Likely to occur in

1. Leached sandy soils where total zinc is low.
2. Just levelled soils where subsoil is exposed for cultivation. Available zinc in surface soil is often double that of the subsoil.
3. Cool wet weather.
4. Soil having heavy and excessive application of phosphate fertilizers, which may hamper use of zinc by the crop.
5. Acid soils having pH below 5.0.
6. Alkaline soils having pH above 7.5.

Integrated nutrient management

1. Get the soil analysed before sowing to estimate the amount of 'available' zinc in the soil.
2. Problematic alkaline or acid soils should be reclaimed.
3. Add organic manures well before sowing.
4. Apply 25–30 kg of zinc sulphate or 10 kg of zinc chelate per hectare once every 2 years in zinc-deficient soils.
5. Zinc fertilizers should not be mixed with phosphate fertilizers.
6. In fields with known zinc deficiency, if deficiency appears in the standing crop, apply foliar application of a 0.5% w/v solution of a soluble zinc salt, Zinc sulphate with 0.25% w/v solution of unsoaked lime.

Further reading

Grundon, N.J. (1987) *Hungry Crops: A Guide to Nutrient Deficiencies in Field Crops*. Queensland Department of Primary Industries, Brisbane, Australia.

Halim, A.H., Wassom, C.E. and Ellis, R. (1968) Zinc deficiency symptoms and zinc and phosphorus interactions in several strains on corn (*Zea mays* L.). *Agronomy Journal* 60, 267–271.

Robson, A.D. (1993) *Zinc in Soils and Plants: Proceedings of the International Symposium on 'Zinc in Soils and Plants' held at The University of Western Australia, 27–28 September 1993*. Kluwer Academic Publishers, Dordrecht, the Netherlands.

Warnok, R.E. (1970) Micronutrient uptake and mobility within corn plants (*Zea mays* L.) in relation to phosphorus induced zinc deficiency. *Soil Science Society of America Journal* 34, 765–769.

Plate 59. Boron-deficient maize plant (Photo by Dr Prakash Kumar.)

Plate 60. Puckering on affected leaf. (Photo by Dr Prakash Kumar.)

Plate 61. Transparent stripes on affected leaf. (Photo by Dr Prakash Kumar.)

Plate 62. Affected cob with few grains. (Photo by Dr Prakash Kumar.)

MAIZE (*Zea mays* Linn.) BORON (B) DEFICIENCY

Symptoms

1. Boron is essential for the normal growth of plants. Boron plays a vital role in many growth-related functions such as cell division, cell elongation, sugar translocation and plant hormone functioning. So, the plant that suffers boron deficiency may suffer with growth-related defects. Boron is also essential for growth of the pollen tube and in this way boron deficiency may also disturb normal pollination and fruit setting. Boron deficiency also reduces root growth.
2. Boron-deficient plants are stunted with stout stems and pale green foliage. Affected plants produce small and bent cobs. Boron deficiency affects pollination by preventing growth of the pollen tube. As a result, boron-deficient plants produce barren ears or ears with few grains (Plate 62).
3. Boron is immobile in plants and under short supply conditions it is not easily mobilized from older to younger leaves. The deficiency symptoms therefore appear first and more severely on younger leaves. Old leaves remain green and healthy.
4. Symptoms appear on new emerging leaves. Affected leaves become pale green, short and held erect.
5. Affected leaves fail to unfold fully and leaf margins show loss in turgidity.
6. If deficiency persists and becomes more severe, puckering appears on affected leaves.
7. In acute deficiency conditions, watery and transparent stripes appear on affected leaves.

Developmental stages

Stage I: Deficient plants produce pale green leaves with signs of restricted leaf growth (Plate 59).
Stage II: If deficiency persists and becomes more severe, affected leaves fail to unfold fully and leaf margins lose turgidity. Puckering appears on the affected leaves (Plate 60).
Stage III: In acute deficiency conditions, watery or transparent stripes develop on affected leaves (Plate 61).

Likely to occur in

1. Drought conditions or low moisture conditions in soil. Drought reduces boron uptake.
2. Soils formed from parent material low in boron, such as acid igneous rocks or freshwater sediments.
3. Sandy soils where boron has been leached by heavy rainfall or excessive irrigation.
4. Alkaline soils having free lime.
5. Soils with excess lime application or disturbed Ca:B balance.
6. Soils low in organic matter.
7. Acid soils having pH below 5.0.
8. Alkaline soils having pH between 7.5 and 9.0.

Integrated nutrient management

1. Get the soil analysed before sowing to estimate the amount of 'available' boron in the soil.
2. Apply analysis-based recommended quantity of boron before sowing by using boron compounds such as borax, boric acid or chelated boron.
3. Add organic manures well before sowing.
4. Foliar application of boric acid or chelated boron is recommended at 30–35 days after seedling emergence or as soon as deficiency symptoms appear.

Further reading

Bell, R.W. and Rerkasem, B. (1997) *Boron in Soils and Plants: Proceedings of the International Symposium on 'Boron in Soils and Plants' held at Chiang Mai, Thailand, 7–11 September 1997*. Kluwer Academic Publishers, Dordrecht, the Netherlands.
Dell, B. and Huang, L. (1997) Physiological response of plants to low boron. *Plant and Soil* 193, 103–120.
Grundon, N.J. (1987) *Hungry Crops: A Guide to Nutrient Deficiencies in Field Crops*. Queensland Department of Primary Industries, Brisbane, Australia.
Lordkaew, L., Dell, B., Jamjod, S. and Rerkasem, B. (2010) Boron deficiency in maize. *Plant and Soil* 342, 207–220.

Plate 63. Nitrogen-deficient rice leaves. (Photo by Dr Prakash Kumar.)

RICE (*Oryza sativa* Linn.)
NITROGEN (N) DEFICIENCY

Plate 64. Deficient pale green rice field (foreground) and a healthy dark field (background). (Photo by Dr Prakash Kumar.)

Plate 65. Nitrogen-deficient crop with a close view. (Photo by Dr Prakash Kumar.)

Plate 66. Nitrogen-deficient leaf with a normal leaf. (Photo by Dr Prakash Kumar.)

Symptoms

1. Deficient plants appear stunted, thin and spindly with pale green to yellowish green leaves. The number of tillers and grain yields are reduced severely.
2. Nitrogen is mobile in plants and under short supply conditions it is easily mobilized from older to younger leaves.
3. The deficiency symptoms appear first and become more severe on older leaves (Plate 63).
4. If deficiency occurs during the young stage of the crop, the whole plant appears uniformly pale green to yellowish green. The deficient rice field gives a clear impression of nitrogen deficiency by providing a yellowish green look to the entire crop (Plates 64 and 65).
5. In later stages of the crop, older leaves become pale yellow while younger leaves remain green.
6. A pale yellow chlorosis develops at the tip of old leaves and proceeds in a broad front towards the leaf base.
7. Yellow chlorotic leaves turn pale brown and die.

Developmental stages

Stage I: In mild deficiency, the entire plant appears uniformly light green to yellowish green in colour (Plate 64).
Stage II: If the deficiency becomes severe, a pale yellow chlorosis begins at the tip of old leaves and proceeds in a broad front towards the base (Plates 63, 65 and 66).
Stage III: As the symptoms advance, the whole leaf becomes pale brown, withers and dies (Plate 63).
Stage IV: In the later stage, the dead leaves hang down making a thatch of dry leaves near the base of the plant.

Likely to occur in

1. Soils having low organic matter.
2. Cases of insufficient application of nitrogenous fertilizers.
3. Soils exhausted by intensive cropping.
4. Soils with nitrogen losses due to heavy rainfall.
5. Permanently submerged conditions.
6. Temporary drying out of the soil during the growing period.
7. Poor biological nitrogen fixation.
8. Acid soils having pH below 6.0.
9. Alkaline soils having pH above 8.0.

Integrated nutrient management

1. Get the soil analysed before sowing to measure the amount of 'available' nitrogen in the soil.
2. Apply analysis-based recommended quantity of nitrogen as basal by using:
- **a.** organic manures;
- **b.** bio-fertilizers;
- **c.** nitrogenous fertilizers.

3. Use slow-release nitrogenous fertilizers such as sulphur-coated urea or urea supergranules in a basal dressing before planting.
4. Top-dress soluble nitrogenous fertilizers such as urea in two or three split doses.
5. For quick recovery, apply urea (2% w/v solution) as a foliar spray in standing crops. Foliar sprays are required to be repeated every 10–15 days.
6. Incorporate legume crops in rotation.

Further reading

Cibes, H.R. and Gaztambide, S. (1980) Mineral deficiency symptoms displayed by the rice plants grown under controlled conditions in the green house. *Journal of Agriculture of the University of Puerto Rico* 64, 369–378.

De Datta, S.K. and Buresh, R.J. (1989) Integrated nitrogen management in irrigated rice. *Advances in Soil Science* 10, 143–169.

Dobermann, A. and Fairhurst, T. (2000) *Rice: Nutrient Disorders and Nutrient Management*. Potash & Phosphate Institute (PPI), Potash & Phosphate Institute of Canada (PPIC) and International Rice Research Institute (IRRI), Los Baños, Philippines.

Tanaka, A., Patnaik, S. and Abichandani, C.T. (1958) Studies on the rice plant (*Oryza sativa* L.): Part I. Influence of nitrogen level on growth and nutrient uptake by rice plant (*Oryza sativa* var. *indica*). *Plant Sciences* 47, 140–154.

Plate 67. Phosphorus-deficient rice crop. (Photo by Dr Prakash Kumar.)

RICE (*Oryza sativa* Linn.) PHOSPHORUS (P) DEFICIENCY

Plate 68. Leaf showing purpling on the margins. (Photo by Dr Manoj Kumar Sharma.)

Plate 69. Close-up of a leaf showing purpling. (Photo by Dr Manoj Kumar Sharma.)

Plate 70. Leaf showing advanced stage of pigmentation. (Photo by Dr Manoj Kumar Sharma.)

Symptoms

1. Phosphorus-deficient rice plants appear dark green with narrow, short and erect leaves (Plate 67). Plants are stunted with reduced tillering. Stems are thin and spindly with retarded growth. The number of panicles and grains per panicle are drastically reduced. Phosphorus deficiency delays crop maturity. Nitrogen application gives no response, if phosphorus is deficient.
2. Phosphorus is mobile in plants and under short supply conditions it is easily mobilized from older to younger leaves.
3. The deficiency symptoms appear first and become more severe on older leaves.
4. Symptoms begin with a dark green to bluish green coloration on older leaves. Younger leaves remain green and healthy.
5. In some rice varieties having the tendency to produce rich amounts of anthocyanin pigment, red and purple colours may develop on affected older leaves.
6. Purpling starts from leaf margins and later covers the entire leaf (Plates 68, 69 and 70).
7. If deficiency persists and becomes more severe, older leaves turn brown and die.

Developmental stages

Stage I: In mild deficiency conditions and during initial stages of crop growth, the entire plant appears dark green (Plate 67).
Stage II: If deficiency persists and becomes more severe, the dark green to bluish green colour develops on older leaves. Younger leaves remain green and healthy. Some rice varieties develop purpling, which starts from the margins of older leaves and proceeds towards the base (Plates 68, 69 and 70).
Stage III: In acute deficiency conditions, affected older leaves turn brown and die.

Likely to occur in

1. Soils exhausted by intensive cropping.
2. Soils with excess liming, resulting in immobilization of phosphorus in calcium phosphates.
3. Coarse-textured soils having low organic matter and small phosphorus reserves.
4. Highly weathered, clayey, acid upland soils, where phosphorus-fixation capacity is very high.
5. Calcareous, saline and sodic soils.
6. Excess use of nitrogenous fertilizer with insufficient application of phosphatic fertilizer.
7. Acid soils having pH below 6.0.
8. Alkaline soils having pH between 7.5 and 8.5.

Integrated nutrient management

1. Get the soil analysed before sowing to measure the amount of 'available' phosphate in the soil.
2. Apply analysis-based recommended quantity of phosphorus as basal by using:
- **a.** organic manures;
- **b.** phosphate-solubilizing microbial cultures; or
- **c.** phosphatic fertilizers.

3. In deficient standing crops apply soluble phosphatic fertilizers such as ammonium phosphate with irrigation water.

Further reading

Cibes, H.R. and Gaztambide, S. (1980) Mineral deficiency symptoms displayed by the rice plants grown under controlled conditions in the green house. *Journal of Agriculture of the University of Puerto Rico* 64, 369–378.

Dobermann, A. and Fairhurst, T. (2000) *Rice: Nutrient Disorders and Nutrient Management*. Potash & Phosphate Institute (PPI), Potash & Phosphate Institute of Canada (PPIC) and International Rice Research Institute (IRRI), Los Baños, Philippines.

Sanyal, S.K. and De Datta, S.K. (1991) Chemistry of phosphorus transformation in soil. *Advances in Soil Science* 16, 1–120.

Tanaka, A., Patnaik, S. and Abichandani, C.T. (1959) Studies on the rice plant (*Oryza sativa* L.): Part VII. Influence of increasing levels of phosphate and potash on growth, yield and nutrient uptake by rice plant (*Oryza sativa* var. *indica*). *Plant Sciences* 50, 305–318.

Plate 71. Potassium-deficient rice leaf showing marginal chlorosis. (Photo by Dr Manoj Kumar Sharma.)

RICE (*Oryza sativa* Linn.) POTASSIUM (K) DEFICIENCY

Plate 72. Potassium-deficient lower leaves. (Photo by Dr Prakash Kumar.)

Plate 73. Leaves with yellowish brown necrosis. (Photo by Dr Prakash Kumar.)

Plate 74. Bronzing of a leaf, a characteristic symptom of potassium deficiency. (Photo by Dr Prakash Kumar.)

Symptoms

1. Rice crop has strong hidden hunger to potassium deficiency. Visual deficiency symptoms appear only in severe deficiency conditions and mostly during later stages of crop growth.
2. Hybrid rice varieties are more sensitive to potassium deficiency than modern inbred improved varieties.
3. Potassium moves readily from old to young leaves, therefore deficiency symptoms appear first on old leaves.
4. Yellowish brown marginal necrosis begins from the tip of the leaf and advances down the margins towards the base (Plate 71).
5. Dark rust brown spots appear on the leaf surface in some rice varieties. Bronzing of older leaves is also a characteristic symptom of potassium deficiency (Plate 74).

Developmental stages

Stage I: Deficiency of potassium causes stunted growth of smaller leaves, thin stems and droopy older leaves.
Stage II: When deficiency persists and becomes more severe, marginal yellowish brown to yellowish orange discoloration followed by necrosis develops on older leaves starting from leaf tips (Plate 71).
Stage III: Necrosis advances down the margins towards the base, leaving the mid-vein and surrounding tissue green (Plates 72 and 73).
Stage IV: Eventually the affected leaf dies.

Likely to occur in

1. Soils formed from parent material low in potassium.
2. Soils low in organic matter.
3. Soils with wide Na:K, Mg:K or Ca:K ratio.
4. Excessive use of nitrogenous or nitrogenous plus phosphatic fertilizers with insufficient application of potassium fertilizers.
5. Direct sown rice during early growth stages.
6. Highly weathered acid soils having pH below 6.0.
7. Poorly drained soils where potassium uptake is hampered by H_2S, organic acids and excessive concentration of Fe^{2+}.
8. Susceptible hybrid rice cultivars.

Integrated nutrient management

1. Get the soil analysed before planting to measure the amount of 'available' potassium in the soil.
2. Add organic manures well before planting.
3. Apply potassium nitrate, potassium sulphate, potassium chloride or a compound fertilizer such as N:P:K to the soil at planting as per soil testing recommendations.
4. Potassium can also be applied in two or three split applications for better use in the rice crop: the first dose at planting, the second dose at panicle initiation (40–50 days after transplant) and the third at first flowering (60–70 days after transplanting).
5. If potassium deficiency appears on the standing crop, apply soluble potassium nitrate, potassium sulphate or potassium chloride with irrigation water. Foliar sprays of these salts are usually not recommended because a number of sprays are needed to fulfil crop requirements.

Further reading

Cibes, H.R. and Gaztambide, S. (1980) Mineral deficiency symptoms displayed by the rice plants grown under controlled conditions in the green house. *Journal of Agriculture of the University of Puerto Rico* 64, 369–378.

De Datta, S.K. and Mikkelsen, D.S. (1985) Potassium nutrition in rice. In: Munson, R.D., Sumer, M.E. and Bishop, W.D. (eds) *Potassium in Agriculture*. American Society of Agronomy, Crop Science Society of America and Soil Science Society of America, Madison, Wisconsin, pp. 665–669.

Dobermann, A. and Fairhurst, T. (2000) *Rice: Nutrient Disorders and Nutrient Management*. Potash & Phosphate Institute (PPI), Potash & Phosphate Institute of Canada (PPIC) and International Rice Research Institute (IRRI), Los Baños, Philippines.

Mghase, J.J., Shivachi, H., Takahashi, H. and Irie, K. (2011) Nutrient deficiencies and their symptoms in upland rice. *ISSAAS Journal* 7, 59–67.

Plate 75. Sulphur-deficient rice plants. (Photo by Dr Prakash Kumar.)

Plate 76. Sulphur-deficient rice crop. (Photo by Dr Prakash Kumar.)

Plate 77. Deficient plant showing the pale yellow youngest leaf. (Photo by Dr Prakash Kumar.)

Plate 78. Sulphur-deficient leaf (left) compared with a healthy leaf (right). (Photo by Dr Prakash Kumar.)

RICE (*Oryza sativa* Linn.) SULPHUR (S) DEFICIENCY

Symptoms

1. Deficient plants appear stunted, thin and spindly with pale yellow leaves. The numbers of tillers and grains per plant are reduced. Sulphur deficiency also delays crop maturity.
2. At initial stages of growth sulphur deficiency symptoms resemble nitrogen deficiency symptoms, as in both cases the whole plant appears pale green. A close observation is required to differentiate between the two deficiencies. In contrast to nitrogen deficiency where older leaves are more pale yellow than the younger leaves, sulphur deficiency produces more pronounced chlorosis on the young leaves with comparatively darker old leaves (Plate 75).
3. Sulphur is less mobile in the plant than nitrogen, so under short supply conditions deficiency symptoms tend to appear first on younger leaves (Plate 76).
4. The young leaves become dull or bright yellow while the old leaves usually remain green (Plate 77).
5. The pattern of yellowing on the entire leaf appears uniform, covering both veins and interveinal tissues uniformly (Plate 78).
6. In acute deficiency conditions, necrosis may develop on the tips of affected younger leaves. The youngest leaves are most severely affected by tip necrosis. However older leaves remain normal, without any tip necrosis.

Developmental stages

Stage I: In mild deficiency and during younger stages of the crop, all leaves on the whole plant become pale green.
Stage II: If the deficiency persists and becomes more severe, the younger leaves turn pale yellow (Plate 75).
Stage III: In acute deficiency conditions, tip necrosis appears on younger leaves; older leaves remain normal without any tip necrosis.

Likely to occur in

1. Soils having low organic matter.
2. Light-textured sandy soils that have been leached by heavy rainfall or excessive irrigation.
3. Soils exhausted by intensive cropping.
4. Soils derived from parent material which is inherently low in sulphur.
5. Acid soils having pH below 6.0.

Integrated nutrient management

1. Get the soil analysed before sowing to measure the amount of 'available' sulphate in the soil.
2. Apply analysis-based recommended quantity of sulphur by mixing in the surface soil, well before sowing, either:
  a. elemental sulphur; or
  b. gypsum.
3. In deficient standing crops apply ammonium sulphate, magnesium sulphate or potassium sulphate with irrigation water.

Further reading

Blair, G.J., Momuat, E.O. and Mamaril, C.P. (1979) Sulphur nutrition in rice. II. Effect of source and rate of S on growth and yield under flooded conditions. *Agronomy Journal* 71, 477–480.

Cibes, H.R. and Gaztambide, S. (1980) Mineral deficiency symptoms displayed by the rice plants grown under controlled conditions in the green house. *Journal of Agriculture of the University of Puerto Rico* 64, 369–378.

Dobermann, A. and Fairhurst, T. (2000) *Rice: Nutrient Disorders and Nutrient Management*. Potash & Phosphate Institute (PPI), Potash & Phosphate Institute of Canada (PPIC) and International Rice Research Institute (IRRI), Los Baños, Philippines.

Yoshida, S. and Chaudhry, M.R. (1979) Sulphur nutrition in rice. *Soil Science and Plant Nutrition* 25, 121–134.

Plate 79. Iron-deficient rice crop. (Photo by Dr Prakash Kumar.)

Plate 80. Interveinal chlorosis on a young leaf. (Photo by Dr Prakash Kumar.)

Plate 81. Leaves completely devoid of chlorophyll. (Photo by Dr Prakash Kumar.)

Plate 82. Leaf showing papery white appearance. (Photo by Dr Prakash Kumar.)

RICE (*Oryza sativa* Linn.)
IRON (Fe) DEFICIENCY

Symptoms

1. Iron deficiency is mainly a problem of rice grown on high-pH, aerobic, upland soils. Iron deficiency does not occur widely in flooded lowland soils. Iron availability usually increases after flooding.

2. Iron is immobile in plants. So, deficiency symptoms appear first and more severely on the younger leaves. The older leaves remain normal and apparently healthy.

3. In mild deficiency, the young leaves of the plants develop temporary fading of interveinal tissues with prominent green veins. Plants recover and regain normal condition after some time.

4. If the deficiency persists and becomes more severe, a bright pale yellow chlorosis develops in interveinal tissues (tissues between the veins), leaving the veins green and prominent. The symptom is described as 'interveinal chlorosis' and is a specific symptom of iron deficiency (Plate 80).

5. As the symptoms advance, the younger leaves become uniformly chlorotic with yellowish white colour.

6. In acute deficiency conditions, the new emerging leaf fails to grow and the younger leaves bleach to papery white. The papery white leaves soon turn necrotic and die.

Developmental stages

Stage I: In mild deficiency the young leaves show temporary fading of interveinal tissues with prominent green veins (Plate 79).

Stage II: If deficiency persists and becomes more severe, interveinal tissues of emerging leaves turn pale yellow with prominent green veins (Plate 80).

Stage III: As symptoms advance, young leaves become pale yellow to yellowish white and emerging leaves fail to grow (Plate 81).

Stage IV: In acute deficiency conditions, affected leaves bleach to papery white and soon turn necrotic (Plate 82).

Likely to occur in

1. Upland soils with a low concentration of soluble iron (Fe^{2+}).

2. Soils with low organic matter status.

3. High-pH alkaline or calcareous soils where large bicarbonate concentrations decrease the solubility and uptake of iron.

4. Soils with a high P:Fe ratio, where iron is bound in iron phosphates and becomes unavailable. This may be due to excessive application of phosphatic fertilizers.

Integrated nutrient management

1. Get the soil analysed before sowing. Problematic alkaline soils should be reclaimed. Add organic manures well before sowing.

2. In high-pH soils use ammonium sulphate fertilizer instead of urea to supply nitrogen. Ammonium sulphate is an acidifying fertilizer that increases iron availability.

3. Apply solid ferrous sulphate ($FeSO_4$) fertilizer as broadcast as per soil test recommendations.

4. If deficiency appears on standing crops, apply foliar application of ferrous sulphate in 2% w/v solution or iron chelates. As iron is immobile in plants, two or three foliar sprays (starting at tillering) are required at 15-day intervals.

Further reading

Anderson, W.B. (1982) Diagnosis and correction of iron deficiency in field crops – an overview. *Journal of Plant Nutrition* 5, 785–795.

Cibes, H.R. and Gaztambide, S. (1980) Mineral deficiency symptoms displayed by the rice plants grown under controlled conditions in the green house. *Journal of Agriculture of the University of Puerto Rico* 64, 369–378.

Dobermann, A. and Fairhurst, T. (2000) *Rice: Nutrient Disorders and Nutrient Management*. Potash & Phosphate Institute (PPI), Potash & Phosphate Institute of Canada (PPIC) and International Rice Research Institute (IRRI), Los Baños, Philippines.

Singh, B.P., Sinha, M.K., Singh, R.A. and Singh, B.N. (1986) Reaction of genotypes of rice (*Oryza sativa*) to iron chlorosis in calcareous soil. *Experimental Agriculture* 22, 75–78.

Plate 83. Zinc-deficient rice plant. (Photo by Dr Prakash Kumar.)

Plate 84. Brown blotches on leaves. (Photo by Dr Prakash Kumar.)

Plate 85. Plant showing deficiency symptoms on lower leaves. (Photo by Dr Manoj Kumar Sharma.)

Plate 86. Close-up of a deficient leaf at advanced stage. (Photo by Dr Manoj Kumar Sharma.)

RICE (*Oryza sativa* Linn.) ZINC (Zn) DEFICIENCY

Symptoms

1. Rice is very sensitive to zinc deficiency. If severe, zinc deficiency may destroy the entire crop. Rice disorders caused by zinc deficiency are known by various popular local names in different rice-growing areas. 'Khaira disease' in India, 'Hadda disease' in Pakistan and 'Akagare Type II disorder' in Japan are a few popular names for zinc-deficiency disorders in rice.
2. Zinc deficiency in rice causes stunted plant growth, reduced tillering, restricted root growth and sterility of spikelets, resulting in very poor crop yield.
3. Zinc deficiency symptoms in rice occur between 2 and 4 weeks after transplanting. Brown blotches and streaks appear on the lower leaves (Plate 85).
4. When deficiency persists, the blotches and streaks form a bigger area and ultimately cover the whole leaf (Plate 86).
5. The symptoms become more severe when bright sunny days prevail.
6. In later stages the entire leaf turns brown or bronze and dries.

Developmental stages

Stage I: Deficiency symptoms appear on the lower leaves of the plant as brown blotches or streaks on the leaves (Plate 83 and 84).
Stage II: When deficiency persists and becomes more severe, the blotches enlarge and coalesce to cover the whole leaf.
Stage III: Eventually the affected leaf turns bronze and dries.

Likely to occur in

1. Leached sandy soils where total zinc is low.
2. Just levelled soils where the subsoil is exposed for cultivation. Available zinc in surface soil is often double that of the subsoil.
3. Soils having heavy and excessive application of phosphate fertilizers, which hampers use of zinc because of the formation of zinc phosphates.
4. Soils having high bicarbonate (HCO_3^-) concentration.
5. Depressed zinc uptake because of an increase in the availability of iron, calcium, magnesium, copper, manganese and phosphorus after flooding.
6. Soils having heavy and excessive liming.
7. Acid soils having pH below 5.0.
8. Alkaline soils having pH above 7.5.

Integrated nutrient management

1. Get the soil analysed before sowing to estimate the amount of available zinc in the soil.
2. Problematic alkaline soils should be reclaimed.
3. Use 2 kg zinc sulphate/ha in the nursery.
4. Apply 25–30 kg zinc sulphate/ha in zinc-deficient soils.
5. Zinc fertilizers should not be mixed with phosphate fertilizers.
6. If deficiency appears in the standing crop, spray 5 kg of zinc sulphate plus 2.5 kg of unslaked lime in 800 l of water at 20, 30 and 40 days after planting.

Further reading

Dobermann, A. and Fairhurst, T. (2000) *Rice: Nutrient Disorders and Nutrient Management*. Potash & Phosphate Institute (PPI), Potash & Phosphate Institute of Canada (PPIC) and International Rice Research Institute (IRRI), Los Baños, Philippines.

Nene, Y.L. (1966) Symptoms, cause and control of Khaira Disease of paddy. *Bulletin of the Indian Phytopathological Society* 3, 97–101.

Tanaka, A. and Yoshida, S. (1970) Nutrient disorders of rice plants in Asia. *IRRI Technical Bulletin* 10, 35–40.

Yoshida, S., Ahn, J.S. and Forno, D.A. (1973) Occurrence, diagnosis and correction of zinc deficiency of low land rice. *Soil Science and Plant Nutrition* 19, 83–95.

Plate 87. Boron-deficient rice plant. (Photo by Dr Manoj Kumar Sharma.)

RICE (*Oryza sativa* Linn.) BORON (B) DEFICIENCY

Plate 88. Leaf tip tends to roll. (Photo by Dr Prakash Kumar.)

Plate 89. Rolled white tip of affected rice leaf. (Photo by Dr Prakash Kumar.)

Plate 90. Advanced stage of leaf damage by boron deficiency. (Photo by Dr Manoj Kumar Sharma.)

Symptoms

1. Boron is essential for the normal growth of plants. Boron plays a vital role in many growth-related functions such as cell division, cell elongation and sugar translocation. So, the plant that suffers boron deficiency may suffer with growth-related defects. Boron-deficient rice crops are stunted with retarded growth and poor yields. Boron deficiency results in reduced viability of pollen and in this way may disturb the grain formation in rice.
2. Boron is immobile in plants and under short supply conditions it is not easily mobilized from older to younger leaves. So, the deficiency symptoms appear first and more severely on the younger leaves. Old leaves remain green and healthy.
3. Initial symptoms appear on the tips of emerging leaves. The affected leaf becomes white on the tip and shows rolling of the leaf tip (Plates 87 and 88).
4. If deficiency persists and becomes more severe, growing points fail to grow and die. Still, new tillers continue to be produced.
5. If deficiency occurs during the panicle formation stage, plants fail to form panicles.

Developmental stages

Stage I: The affected leaf becomes white on the tip and shows rolling of the leaf tip (Plate 88).
Stage II: If deficiency persists and becomes more severe, growing points fail to grow further and die (Plates 88, 89 and 90).

Likely to occur in

1. Drought conditions or low moisture conditions in soil. Drought reduces boron uptake.
2. Soils formed from parent material low in boron, such as acid igneous rocks or freshwater sediments.
3. Sandy soils where boron has been leached by heavy rainfall or excessive irrigation.
4. Alkaline soils having free lime.
5. Soils with excess lime application or disturbed Ca:B balance.
6. Acid soils having pH below 5.0.
7. Alkaline soils having pH between 7.5 and 9.0.

Integrated nutrient management

1. Get the soil analysed before sowing to estimate the amount of available boron in the soil.
2. Apply analysis-based recommended quantity of boron before sowing by using boron compounds such as borax, boric acid or chelated boron. In boron-deficient fields, soil application of boron at the rate of 2–3 kg/ha in clayey soils and 3–5 kg/ha in sandy soils is sufficient to fulfil crop requirement. In rice–wheat rotation, soil application of boron once in a year is sufficient. Excess amounts of boron may induce toxicity.
3. Borax and fertilizer borates should not be mixed with ammonium fertilizers as this may cause NH_3 volatilization and rapid loss of nitrogen.
4. Foliar application of boric acid or chelated boron is recommended during vegetative growth or as soon as deficiency symptoms appear.

Further reading

Cibes, H.R. and Gaztambide, S. (1980) Mineral deficiency symptoms displayed by the rice plants grown under controlled conditions in the green house. *Journal of Agriculture of the University of Puerto Rico* 64, 369–378.

Dobermann, A. and Fairhurst, T. (2000) *Rice: Nutrient Disorders and Nutrient Management*. Potash & Phosphate Institute (PPI), Potash & Phosphate Institute of Canada (PPIC) and International Rice Research Institute (IRRI), Los Baños, Philippines.

Shorrocks, V.M. (1997) The occurrence and correction of boron deficiency. *Plant and Soil* 193, 121–197.

Xiaohe, Y. and Bell, P.F. (1998) Nutrient deficiency symptoms and boron uptake mechanisms of rice. *Journal of Plant Nutrition* 21, 2077–2088.

Plate 91. Nitrogen-deficient sorghum plant. (Photo by Dr Prakash Kumar.)

Plate 92. Deficient pale yellow crop (foreground) and a healthy green crop (background). (Photo by Dr Prakash Kumar and Dr Manoj Kumar Sharma.)

Plate 93. Chlorosis on leaf in a V-shaped pattern. (Photo by Dr Prakash Kumar.)

Plate 94. Leaf showing yellow chlorosis and brown necrosis. (Photo by Dr Prakash Kumar.)

SORGHUM (*Sorghum vulgare* Pers.) NITROGEN (N) DEFICIENCY

Symptoms

1. Being a fast-growing grass species, sorghum is very sensitive to nitrogen deficiency. Deficiency symptoms appear even in mild deficiency. Nitrogen deficiency during any stage of crop growth may cause severe losses to the crop in terms of fodder production or grain yield.

2. Nitrogen-deficient plants are stunted with spindly stems and pale green to yellow leaves. The deficient plant produces fewer tillers.

3. Nitrogen is mobile in plants and under short supply conditions it is easily mobilized from older to younger leaves.

4. The deficiency symptoms appear first and become more severe on the older leaves (Plate 91).

5. A pale yellow chlorosis develops at the tip of old leaves and proceeds towards the leaf base along the midrib in a V-shaped pattern. This is the specific symptom of nitrogen deficiency in sorghum.

6. In mature deficient plants, pale green young leaves, pale yellow to pale brown middle leaves and dry old leaves can appear simultaneously (Plate 91).

Developmental stages

Stage I: In mild deficiency, the entire plant appears uniformly light green in colour (Plate 92).

Stage II: If the deficiency becomes severe, a pale yellow chlorosis begins at the tip of old leaves and proceeds towards the base along the midrib in a V-shaped pattern (Plate 93).

Stage III: As the symptoms advance, the pale yellow chlorosis is followed by a pale brown necrosis (Plate 94).

Stage IV: In the most advanced stage, the affected leaves eventually burn and die. The dead leaves hang down, making a thatch of dry leaves near the base of the plant.

Likely to occur in

1. Soils having low organic matter.

2. Light-textured sandy soils that have been leached by heavy rainfall or excessive irrigation.

3. Soils exhausted by intensive cropping.

4. Waterlogged conditions.

5. Acid soils having pH below 6.0.

6. Alkaline soils having pH above 8.0.

Integrated nutrient management

1. Get the soil analysed before sowing to measure the amount of 'available' nitrogen in the soil.

2. Apply analysis-based recommended quantity of nitrogen as basal by using:

- **a.** organic manures;
- **b.** bio-fertilizers;
- **c.** nitrogenous fertilizers.

3. Top-dress soluble nitrogenous fertilizers such as urea in split doses.

4. For quick recovery, apply urea (2% w/v solution) as a foliar spray in the standing crop. Foliar sprays are required to be repeated every 10–15 days.

5. Incorporate legume crops in rotation.

Further reading

Asher, C.J. and Cowie, A.M. (1986) Effects of nitrogen supply on growth, nitrogen uptake and nitrogen distribution in hybrid grain sorghum. In: Foale, M.A. and Henzell, R.G. (eds) *Proceedings of the First Australian Sorghum Conference held at Queensland Agricultural College, Gatton, February 1986*. Australian Sorghum Conference Organising Committee, Gatton, Australia, pp. 6.68–6.78.

Grundon, N.J., Edwards, D.G., Takkar, P.N., Asher, C.J. and Clark, R.B. (1987) *Nutritional Disorders of Grain Sorghum*. Australian Centre for International Agriculture Research, Canberra, pp. 26–29.

Krantz, B.A. and Melsted, S.W. (1964) Nutritional deficiencies in corn, sorghum and small grains. In: Sprangue, H.B. (ed.) *Hunger Signs of Crops – A Symposium*, 3rd edn. David McKay Company, New York, pp. 25–57.

Zhao, D., Reddy, K.R., Kakani, V.G. and Reddy, V.R. (2005) Nitrogen deficiency effects on plant growth, leaf photosynthesis and hyperspectral reflectance properties of sorghum. *European Journal of Agronomy* 22, 391–403.

Plate 95. Phosphorus-deficient sorghum leaf. (Photo by Dr Prakash Kumar.)

SORGHUM (*Sorghum vulgare* Pers.) PHOSPHORUS (P) DEFICIENCY

Plate 96. Purpling on a lower leaf. (Photo by Dr Prakash Kumar.)

Plate 97. Purpling starts from margins of the leaf. (Photo by Dr Prakash Kumar.)

Plate 98. Entire leaf suffused with purple colour. (Photo by Dr Prakash Kumar.)

Symptoms

1. Sorghum is one of the good indicator plants for phosphorus deficiency. Deficiency symptoms appear even in mild deficiency conditions. Deficient plants appear stunted and spindly with dark green leaves.
2. Phosphorus deficiency delays crop maturity. Deficient plants produce small heads and grain size, resulting in poor grain yields.
3. Phosphorus is mobile in plants and under short supply conditions it is easily mobilized from older to younger leaves.
4. The deficiency symptoms appear first and become more severe on older leaves, while young leaves usually remain normal.
5. Old leaves develop a characteristic dark green to bluish green coloration (Plate 95).
6. Under severe deficient conditions, purple or purple–red colours develop on older dark green leaves. Purpling usually starts from the margins of older leaves (Plates 96 and 97).
7. In acute deficiency conditions, the purpling may cover the entire plant. In the most advanced stage, affected leaves burn and die.

Developmental stages

Stage I: In mild deficiency conditions, older leaves develop a characteristic dark green coloration.
Stage II: If deficiency persists, a dark purple colour develops on the margins of old leaves, usually beginning from the tip and proceeding towards the base of the leaves (Plates 96 and 97).
Stage III: If the deficiency becomes severe, the entire leaf becomes reddish brown or purple (Plate 98).
Stage IV: In acute deficiency conditions, a brown necrosis develops on the tip and margins of the affected leaves and proceeds along the margins towards the base of the leaves. The affected lower leaves eventually burn and die.

Likely to occur in

1. Soils having low organic matter.
2. Alkaline and calcareous soils.
3. Soils exhausted by intensive cropping.
4. Acid soils and highly weathered soils.
5. Soils where topsoil has been removed by erosion.
6. Acid soils having pH below 6.0.
7. Alkaline soils having pH between 7.5 and 8.5.

Integrated nutrient management

1. Get the soil analysed before sowing to measure the amount of 'available' phosphate in soil.
2. Apply analysis-based recommended quantity of phosphorus as basal by using:
- **a.** organic manures;
- **b.** phosphate-solubilizing microbial cultures;
- **c.** phosphatic fertilizers.

3. In deficient standing crops apply soluble phosphatic fertilizers such as ammonium phosphate with irrigation water.

Further reading

Brown, J.C., Clark, R.B. and Jones, W.E. (1977) Efficient and inefficient use of phosphorus by sorghum. *Soil Science Society of America Journal* 41, 747–750.

Grundon, N.J., Edwards, D.G., Takkar, P.N., Asher, C.J. and Clark, R.B. (1987) *Nutritional Disorders of Grain Sorghum*. Australian Centre for International Agriculture Research, Canberra.

Krantz, B.A. and Melsted, S.W. (1964) Nutritional deficiencies in corn, sorghum and small grains. In: Sprangue, H.B. (ed.) *Hunger Signs of Crops – A Symposium*, 3rd edn. David McKay Company, New York, pp. 25–57.

Reuter, J.R.B., Jones, G.D. and Friendericks, J.B. (1983) Effect of P and K on yield and chemical composition of forage sorghum. *Agronomy Journal* 75, 5–8.

Plate 99. Potassium-deficient sorghum crop. (Photo by Dr Prakash Kumar.)

Plate 100. Potassium-deficient leaf showing marginal chlorosis. (Photo by Dr Prakash Kumar.)

Plate 101. Deficient leaf showing chlorosis and necrosis. (Photo by Dr Prakash Kumar.)

Plate 102. Close-up of a deficient leaf showing burning. (Photo by Dr Prakash Kumar.)

SORGHUM (*Sorghum vulgare* Pers.) POTASSIUM (K) DEFICIENCY

Symptoms

1. Potassium deficiency causes shortening of the internodes and dwarfing of plants, with a general loss of healthy growth indicated by dark green colour of the foliage. Deficient plants produce pale green to bronze–yellow foliage. Affected plants produce small heads that set few grains.

2. Potassium deficiency symptoms in sorghum usually appear during the later stages of crop growth.

3. Potassium moves readily from old to young leaves; therefore deficiency symptoms appear first and more severely on old leaves, while young leaves usually remain normal and healthy.

4. Symptoms begin as marginal chlorosis and necrosis on old leaves. The pale brown necrosis develops at the tip of the leaf and advances down the margins, leaving the midrib and surrounding tissues green.

5. In severe deficient conditions, affected leaves burn and die and deficiency symptoms move on to upper leaves. Marginal chlorosis and necrosis of older leaves is the specific symptom of potassium deficiency (Plate 99).

Developmental stages

Stage I: Mild deficiency of potassium causes stunted growth, thin stems and pale green to bronze–yellow foliage.

Stage II: If deficiency persists and becomes more severe, marginal chlorosis develops on older leaves starting from the tips of the leaves (Plate 100).

Stage III: Chlorosis is followed by pale brown necrosis. Both chlorosis and necrosis advance down the margins towards the base, leaving the mid-vein and surrounding tissue pale green (Plate 101).

Stage IV: In acute deficiency conditions the entire affected leaf burns and dies (Plate 102).

Likely to occur in

1. Soils formed from parent material low in potassium.

2. Light-textured soils where potassium has been leached by heavy rainfall or excessive irrigation.

3. Soils low in organic matter.

4. Soils with wide Na:K, Mg:K or Ca:K ratio.

5. Acute sodic/saline/acid soil conditions.

6. Large bicarbonate concentration in irrigation water.

7. Highly acid soils having pH below 6.0.

Integrated nutrient management

1. Get the soil analysed before sowing to measure the amount of 'available' potassium in the soil.

2. Problematic acid/alkaline/saline soils should be reclaimed.

3. Add organic manures well before sowing.

4. Apply potassium nitrate, potassium sulphate or potassium chloride to the soil at or before sowing as per soil testing recommendations.

5. If potassium deficiency appears on the standing crop, apply soluble potassium salts such as potassium nitrate, potassium sulphate or potassium chloride with irrigation water. Foliar sprays of these salts are usually not recommended because a number of sprays are needed to fulfil crop requirements.

Further reading

Grundon, N.J., Edwards, D.G., Takkar, P.N., Asher, C.J. and Clark, R.B. (1987) *Nutritional Disorders of Grain Sorghum*. Australian Centre for International Agriculture Research, Canberra.

Krantz, B.A. and Melsted, S.W. (1964) Nutritional deficiencies in corn, sorghum and small grains. In: Sprangue, H.B. (ed.) *Hunger Signs of Crops – A Symposium*, 3rd edn. David McKay Company, New York, pp. 25–57.

Reuter, J.R.B., Jones, G.D. and Friendericks, J.B. (1983) Effect of P and K on yield and chemical composition of forage sorghum. *Agronomy Journal* 75, 5–8.

Robson, A.D. and Snowball, K. (1986) Nutrient deficiency and toxicity symptoms. In: Reuter, D.J. and Robinson, J.B. (eds) *Plant Analysis: An Interpretation Manual*. Inkata Press Pty Ltd, Melbourne, Australia, pp. 13–19.

Plate 103. Calcium-deficient sorghum plant. (Photo by Dr Prakash Kumar.)

SORGHUM (*Sorghum vulgare* Pers.) CALCIUM (Ca) DEFICIENCY

Plate 104. Plant with rolled younger leaves. (Photo by Dr Prakash Kumar.)

Plate 105. Plant with 'ladder-like' appearance. (Photo by Dr Prakash Kumar.)

Plate 106. Plant with a 'bull whip-like' structure. (Photo by Dr Prakash Kumar.)

Symptoms

1. Calcium deficiency, if severe, may destroy the entire sorghum crop. Deficient plants appear stunted with distorted, torn and ragged foliage. Calcium deficiency hampers plant growth. Severely affected plants fail to grow and die before maturity.
2. Calcium is immobile in plants and under short supply conditions it is not easily mobilized from older to younger leaves.
3. The deficiency symptoms appear first and become more severe on younger leaves while older leaves usually remain green and healthy.
4. Calcium deficiency symptoms begin with yellow to white interveinal lesions on young leaves.
5. If the deficiency persists and become mores severe, the new emerging leaves fail to unroll and make a 'bull whip-like' structure.
6. As the symptoms advance, the new leaves develop holes in the lamina. The torn and malformed leaves give the plant a ragged appearance.
7. Under acute deficiency conditions, the tips of the youngest leaves are joined together and do not separate from the whorl.

Developmental stages

Stage I: In the early stage of deficiency, the young leaves become pale green and then develop yellow to white lesions in interveinal tissues.
Stage II: If the deficiency persists and becomes more severe, the youngest leaves remain rolled (Plate 104).
Stage III: As the symptoms advance, the new leaves develop holes in the lamina. The torn and malformed leaves give the plant a ragged appearance (Plate 103).
Stage IV: In acute deficiency conditions, the tips of the growing youngest leaves are joined together and do not separate from the whorl. This gives the plant a 'ladder-like' appearance (Plates 105 and 106).

Likely to occur in

1. Acid sandy soils that have been leached by heavy rainfall.
2. Sodic soils that are rich in exchangeable sodium.
3. Soils having high soluble aluminium and low exchangeable calcium.
4. Strongly acid peat and muck soils that are low in calcium.
5. Acid soils having pH below 6.5.
6. Alkaline soils having pH above 8.5.

Integrated nutrient management

1. Get the soil analysed before sowing to measure the lime or calcium requirement of the soil.
2. Apply analysis-based recommended quantity of calcium-containing fertilizers well before sowing. Suitable calcium-containing fertilizers may be gypsum (calcium sulphate), calcium nitrate or calcium chloride.
3. In acid soils, lime or limestone (calcium carbonate) and dolomite (a mixture of calcium carbonate and magnesium carbonate) are more suitable calcium supplements.
4. The foliar application of 2% w/v calcium sulphate solution (twice) is recommended in standing crops.

Further reading

Grundon, N.J., Edwards, D.G., Takkar, P.N., Asher, C.J. and Clark, R.B. (1987) *Nutritional Disorders of Grain Sorghum*. Australian Centre for International Agriculture Research, Canberra.

Kawaski, T. and Moritsugu, M. (1979) A characteristic symptom of calcium deficiency in maize and sorghum. *Communications in Soil Science and Plant Analysis* 10, 42–56.

Krantz, B.A. and Melsted, S.W. (1964) Nutritional deficiencies in corn, sorghum and small grains. In: Sprangue, H.B. (ed.) *Hunger Signs of Crops – A Symposium*, 3rd edn. David McKay Company, New York, pp. 25–57.

Robson, A.D. and Snowball, K. (1986) Nutrient deficiency and toxicity symptoms. In: Reuter, D.J. and Robinson, J.B. (eds) *Plant Analysis: An Interpretation Manual*. Inkata Press Pty Ltd, Melbourne, Australia, pp. 13–19.

Plate 107. Sulphur-deficient sorghum plant. (Photo by Dr Prakash Kumar.)

Plate 108. Plant showing affected younger and healthy older leaves. (Photo by Dr Prakash Kumar.)

Plate 109. Sulphur-deficient pale green leaf. (Photo by Dr Prakash Kumar and Dr Manoj Kumar Sharma.)

Plate 110. Sulphur-deficient pale yellow leaf. (Photo by Dr Prakash Kumar and Dr Manoj Kumar Sharma.)

SORGHUM (*Sorghum vulgare* Pers.) SULPHUR (S) DEFICIENCY

Symptoms

1. Sulphur-deficient sorghum plants are stunted with pale green to yellow foliage. Deficient plants lack vigour, produce small heads and fewer grains, resulting in poor crop yields.
2. At the initial crop stage, sulphur deficiency symptoms are often confused with those caused by nitrogen deficiency.
3. A close observation is required to see whether the older leaves are more dark green and the younger ones are more pale (case of sulphur deficiency) or whether the younger leaves are more dark green and the older ones are more pale (case of nitrogen deficiency).
4. Sulphur deficiency, even if severe, does not produce any type of necrosis or burning of leaves. The change in leaf colour is the only indicative symptom of sulphur deficiency.
5. Sulphur is immobile in plants and under short supply conditions it is not easily mobilized from older to younger leaves. So, the deficiency symptoms appear first and become more severe on younger leaves.
6. Deficiency symptoms appear as an even and uniform pale green to pale yellow chlorosis across the lamina of young leaves (Plates 109 and 110).
7. The pattern of yellowing on the entire leaf appears uniform, affecting both veins and interveinal tissues uniformly.
8. The youngest leaves are the palest.

Developmental stages

Stage I: In mild deficiency or in the young plant stage, all leaves on the plant become pale green to yellow, although older leaves remain comparatively darker.
Stage II: As the symptoms advance in the growing crop, the younger leaves turn pale yellow with the older leaves green and normal (Plates 107 and 108).
Stage III: In severe deficiency conditions, the youngest leaves turn pale yellow.
Stage IV: If deficiency persists and becomes more severe, the deficiency symptoms move downwards covering more leaves.

Likely to occur in

1. Soils having low organic matter.
2. Light-textured sandy soils that have been leached by heavy rainfall or excessive irrigation.
3. Soils exhausted by intensive cropping.
4. Soils derived from parent material that is inherently low in sulphur.
5. Acid soils having pH below 6.0.

Integrated nutrient management

1. Get the soil analysed before sowing to measure the amount of 'available' sulphate in the soil and to predict the amount of fertilizer required.
2. Apply analysis-based recommended quantity of sulphur by mixing into the soil, well before sowing, either:
   a. elemental sulphur (flowers of sulphur); or
   b. gypsum (calcium sulphate).
3. In deficient standing crops apply ammonium sulphate, magnesium sulphate or potassium sulphate with irrigation water.
4. Problematic acid soils should be reclaimed.

Further reading

Bennett, W.F. (1993) *Nutrient Deficiencies and Toxicities in Crop Plants*. American Phytopathological Society, St Paul, Minnesota.
Grundon, N.J., Edwards, D.G., Takkar, P.N., Asher, C.J. and Clark, R.B. (1987) *Nutritional Disorders of Grain Sorghum*. Australian Centre for International Agriculture Research, Canberra.
Krantz, B.A. and Melsted, S.W. (1964) Nutritional deficiencies in corn, sorghum and small grains. In: Sprangue, H.B. (ed.) *Hunger Signs of Crops – A Symposium*, 3rd edn. David McKay Company, New York, pp. 25–57.
McLachan, K.D. (1978) *An Atlas of Sulphur Deficiency in Commercial Plants*. CSIRO Publishing, Melbourne, Australia.

Plate 111. Iron-deficient sorghum plant. (Photo by Dr Prakash Kumar.)

Plate 112. Plant showing interveinal chlorosis. (Photo by Dr Prakash Kumar.)

Plate 113. Iron deficiency in sorghum crop. (Photo by Dr Prakash Kumar and Dr Manoj Kumar Sharma.)

Plate 114. Crop showing papery white leaves with necrosis. (Photo by Dr Prakash Kumar and Dr Manoj Kumar Sharma.)

SORGHUM (*Sorghum vulgare* Pers.) IRON (Fe) DEFICIENCY

Symptoms

1. Sorghum is highly sensitive to iron deficiency. Iron-deficiency chlorosis is a most common disorder of sorghum grown on alkaline calcareous soils.
2. Iron is immobile in plants. So, deficiency symptoms appear first and more severely on the younger leaves. The older leaves remain normal and apparently healthy (Plate 111).
3. In mild deficiency the topmost leaves of the plants develop temporary fading of interveinal tissues with prominent green veins (Plate 111).
4. If the deficiency persists and becomes more severe, a bright pale yellow chlorosis develops in interveinal tissues (tissues between the veins), leaving the veins green and prominent (Plates 112 and 113).
5. As the symptoms advance, the prominent green veins also fade and become light green to pale yellow. In acute deficiency conditions, the entire leaf bleaches to papery white and burns (Plate 114).

Developmental stages

Stage I: The topmost leaves of the plants develop temporary fading of interveinal tissues with prominent green veins.
Stage II: Interveinal chlorosis develops on younger leaves (Plate 112). Interveinal chlorosis extends uniformly along the full length of the leaf.
Stage III: The prominent green veins also fade and become light green to pale yellow (Plate 113).
Stage IV: The affected leaves bleach to papery white and burn (Plate 114).

Likely to occur in

1. Sandy soils having low total iron.
2. Calcareous soils where solubility of iron is very low.
3. Peat and muck soils where organic matter ties up iron and reduces its availability in soil solution.
4. Acid soils where, despite high availability of iron in soil solution, excessively high levels of soluble zinc, manganese, copper or nickel hamper the uptake of iron by the plant.
5. Waterlogged soils.
6. Alkaline soils having pH above 7.5.

Integrated nutrient management

1. Get the soil analysed before sowing to measure the amount of 'available' iron in the soil.
2. Inorganic soluble salts of iron such as iron sulphates or chlorides may be applied, but these forms of iron quickly become insoluble. Soil dressing with organic forms of iron such as iron chelates (10 kg/ha) is more effective than with inorganic forms of iron. These organic iron forms should be used as per soil pH. If the soil is neutral (around pH 7.0), the most suitable chelates are FeHEDTA and FeDTPA. For acid soils FeEDTA is the most effective chelate, while on alkaline calcareous soils FeEDDHA is recommended.
3. If deficiency appears on the standing crop, apply organic iron chelates or inorganic soluble salts such as iron sulphates or chlorides (0.5 to 1.0% w/v solution) as a foliar spray. Apply two or three sprays at intervals of 10–15 days.

Further reading

Clark, R.B., Pier, P.A., Knudsen, D. and Maranville, J.W. (2008) Effect of trace element deficiencies and excesses on mineral nutrients in sorghum. In: Special issue on Trace Element Stress in Plants: Effects and Methodology. *Journal of Plant Nutrition* 3(1–4), 357–374.

Grundon, N.J., Edwards, D.G., Takkar, P.N., Asher, C.J. and Clark, R.B. (1987) *Nutritional Disorders of Grain Sorghum*. Australian Centre for International Agriculture Research, Canberra.

Krantz, B.A. and Melsted, S.W. (1964) Nutritional deficiencies in corn, sorghum and small grains. In: Sprangue, H.B. (ed.) *Hunger Signs of Crops – A Symposium*, 3rd edn. David McKay Company, New York, pp. 25–57.

Plate 115. Zinc-deficient sorghum plant. (Photo by Dr Prakash Kumar and Dr Manoj Kumar Sharma.)

SORGHUM (*Sorghum vulgare* Pers.) ZINC (Zn) DEFICIENCY

Plate 116. White to yellow bands begin at the base of the leaf. (Photo by Dr Prakash Kumar.)

Symptoms

1. Sorghum is a sensitive crop to zinc deficiency similar to maize.
2. Deficiency prevents elongation of internodes, resulting in stunted growth. The affected crop lacks vigour and yields poorly.
3. Zinc deficiency symptoms in sorghum occur within 2–3 weeks after seedling emergence (Plate 115).
4. Because zinc is not readily transferred from old to young parts in the plant, symptoms develop first and more severely on young leaves. Younger leaves are most affected while older leaves remain green and apparently healthy.
5. White to yellow bands or streaks of bleached tissues appear on each side of the midrib beginning at the base of the leaf (Plate 116). The midrib and the leaf margins remain green. This is the specific symptom of zinc deficiency in sorghum (Plates 117 and 118).
6. In acute deficiency conditions, the affected tissue eventually dies and turns pale grey, leaving the midrib and margins green.

Plate 117. Close-up of a leaf showing broad white bands. (Photo by Dr Prakash Kumar.)

Developmental stages

Stage I: In mild deficiency young leaves turn to pale green.
Stage II: When deficiency persists and becomes more severe, pale yellow to white broad bands or streaks appear between the midrib and margin in the basal half of the leaf. Pale yellow to white broad bands on the upper leaves is the specific symptom of zinc deficiency in sorghum.
Stage III: In acute deficiency conditions, deficiency symptoms also develop on some of the older leaves down the stem.

Likely to occur in

1. Leached sandy soils where total zinc is low.
2. Just-levelled soils where the subsoil is exposed for cultivation. Available zinc in surface soil is often double that of the subsoil.
3. Cool, wet weather.
4. Soils having heavy and excessive application of phosphate fertilizers, which may hamper use of zinc by the crop.
5. Acid soils having pH below 5.0.
6. Alkaline soils having pH above 7.5.

Plate 118. Both streaks and white bands on a leaf. (Photo by Dr Prakash Kumar.)

Integrated nutrient management

1. Get the soil analysed before sowing to estimate the amount of 'available' zinc in the soil.
2. Problematic alkaline or acid soils should be reclaimed.
3. Add organic manures well before sowing.
4. Apply 25–30 kg of zinc sulphate or 10 kg of zinc chelate per hectare once every 2 years in zinc-deficient soils.
5. Zinc fertilizers should not be mixed with phosphate fertilizers.
6. In fields with known zinc deficiency, if deficiency appears in a standing crop, apply foliar application of a 0.5% w/v solution of a soluble zinc salt, Zinc sulphate with 0.25% w/v solution of unsoaked lime.

Further reading

Grundon, N.J., Edwards, D.G., Takkar, P.N., Asher, C.J. and Clark, R.B. (1987) *Nutritional Disorders of Grain Sorghum*. Australian Centre for International Agriculture Research, Canberra.

Krantz, B.A. and Melsted, S.W. (1964) Nutritional deficiencies in corn, sorghum and small grains. In: Sprangue, H.B. (ed.) *Hunger Signs of Crops – A Symposium*, 3rd edn. David McKay Company, New York, pp. 25–57.

Ohki, K. (1984) Zinc nutrition related to critical deficiency and toxicity levels for sorghum. *Agronomy Journal* 76, 253–256.

Robson, A.D. (1993) *Zinc in Soils and Plants: Proceedings of the International Symposium on 'Zinc in Soils and Plants' held at The University of Western Australia, 27–28 September 1993*. Kluwer Academic Publishers, Dordrecht, the Netherlands.

Plate 119. Manganese-deficient sorghum plant. (Photo by Dr Prakash Kumar.)

Plate 120. 'White flecks' developing on a leaf. (Photo by Dr Prakash Kumar.)

Plate 121. Reddish interveinal necrosis at initial stage. (Photo by Dr Prakash Kumar and Dr Manoj Kumar Sharma.)

Plate 122. Necrosis covering more leaf area. (Photo by Dr Prakash Kumar and Dr Manoj Kumar Sharma.)

SORGHUM (*Sorghum vulgare* Pers.) MANGANESE (Mn) DEFICIENCY

Symptoms

1. Manganese-deficient sorghum plants appear stunted with thin, spindly stems and pale green to yellow leaves.
2. Manganese is immobile in plants and under short supply conditions it is not easily mobilized from older to younger leaves.
3. The deficiency symptoms appear first and become more severe on the younger leaves, the older leaves remain normal.
4. The symptoms begin with a faint interveinal chlorosis on young leaves (Plate 119).
5. As the symptoms advance, interveinal tissue of affected leaves turns into interveinal 'white flecks'. The white flecks are somewhat similar to the 'white flecks' of maize in manganese deficiency (Plate 120).
6. The white flecks develop initially only a short length up the leaf blade, but may extend to almost the full length of the blade in advanced stages of deficiency.
7. If deficiency persists and becomes more severe, reddish or brownish necrotic lesions develop within interveinal areas. These necrotic lesions become darker and cover large sections of the leaf blade, in advanced stage of deficiency.

Developmental stages

Stage I: In mild deficiency, the younger leaves develop temporary fading of interveinal tissues. The plant recovers soon (Plate 119).
Stage II: As the symptoms advance, the younger leaves develop white flecks in interveinal areas.
Stage III: If the deficiency persists and becomes more severe, reddish or brownish necrotic lesions develop within interveinal areas (Plate 121).
Stage IV: These necrotic lesions become darker and cover larger sections of the leaf blade in the advanced stage of deficiency (Plate 122).

Likely to occur in

1. Light-textured sandy soils that have been leached by heavy rainfall or excessive irrigation.
2. Calcareous and alkaline soils where solubility of manganese is very low.
3. Waterlogged peaty soils where organic matter ties up manganese and reduces its availability in solution.
4. Soils derived from parent material that is inherently low in manganese.
5. Acid soils having pH below 5.0.
6. Alkaline soils having pH above 7.5.

Integrated nutrient management

1. Get the soil analysed before sowing to measure the amount of 'available' manganese in the soil.
2. Problematic alkaline or acid soils should be reclaimed.
3. Add organic manures well before sowing.
4. Apply soluble salts of manganese such as manganese sulphate as basal.
5. In deficient standing crops apply manganese sulphate (0.2 to 0.3% w/v solution) as a foliar spray.

Further reading

Clark, R.B., Pier, P.A., Knudsen, D. and Maranville, J.W. (2008) Effect of trace element deficiencies and excesses on mineral nutrients in sorghum. In: Special Issue on Trace Element Stress in Plants: Effects and Methodology. *Journal of Plant Nutrition* 3(1–4), 357–374.

Grundon, N.J., Edwards, D.G., Takkar, P.N., Asher, C.J. and Clark, R.B. (1987) *Nutritional Disorders of Grain Sorghum*. Australian Centre for International Agriculture Research, Canberra.

Krantz, B.A. and Melsted, S.W. (1964) Nutritional deficiencies in corn, sorghum and small grains. In: Sprangue, H.B. (ed.) *Hunger Signs of Crops – A Symposium*, 3rd edn. David McKay Company, New York, pp. 25–57.

Ohki, K. (1974) Early growth of grain sorghum as related to manganese nutrition. *Agronomy Journal* 66, 328–330.

Plate 123. Nitrogen-deficient pearl millet plant. (Photo by Dr Prakash Kumar.)

Plate 124. Demonstration field with (left) and without (right) nitrogen application. (Photo by Dr Prakash Kumar.)

Plate 125. Chlorosis on leaf in a V-shaped pattern. (Photo by Dr Prakash Kumar.)

Plate 126. Leaf showing yellow chlorosis and brown necrosis. (Photo by Dr Prakash Kumar.)

PEARL MILLET (*Pennisetum typhoides* (Burm.f) Stapf & C.E. Hubb.) NITROGEN (N) DEFICIENCY

Symptoms

1. Pearl millet is one of the most sensitive crops to nitrogen deficiency. There are numerous reports of crop failure due to heavy rains in light-textured soils where pearl millet is grown as a rain-fed crop. The nitrogen deficiency is one of the major reasons behind such failures. Nitrogen deficiency during any stage of crop growth may cause severe losses to the crop in terms of fodder production or grain yield.
2. Nitrogen-deficient plants are stunted with spindly stems and pale green to yellow leaves. The deficient plant produces fewer tillers.
3. In severe deficiency conditions, the plant does not produce any tillers. The height of the plant is reduced to half. The plant produces a tiny ear head with few grains or without any grains.
4. Nitrogen is mobile in plants and under short supply conditions it is easily mobilized from older to younger leaves.
5. The deficiency symptoms appear first and become more severe on older leaves (Plate 123).
6. A pale yellow chlorosis develops at the tip of old leaves and proceeds towards the leaf base along the midrib in a V-shaped pattern. This is the specific symptom of nitrogen deficiency in pearl millet.
7. The pale yellow chlorosis is followed by a pale brown necrosis. The deficient leaves eventually die.

Developmental stages

Stage I: In mild deficiency, the entire plant appears uniformly light green in colour (Plate 124).
Stage II: If the deficiency becomes severe, a pale yellow chlorosis begins at the tip of old leaves and proceeds towards the base along the midrib in a V-shaped pattern (Plate 125).
Stage III: As the symptoms advance, the pale yellow chlorosis is followed by a pale brown necrosis (Plate 126).
Stage IV: In the most advanced stage, the affected leaves eventually burn and die. The dead leaves hang down, making a thatch of dry leaves near the base of the plant.

Likely to occur in

1. Soils having low organic matter.
2. Light-textured sandy soils that have been leached by heavy rainfall or excessive irrigation.
3. Soils exhausted by intensive cropping.
4. Waterlogged conditions.
5. Acid soils having pH below 6.0.
6. Alkaline soils having pH above 8.0.

Integrated nutrient management

1. Get the soil analysed before sowing to measure the amount of 'available' nitrogen in the soil.
2. Apply analysis-based recommended quantity of nitrogen as basal by using:
- **a.** organic manures;
- **b.** bio-fertilizers;
- **c.** nitrogenous fertilizers.

3. Top-dress soluble nitrogenous fertilizers such as urea in split doses.
4. Incorporate legume crops in rotation.

Further reading

Gregory, P.J. (1978) Uptake of N, P and K by irrigated and unirrigated pearl millet (*Pennisetum typhoides*). *Experimental Agriculture* 15, 217–223.

Hafner, H. (1992) Effect of organic and inorganic fertilizer application on growth and mineral nutrient uptake of pearl millet (*Pennisetum glaucum* L.) and ground nut (*Arachis hypogaea* L.) in acid sandy soils of Niger. Thesis published by Verlag Ulrich E. Grauer, Wendlingen, Germany.

Khairwal, I.S., Rai, K.N., Divwakar, B., Sharma, Y.K., Rajpurohit, B.S., Nirwan, B. and Bhattacharjee, R. (2007) *Pearl Millet Crop and Seed Production Manual*. International Crops Research Institute for the Semi-Arid Tropics (ICRISAT), Patancheru, India, pp. 23–37.

Voortman, R.L. and Brouwer, J. (2003) An empirical analysis of the simultaneous effects of nitrogen, phosphorus and potassium in millet production on spatially variable fields in SW Niger. *Nutrient Cycling in Agroecosystems* 66, 143–164.

Plate 127. Phosphorus-deficient pearl millet leaf. (Photo by Dr Prakash Kumar.)

Plate 128. Leaf showing marginal purpling. (Photo by Dr Prakash Kumar.)

Plate 129. Reddish pigmentation covering margins of leaf. (Photo by Dr Prakash Kumar.)

Plate 130. Close-up of a phosphorus-deficient leaf. (Photo by Dr Prakash Kumar.)

PEARL MILLET (*Pennisetum typhoides* (Burm.f) Stapf & C.E. Hubb.) PHOSPHORUS (P) DEFICIENCY

Symptoms

1. Pearl millet growth and root development are severely affected by phosphorus deficiency. Deficient plants appear stunted and spindly with dark green leaves.
2. Phosphorus deficiency delays crop maturity. Deficient plants produce small ear heads and grain size, resulting in poor grain yields.
3. Phosphorus is mobile in plants and under short supply conditions it is easily mobilized from older to younger leaves.
4. The deficiency symptoms appear first and become more severe on older leaves while young leaves usually remain normal.
5. The symptoms begin as a dark green appearance of the plant. As symptoms advance lower leaves become darker.
6. Under severe deficient conditions, purple or purple–red colours develop on older dark green leaves. Purpling usually starts from the margins of older leaves (Plates 127 and 128).
7. In acute deficiency conditions, the purpling may cover the entire plant (Plate 130). In the most advanced stage, affected leaves burn and die.

Developmental stages

Stage I: In mild deficiency conditions, the plant appears dark green and stunted.
Stage II: If deficiency persists, the lower leaves become dark bluish green.
Stage III: If the deficiency becomes more severe, a reddish to dark purple colour develops on the margins of old leaves, usually beginning from the tip and proceeding towards the base of the leaves (Plates 127 and 129).
Stage IV: In acute deficiency conditions, a brown necrosis develops on the affected leaves.

Likely to occur in

1. Soils having low organic matter.
2. Alkaline and calcareous soils.
3. Soils exhausted by intensive cropping.
4. Acid soils and highly weathered soils.
5. Soils where the topsoil has been removed by erosion.
6. Acid soils having pH below 6.0.
7. Alkaline soils having pH between 7.5 and 8.5.

Integrated nutrient management

1. Get the soil analysed before sowing to measure the amount of 'available' phosphate in the soil.
2. Apply analysis-based recommended quantity of phosphorus as basal by using:
  a. organic manures;
  b. phosphate-solubilizing microbial cultures; or
  c. phosphatic fertilizers.
3. No recommendation is available as yet to correct phosphorus deficiency in a standing crop of pearl millet.

Further reading

Faye, I., Diouf, O., Guisse, A., Sene, M. and Diallo, N. (2006) Characterising root responses to low phosphorus in pearl millet [(*Pennisetum glaucum* (L.) R. Br.]. *Agronomy Journal* 98, 1187–1194.

Gregory, P.J. (1978) Uptake of N, P and K by irrigated and unirrigated pearl millet (*Pennisetum typhoides*). *Experimental Agriculture* 15, 217–223.

Khairwal, I.S., Rai, K.N., Divwakar, B., Sharma, Y.K., Rajpurohit, B.S., Nirwan, B. and Bhattacharjee, R. (2007) *Pearl Millet Crop and Seed Production Manual*. International Crops Research Institute for the Semi-Arid Tropics (ICRISAT), Patancheru, India, pp. 23–37.

Payne, W.A., Lascano, R.J., Hossner, L.R., Wendt, C.W. and Onken, A.B. (1991) Pearl millet growth as affected by phosphorus and water. *Agronomy Journal* 83, 942–948.

Plate 131. Potassium-deficient pearl millet crop. (Photo by Dr Prakash Kumar.)

Plate 132. Young-stage crop suffering with potassium deficiency. (Photo by Dr Prakash Kumar.)

Plate 133. Severely deficient crop with chlorosis and necrosis. (Photo by Dr Prakash Kumar.)

Plate 134. Deficient leaves showing stages of deficiency. (Photo by Dr Prakash Kumar.)

PEARL MILLET (*Pennisetum typhoides* (Burm.f) Stapf & C.E. Hubb.) POTASSIUM (K) DEFICIENCY

Symptoms

1. Potassium deficiency symptoms in pearl millet are more prominent in high-yielding hybrids than in local cultivars.
2. Potassium deficiency causes shortening of the internodes and dwarfing of the plant, with a general loss of healthy growth indicated by dark green colour of the foliage. Affected plants produce small ear heads.
3. Potassium moves readily from old to young leaves; therefore deficiency symptoms appear first and more severely on old leaves, while young leaves usually remain normal and healthy (Plate 132).
4. Symptoms begin as marginal chlorosis and necrosis on old leaves. The pale brown necrosis develops at the tip of the leaf and advances down the margins, leaving the midrib and surrounding tissues green.
5. In severe deficient conditions, affected leaves burn and die and deficiency symptoms move on to upper leaves. Marginal chlorosis and necrosis of older leaves is the specific symptom of potassium deficiency (Plate 133).

Developmental stages

Stage I: Mild deficiency of potassium causes stunted growth, thin stems and pale green foliage (Plate 131).
Stage II: If deficiency persists and becomes more severe, marginal chlorosis develops on older leaves starting from the tips of the leaves (Plate 134).
Stage III: Chlorosis is followed by pale brown necrosis. Both chlorosis and necrosis advance down the margins towards the base, leaving the mid-vein and surrounding tissue pale green (Plates 133 and 134).
Stage IV: In acute deficiency conditions the entire affected leaf burns and dies.

Likely to occur in

1. Soils formed from parent material low in potassium.
2. Light-textured soils where potassium has been leached by heavy rainfall or excessive irrigation.
3. Soils low in organic matter.
4. Soils with wide Na:K, Mg:K or Ca:K ratio.
5. Acute sodic/saline/acid soil conditions.
6. Large bicarbonate concentration in irrigation water.
7. Highly acid soils having pH below 6.0.

Integrated nutrient management

1. Get the soil analysed before sowing to measure the amount of 'available' potassium in the soil.
2. Problematic acid/alkaline/saline soils should be reclaimed.
3. Add organic manures well before sowing.
4. Apply potassium nitrate, potassium sulphate or potassium chloride to the soil at or before sowing as per soil testing recommendations.
5. No recommendation is available as yet to correct potassium deficiency in a standing crop of pearl millet.

Further reading

Ashraf, M., Zafer, Z.U. and Cheema, Z.A. (1994) Effect of low potassium regimes on some salt and drought tolerant line of pearl millet. *Phyton (Horn, Austria)* 34, 219–227.

Gregory, P.J. (1978) Uptake of N, P and K by irrigated and unirrigated pearl millet (*Pennisetum typhoides*). *Experimental Agriculture* 15, 217–223.

Khairwal, I.S., Rai, K.N., Divwakar, B., Sharma, Y.K., Rajpurohit, B.S., Nirwan, B. and Bhattacharjee, R. (2007) *Pearl Millet Crop and Seed Production Manual*. International Crops Research Institute for the Semi-Arid Tropics (ICRISAT), Patancheru, India, pp. 23–37.

Voortman, R.L. and Brouwer, J. (2003) An empirical analysis of the simultaneous effects of nitrogen, phosphorus and potassium in millet production on spatially variable fields in SW Niger. *Nutrient Cycling in Agroecosystems* 66, 143–164.

Plate 135. Calcium-deficient pearl millet plant. (Photo by Dr Prakash Kumar.)

Plate 136. Leaf with yellow and white lesions. (Photo by Dr Prakash Kumar.)

Plate 137. Torn and malformed leaf. (Photo by Dr Prakash Kumar.)

Plate 138. Plant with 'bull whip-like' structure. (Photo by Dr Prakash Kumar.)

PEARL MILLET (*Pennisetum typhoides* (Burm.f) Stapf & C.E. Hubb.) CALCIUM (Ca) DEFICIENCY

Symptoms

1. Calcium deficiency usually appears in early stages of crop growth. Deficient plants appear stunted with distorted, torn and ragged foliage. Calcium deficiency hampers plant growth. Severely affected plants fail to grow and die before maturity.
2. Calcium is immobile in plants and under short supply conditions it is not easily mobilized from older to younger leaves.
3. The deficiency symptoms appear first and become more severe on younger leaves while older leaves usually remain green and healthy.
4. The new emerging leaves fail to unroll and may develop yellow to white lesions on the youngest affected leaves.
5. As the symptoms advance, the new leaves develop holes in the lamina. The torn and malformed leaves give the plant a ragged appearance.
6. Under acute deficiency conditions, the tips of the youngest leaves are joined together and do not separate from the whorl.

Developmental stages

Stage I: The youngest leaves remain rolled (Plate 135) and yellow to white lesions may develop on the youngest affected leaves (Plate 136).
Stage II: As the symptoms advance, the new leaves develop holes in the lamina. The torn and malformed leaves give the plant a ragged appearance (Plate 137).
Stage III: In acute deficiency conditions, the tips of the growing youngest leaves are joined together and do not separate from the whorl. The young emerging leaves make a 'bull whip-like' structure (Plate 138).

Likely to occur in

1. Acid sandy soils that have been leached by heavy rainfall.
2. Sodic soils that are rich in exchangeable sodium.
3. Soils having high soluble aluminium and low exchangeable calcium.
4. Strongly acid peat and muck soils which are low in calcium.
5. Acid soils having pH below 6.5.
6. Alkaline soils having pH above 8.5.

Integrated nutrient management

1. Get the soil analysed before sowing to measure the lime or calcium requirement of the soil.
2. Apply analysis-based recommended quantity of calcium-containing fertilizers well before sowing. Suitable calcium-containing fertilizers may be gypsum (calcium sulphate), calcium nitrate or calcium chloride.
3. In acid soils, lime or limestone (calcium carbonate) and dolomite (a mixture of calcium carbonate and magnesium carbonate) are more suitable calcium supplements.
4. No recommendation is available as yet to correct calcium deficiency in a standing crop of pearl millet.

Further reading

Grundon, N.J., Edwards, D.G., Takkar, P.N., Asher, C.J. and Clark, R.B. (1987) *Nutritional Disorders of Grain Sorghum*. Australian Centre for International Agriculture Research, Canberra.

Kawaski, T. and Moritsugu, M. (1979) A characteristic symptom of calcium deficiency in maize and sorghum. *Communications in Soil Science and Plant Analysis* 10, 42–56.

Krantz, B.A. and Melsted, S.W. (1964) Nutritional deficiencies in corn, sorghum and small grains. In: Sprague, H.B. (ed.) *Hunger Signs of Crops – A Symposium*, 3rd edn. David McKay Company, New York, pp. 25–57.

Maas, E.V. and Grieve, C.M. (1987) Sodium-induced calcium deficiency in salt-stressed corn. *Plant, Cell and Environment* 10, 559–564.

Plate 139. Sulphur-deficient pearl millet plant. (Photo by Dr Prakash Kumar and Dr Manoj Kumar Sharma.)

PEARL MILLET (*Pennisetum typhoides* (Burm.f) Stapf & C.E. Hubb.) SULPHUR (S) DEFICIENCY

Plate 140. Plant showing pale green young leaves. (Photo by Dr Prakash Kumar.)

Plate 141. Plant showing pale yellow young leaves. (Photo by Dr Prakash Kumar and Dr Manoj Kumar Sharma.)

Plate 142. Sulphur-deficient leaf (left) and normal leaf (right). (Photo by Dr Prakash Kumar.)

Symptoms

1. Pearl millet is mostly grown in light-textured soils under rainfed conditions, where the temporary depletion of sulphur due to leaching with rain water is most common. Sulphur deficiency symptoms are mostly found during the early stages of crop growth. Sulphur-deficient pearl millet plants are stunted with pale green to yellow foliage. Deficient plants lack vigour, produce small ear heads and lower number of grains, resulting in poor crop yields.
2. At the initial stage of the crop, the sulphur deficiency symptoms are often confused with those caused by nitrogen deficiency. A close observation is required to see whether the older leaves are more dark green and the younger ones are more pale (case of sulphur deficiency) or whether the younger leaves are more dark green and the older ones are more pale (case of nitrogen deficiency).
3. Sulphur deficiency, even if severe, does not produce any type of necrosis or burning of leaves. The change in leaf colour is the only indicative symptom of sulphur deficiency (Plate 142).
4. Sulphur is immobile in plants and under short supply conditions it is not easily mobilized from older to younger leaves. So, the deficiency symptoms appear first and become more severe on younger leaves.
5. Deficiency symptoms appear as an even and uniform pale green to pale yellow chlorosis across the lamina of young leaves. The pattern of yellowing on the entire leaf appears uniform, affecting both the veins and interveinal tissues uniformly. The youngest leaves are the palest.

Developmental stages

Stage I: In mild deficiency or in the young plant stage, the new emerging younger leaves become light green to yellowish green.
Stage II: As the symptoms advance with the growing crop, the younger leaves turn pale green with the older leaves green and normal (Plates 139 and 140).
Stage III: If deficiency persists and becomes more severe, the deficiency symptoms move down the stem covering more leaves (Plate 141).

Likely to occur in

1. Soils having low organic matter.
2. Light-textured sandy soils that have been leached by heavy rainfall or excessive irrigation.
3. Soils exhausted by intensive cropping.
4. Soils derived from parent material that is inherently low in sulphur.
5. Acid soils having pH below 6.0.

Integrated nutrient management

1. Get the soil analysed before sowing to measure the amount of 'available' sulphate in the soil and to predict the amount of fertilizer required.
2. Apply analysis-based recommended quantity of sulphur by mixing into the soil, well before sowing, either:
  a. elemental sulphur (flowers of sulphur); or
  b. gypsum (calcium sulphate).
3. Problematic acid soils should be reclaimed.

Further reading

Grundon, N.J., Edwards, D.G., Takkar, P.N., Asher, C.J. and Clark, R.B. (1987) *Nutritional Disorders of Grain Sorghum*. Australian Centre for International Agriculture Research, Canberra.

Krantz, B.A. and Melsted, S.W. (1964) Nutritional deficiencies in corn, sorghum and small grains. In: Sprague, H.B. (ed.) *Hunger Signs of Crops – A Symposium*, 3rd edn. David McKay Company, New York, pp. 25–57.

Khairwal, I.S., Rai, K.N., Divwakar, B., Sharma, Y.K., Rajpurohit, B.S., Nirwan, B. and Bhattacharjee, R. (2007) *Pearl Millet Crop and Seed Production Manual*. International Crops Research Institute for the Semi-Arid Tropics (ICRISAT), Patancheru, India, pp. 23–37.

McLachan, K.D. (1978) *An Atlas of Sulphur Deficiency in Commercial Plants*. CSIRO Publishing, Melbourne, Australia.

Plate 143. Iron-deficient pearl millet plant. (Photo by Dr Prakash Kumar.)

PEARL MILLET (*Pennisetum typhoides* (Burm.f) Stapf & C.E. Hubb.) IRON (Fe) DEFICIENCY

Plate 144. Leaf showing interveinal chlorosis. (Photo by Dr Prakash Kumar and Dr Manoj Kumar Sharma.)

Symptoms

1. Iron is immobile in plants. So, deficiency symptoms appear first and more severely on the younger leaves. The older leaves remain normal and apparently healthy (Plate 143).
2. In mild deficiency the young leaves of the plants develop temporary fading of interveinal tissues with prominent green veins.
3. If the deficiency persists and becomes more severe, a bright pale yellow chlorosis develops in interveinal tissues (tissues between the veins), leaving the veins green and prominent (Plates 144 and 145).
4. As the symptoms advance, the prominent green veins also fade and become light green to pale yellow. In acute deficiency conditions, the entire leaf bleaches to papery white and burns (Plate 146).

Developmental stages

Stage I: The topmost leaves of the plant develop temporary fading of interveinal tissues with prominent green veins.
Stage II: Interveinal chlorosis develops on younger leaves (Plates 143 and 144). Interveinal chlorosis uniformly extends the full length of the leaf.
Stage III: The prominent green veins also fade and become light green to pale yellow (Plate 145).
Stage IV: The entire leaf bleaches to papery white (Plate 146).

Plate 145. Faded veins in the more advanced stage. (Photo by Dr Prakash Kumar and Dr Manoj Kumar Sharma.)

Likely to occur in

1. Sandy soils having low total iron.
2. Calcareous soils where solubility of iron is very low.
3. Peat and muck soils where organic matter ties up iron and reduces its availability in soil solution.
4. Acid soils where, despite high availability of iron in soil solution, excessively high levels of soluble zinc, manganese, copper or nickel hamper the uptake of iron by the plant.
5. Waterlogged soils.
6. Alkaline soils having pH above 7.5.

Integrated nutrient management

1. Get the soil analysed before sowing to measure the amount of 'available' iron in the soil.
2. Problematic alkaline soils should be reclaimed.
3. Inorganic soluble salts of iron such as iron sulphates or chlorides may be applied, but these forms of iron quickly become insoluble. Soil dressing with organic forms of iron such as iron chelates (10 kg/ha) is more effective than with inorganic forms of iron. These organic iron forms should be used as per soil pH. If the soil is neutral (around pH 7.0), the most suitable chelates are FeHEDTA and FeDTPA. For acid soils FeEDTA is the most effective chelate, while on alkaline calcareous soils FeEDDHA is recommended.
4. If deficiency appears on a standing crop, apply organic iron chelates or inorganic soluble salts such as iron sulphates or chlorides (0.5 to 1.0% w/v solution) as a foliar spray. Apply two or three sprays at intervals of 10–15 days.

Plate 146. Plant showing papery white leaves. (Photo by Dr Prakash Kumar.)

Further reading

Clark, R.B., Pier, P.A., Knudsen, D. and Maranville, J.W. (2008) Effect of trace element deficiencies and excesses on mineral nutrients in sorghum. In: Special issue on Trace Element Stress in Plants: Effects and Methodology. *Journal of Plant Nutrition* 3(1–4), 357–374.

Grundon, N.J., Edwards, D.G., Takkar, P.N., Asher, C.J. and Clark, R.B. (1987) *Nutritional Disorders of Grain Sorghum*. Australian Centre for International Agriculture Research, Canberra.

Krantz, B.A. and Melsted, S.W. (1964) Nutritional deficiencies in corn, sorghum and small grains. In: Sprague, H.B. (ed.) *Hunger Signs of Crops – A Symposium*, 3rd edn. David McKay Company, New York, pp. 25–57.

Plate 147. Zinc-deficient pearl millet plant. (Photo by Dr Prakash Kumar and Dr Manoj Kumar Sharma.)

Plate 148. White to yellow bands begin at the base of the leaf. (Photo by Dr Prakash Kumar.)

Plate 149. Close-up of a leaf showing broad white bands. (Photo by Dr Prakash Kumar and Dr Manoj Kumar Sharma.)

Plate 150. White bands in the advanced stage of deficiency. (Photo by Dr Prakash Kumar.)

PEARL MILLET (*Pennisetum typhoides* (Burm.f) Stapf & C.E. Hubb.) ZINC (Zn) DEFICIENCY

Symptoms

1. Like maize and sorghum, pearl millet is also very sensitive to zinc deficiency.
2. Zinc deficiency prevents elongation of internodes, which results in crowding of the upper leaves, producing a fan-shaped appearance (Plate 147).
3. The affected plants are very stunted. The crop lacks vigour and yields poorly.
4. Zinc deficiency symptoms in pearl millet occur within 2–3 weeks after seedling emergence.
5. Because zinc is not readily transferred from old to young parts in the plant, symptoms develop first and more severely on young leaves. Younger leaves are most affected while older leaves remain green and apparently healthy.
6. White to yellow bands or streaks of bleached tissues appear on each side of the midrib beginning at the base of the leaf. The midrib and the leaf margins remain green. This is the specific symptom of zinc deficiency in pearl millet (Plates 148, 149 and 150).
7. In acute deficiency conditions, the affected tissue eventually dies and turns pale grey, leaving the midrib and margins green.

Developmental stages

Stage I: In mild deficiency young leaves turn to pale green.
Stage II: When deficiency persists and becomes more severe, pale yellow to white broad bands or streaks appear between the midrib and margin in the basal half of the leaf. Pale yellow to white broad bands on upper leaves are the specific symptoms of zinc deficiency in pearl millet.
Stage III: In acute deficiency conditions, deficiency symptoms also develop on some of the older leaves down the stem.
Stage IV: At last, the affected tissue eventually dies and turns pale grey, leaving the midrib and margins green.

Likely to occur in

1. Leached sandy soils where total zinc is low.
2. Just-levelled soils where the subsoil is exposed for cultivation. Available zinc in surface soil is often double that of the subsoil.
3. Cool wet weather.
4. Soil having heavy and excessive application of phosphate fertilizers which may hamper use of zinc by the crop.
5. Acid soils having pH below 5.0.
6. Alkaline soils having pH above 7.5.

Integrated nutrient management

1. Get the soil analysed before sowing to estimate the amount of 'available' zinc in the soil.
2. Problematic alkaline or acid soils should be reclaimed.
3. Add organic manures well before sowing.
4. Apply 25–30 kg of zinc sulphate or 10 kg of zinc chelate per hectare every 2 years in zinc-deficient soils.
5. Zinc fertilizers should not be mixed with phosphate fertilizers.
6. In fields with known zinc deficiency, if deficiency appears in a standing crop, apply foliar application of a 0.5% w/v solution of a soluble zinc salt, Zinc sulphate with 0.25% w/v solution of unsoaked lime.

Further reading

Krantz, B.A. and Melsted, S.W. (1964) Nutritional deficiencies in corn, sorghum and small grains. In: Sprague, H.B. (ed.) *Hunger Signs of Crops – A Symposium*, 3rd edn. David McKay Company, New York, pp. 25–57.

Ohki, K. (1984) Zinc nutrition related to critical deficiency and toxicity levels for sorghum. *Agronomy Journal* 76, 253–256.

Robson, A.D. (1993) *Zinc in Soils and Plants: Proceedings of the International Symposium on 'Zinc in Soils and Plants' held at The University of Western Australia, 27–28 September 1993*. Kluwer Academic Publishers, Dordrecht, the Netherlands.

Shukla, U.C. and Raj, H. (1987) Relative response of corn (*Zea mays* L.), pearl millet (*Pennisetum typhoides* (Burm F.) Stapf and C.E. Hubb), sorghum (*Sorghum vulgare*) and cowpea (*Vigna unguiculata* (L.) Walp) to zinc deficiency in soil. *Journal of Plant Nutrition* 10, 2057–2067.

Plate 151. Manganese-deficient pearl millet plant. (Photo by Dr Prakash Kumar.)

Plate 152. Close-up of a leaf showing interveinal chlorosis. (Photo by Dr Prakash Kumar.)

Plate 153. Necrotic white flecks on a leaf. (Photo by Dr Prakash Kumar)

Plate 154. Close-up of a leaf showing white flecks. (Photo by Dr Prakash Kumar.)

PEARL MILLET (*Pennisetum typhoides* (Burm.f) Stapf & C.E. Hubb.) MANGANESE (Mn) DEFICIENCY

Symptoms

1. The visual symptoms of manganese deficiency in pearl millet resemble iron deficiency symptoms. Both appear on younger leaves and produce interveinal chlorosis. So, a careful observation is required to differentiate between the two. First, in the case of iron deficiency, interveinal chlorosis is more bright and clear than in manganese deficiency. Second, the iron-deficient leaf first bleaches to papery white and then becomes necrotic (there will be no necrosis or burning on the leaf while veins remain green); in the case of manganese deficiency, the interveinal tissues soon become necrotic, turn into 'necrotic white flecks' and the veins remain green and prominent up until the advanced stage of necrosis.
2. The symptoms begin with a faint interveinal chlorosis on young leaves (Plates 151 and 152).
3. As the symptoms advance, the interveinal tissue of affected leaves turns to interveinal white flecks.
4. If deficiency persists and becomes more severe, reddish or brownish necrotic lesions develop within interveinal areas.

Developmental stages

Stage I: The younger leaves develop temporary fading of interveinal tissues (Plates 151 and 152).
Stage II: The younger leaves develop white flecks in interveinal areas (Plates 153 and 154).
Stage III: Reddish or brownish necrotic lesions may develop within interveinal areas.

Likely to occur in

1. Light-textured sandy soils that have been leached by heavy rainfall or excessive irrigation.
2. Calcareous and alkaline soils where solubility of manganese is very low.
3. Waterlogged peat soils where organic matter ties up manganese and reduces its availability in solution.
4. Soils derived from parent material that is inherently low in manganese.
5. Acid soils having pH below 5.0.
6. Alkaline soils having pH above 7.5.

Integrated nutrient management

1. Get the soil analysed before sowing to measure the amount of 'available' manganese in the soil.
2. Problematic alkaline or acid soils should be reclaimed.
3. Add organic manures well before sowing.
4. Apply soluble salts of manganese such as manganese sulphate as basal.
5. In deficient standing crops apply manganese sulphate (0.2 to 0.3% w/v solution) as a foliar spray.

Further reading

Grundon, N.J., Edwards, D.G., Takkar, P.N., Asher, C.J. and Clark, R.B. (1987) *Nutritional Disorders of Grain Sorghum*. Australian Centre for International Agriculture Research, Canberra.

Krantz, B.A. and Melsted, S.W. (1964) Nutritional deficiencies in corn, sorghum and small grains. In: Sprague, H.B. (ed.) *Hunger Signs of Crops – A Symposium*, 3rd edn. David McKay Company, New York, pp. 25–57.

Ohki, K. (1974) Early growth of grain sorghum as related to manganese nutrition. *Agronomy Journal* 66, 328–330.

Plate 155. Nitrogen-deficient wheat plant. (Photo by Dr Prakash Kumar and Dr Manoj Kumar Sharma.)

WHEAT (*Triticum aestivum* Linn.) NITROGEN (N) DEFICIENCY

Plate 156. Chlorosis of older leaves. (Photo by Dr Prakash Kumar and Dr Manoj Kumar Sharma.)

Plate 157. Close-up of a chlorotic leaf. (Photo by Dr Prakash Kumar and Dr Manoj Kumar Sharma.)

Plate 158. Severely deficient plant showing chlorotic and dead leaves. (Photo by Dr Prakash Kumar and Dr Manoj Kumar Sharma.)

Symptoms

1. Nitrogen deficiency is most common and widespread in wheat-grown areas. Wheat is very sensitive to nitrogen deficiency. Deficiency symptoms appear even in mild deficiency conditions.
2. Deficient plants appear pale, stunted, thin and spindly. The number of tillers and the grain yield are reduced severely.
3. Nitrogen is mobile in plants and under short supply conditions it is easily mobilized from older to younger leaves. The deficiency symptoms appear first and become more severe on older leaves.
4. In mild deficiency conditions or during the young stage of the crop, the entire plant becomes pale green to yellow.
5. If deficiency persists and becomes more severe, a pale yellow chlorosis develops at the tip of older leaves and advances in a broad front towards the leaf base.
6. Pale yellow to almost white chlorotic leaves turn pale brown and die.
7. In deficient crops, green youngest leaves, pale green middle leaves and lemon yellow to pale brown older leaves may appear simultaneously (Plate 155).

Developmental stages

Stage I: During the young stage of the crop or in mild deficiency conditions, the entire plant appears uniformly pale green in colour.
Stage II: If the deficiency persists and becomes more severe, a pale yellow chlorosis begins at the tip of older leaves and proceeds in a broad front towards the base (Plate 156).
Stage III: As the symptoms advance, the whole leaf becomes chlorotic, changing from yellow to almost white (Plate 157).
Stage IV: In the later stage, the chlorotic leaf withers and dies. The dead leaves hang down and make a thatch of dry leaves near the base of the plant (Plate 158).

Likely to occur in

1. Soils having low organic matter.
2. Light-textured sandy soils that have been leached by heavy rainfall or excessive irrigation.
3. Soils exhausted by intensive cropping.
4. Waterlogged conditions.
5. Acid soils having pH below 6.0.
6. Alkaline soils having pH above 8.0.

Integrated nutrient management

1. Get the soil analysed before sowing to measure the amount of 'available' nitrogen in the soil.
2. Apply analysis-based recommended quantity of nitrogen as basal by using:
  a. organic manures;
  b. bio-fertilizers;
  c. nitrogenous fertilizers.
3. Top-dress soluble nitrogenous fertilizers such as urea in two split doses.
4. For quick recovery, apply urea with irrigation water or as a foliar spray in the standing crop. Foliar sprays require to be repeated every 10–15 days.
5. Incorporate legume crops in rotation.

Further reading

Grundon, N.J. (1987) *Hungry Crops: A Guide to Nutrient Deficiencies in Field Crops*. Queensland Department of Primary Industries, Brisbane, Australia.

Novoa, R. and Loomis, R.S. (1981) Nitrogen and plant production. *Plant and Soil* 58, 177–204.

Snowball, K. and Robson, A.D. (1991) *Nutrient Deficiencies and Toxicities in Wheat: A Guide for Field Identification*. International Maize and Wheat Improvement Center (CIMMYT), Mexico City.

Stewart, B.A. and Porter, L.K. (1969) Nitrogen–sulphur relationships in wheat (*Triticum aestivum* L.), corn (*Zea mays*), and beans (*Phaseolus vulgaris*). *Agronomy Journal* 61, 267–271.

Plate 159. Phosphorus-deficient wheat plant. (Photo by Dr Prakash Kumar and Dr Manoj Kumar Sharma.)

Plate 160. Older leaves showing purple–red pigmentation. (Photo by Dr Prakash Kumar and Dr Manoj Kumar Sharma.)

Plate 161. Purple–red pigmentation starts from the tip of the leaf. (Photo by Dr Prakash Kumar and Dr Manoj Kumar Sharma.)

Plate 162. Purple–red striping on stem and chlorosis on leaf. (Photo by Dr Prakash Kumar and Dr Manoj Kumar Sharma.)

WHEAT (*Triticum aestivum* Linn.) PHOSPHORUS (P) DEFICIENCY

Symptoms

1. Wheat has strong hidden hunger for phosphorus deficiency. Deficiency symptoms are not as easily and clearly visible on wheat as they are on maize, sorghum and other grasses. Characteristic phosphorus-deficiency symptoms (purple pigmentation on older leaves) may be visible at the very initial stage of the crop (Plate 159). These symptoms disappear very soon with growth of the crop.
2. Deficient plants appear dull dark green, stunted, thin and spindly. The number of tillers, head size and grain yields are reduced severely.
3. Phosphorus deficiency delays crop maturity. Small heads and irregular maturity of the crop are the noticeable features of phosphorus deficiency.
4. Phosphorus is mobile in plants and under short supply conditions it is easily mobilized from older to younger leaves. The deficiency symptoms appear first and become more severe on older leaves.
5. Old leaves develop a dark purple colour on the leaf tips which proceeds in a broad front towards the base (Plates 160 and 161).
6. If deficiency persists, a dark yellow chlorosis develops on the affected leaves. Chlorosis starts from the tip of the leaf and advances towards the leaf base (Plate 162).
7. In severe deficiency conditions, a purple suffusion combines with yellow colour providing an orange–purple hue to the affected leaf.
8. Purple–red striping develops on the stem and leaf sheaths (Plate 162).

Developmental stages

Stage I: Deficient plants appear dull dark green (Plate 159).
Stage II: Purple or purple–red colour develops on the margins of old leaves, beginning from the tip of the leaf and proceeding towards the base (Plates 160 and 161).
Stage III: If the deficiency becomes severe, the affected leaves become orange–yellow to orange–purple. Purple–red striping develops on the stem and the leaf sheaths (Plate 162).
Stage IV: In acute deficiency conditions, the affected leaves die and form a thatch of dead brown leaves around the base.

Likely to occur in

1. Soils having low organic matter.
2. Alkaline and calcareous soils.
3. Soils exhausted by intensive cropping.
4. Acid soils and highly weathered soils.
5. Soils where the topsoil has been removed by erosion.
6. Acid soils having pH below 6.0.
7. Alkaline soils having pH between 7.5 and 8.5.

Integrated nutrient management

1. Get the soil analysed before sowing to measure the amount of 'available' phosphate in the soil.
2. Apply analysis-based recommended quantity of phosphorus as basal by using:
 a. organic manures;
 b. phosphate-solubilizing microbial cultures; or
 c. phosphatic fertilizers.
3. In deficient standing crops apply soluble phosphatic fertilizers such as ammonium phosphate with irrigation water.

Further reading

Grundon, N.J. (1987) *Hungry Crops: A Guide to Nutrient Deficiencies in Field Crops*. Queensland Department of Primary Industries, Brisbane, Australia.

Ozturk, L., Eker, S., Torun, B. and Cakmak, I. (2005) Variation in phosphorus efficiency among 73 bread and durum wheat genotypes grown in a phosphorus-deficient calcareous soil. *Plant and Soil* 269, 69–80.

Rodriguez, D., Pomar, M.C. and Goudriaan, J. (1998) Leaf primordial, leaf emergence and tillering in wheat (*Triticum aestivum* L.) grown under low-phosphorus conditions. *Plant and Soil* 202, 149–157.

Snowball, K. and Robson, A.D. (1991) *Nutrient Deficiencies and Toxicities in Wheat: A Guide for Field Identification*. International Maize and Wheat Improvement Center (CIMMYT), Mexico City.

Plate 163. Potassium-deficient wheat plants. (Photo by Dr Prakash Kumar.)

WHEAT (*Triticum aestivum* Linn.) POTASSIUM (K) DEFICIENCY

Plate 164. Leaf showing initial stage of marginal chlorosis. (Photo by Dr Prakash Kumar.)

Plate 165. Close-up of an older leaf showing advanced stage of deficiency. (Photo by Dr Prakash Kumar.)

Plate 166. Close-up of affected leaves with chlorosis and necrosis. (Photo by Dr Prakash Kumar.)

Symptoms

1. Strong hidden hunger for potassium deficiency is found in wheat. Visual symptoms hardly appear in mild deficiency conditions.
2. Because potassium is readily transferred from older to younger leaves, the deficiency symptoms appear first on old leaves. Younger, actively growing leaves draw potassium from old parts of the plant, therefore young leaves usually remain green and apparently healthy.
3. Deficient plants are stunted with spindly stems. Tillers produce small heads and few grains with reduced grain size, resulting in a drastic reduction in crop yield.
4. Symptoms begin as marginal chlorosis and necrosis on old leaves. The pale brown necrosis develops at the tip of the leaf and advances down the margins, leaving the midrib and surrounding tissues green (Plate 163).
5. In severe deficient conditions, affected leaves burn and die and deficiency symptoms move on to the upper leaves. Marginal chlorosis and necrosis of older leaves is the specific symptom of potassium deficiency.

Developmental stages

Stage I: In mild potassium deficiency, wheat leaves appear pale green, limp and wilted but recognizable symptoms appear only in severe deficiency conditions.
Stage II: When deficiency persists and becomes more severe, marginal chlorosis develops on older leaves starting from the tip of the leaves (Plates 164 and 165).
Stage III: In acute deficiency conditions the entire affected leaf dies, turning brown in colour (Plate 166).

Likely to occur in

1. Soils formed from parent material low in potassium.
2. Light-textured soils where potassium has been leached by heavy rainfall or excessive irrigation.
3. Soils low in organic matter.
4. Soils with wide Na:K, Mg:K or Ca:K ratio.
5. Acute sodic/saline/acid soil conditions.
6. Large bicarbonate concentration in irrigation water.
7. Highly acid soils having pH below 6.0.

Integrated nutrient management

1. Get the soil analysed before sowing to measure the amount of 'available' potassium in the soil.
2. Problematic acid/alkaline/saline soils should be reclaimed.
3. Add organic manures well before sowing.
4. Apply potassium nitrate, potassium sulphate or potassium chloride to the soil at or before sowing as per soil testing recommendations.
5. If potassium deficiency appears on a standing crop, apply soluble potassium salts such as potassium nitrate, potassium sulphate or potassium chloride with irrigation water. Foliar sprays of these salts are usually not recommended because a number of sprays are needed to fulfil crop requirements.

Further reading

Bennett, W.F. (1993) *Nutrient Deficiencies and Toxicities in Crop Plants*. American Phytopathological Society, St Paul, Minnesota.

Grundon, N.J. (1987) *Hungry Crops: A Guide to Nutrient Deficiencies in Field Crops*. Queensland Department of Primary Industries, Brisbane, Australia.

Snowball, K. and Robson, A.D. (1991) *Nutrient Deficiencies and Toxicities in Wheat: A Guide for Field Identification*. International Maize and Wheat Improvement Center (CIMMYT), Mexico City.

Woodend, J.J. and Glass, A.D.M. (1993) Genotype–environment interaction and correlation between vegetative and grain production measures of potassium use efficiency in wheat (*T. aestivum* L.) grown under potassium stress. *Plant and Soil* 151, 39–44.

Plate 167. Sulphur-deficient wheat plant. (Photo by Dr Prakash Kumar and Dr Manoj Kumar Sharma.)

Plate 168. Sulphur-deficient wheat field. (Photo by Dr Prakash Kumar.)

Plate 169. Deficient plant showing pale yellow younger leaves. (Photo by Dr Prakash Kumar and Dr Manoj Kumar Sharma.)

Plate 170. Sulphur-deficient leaf (bottom), compared with a healthy leaf (top). (Photo by Dr Prakash Kumar and Dr Manoj Kumar Sharma.)

WHEAT (*Triticum aestivum* Linn.) SULPHUR (S) DEFICIENCY

Symptoms

1. Sulphur deficiency in wheat, even if severe, does not produce any type of necrosis or burning of leaves. The change in leaf colour is the only indicative symptom of sulphur deficiency in wheat. Deficient plants appear stunted, thin and spindly with pale yellow foliage (Plate 168). The number of tillers and grains per plant are reduced. Sulphur deficiency also delays crop maturity.

2. At the initial stages of growth sulphur-deficiency symptoms resemble nitrogen deficiency symptoms, as in both cases the whole plant appears pale green. A close observation is required to differentiate between the two deficiencies. In contrast to nitrogen deficiency where older leaves are more pale yellow than younger leaves, sulphur deficiency produces chlorosis that is more pronounced on young leaves with comparatively darker old leaves (Plates 167 and 169).

3. Sulphur is less mobile in the plant than nitrogen, so under short supply conditions deficiency symptoms tend to appear first on younger leaves.

4. The young leaves become dull or bright yellow while the old leaves usually remain green (Plate 170).

5. The pattern of yellowing on the entire leaf appears uniform, covering both veins and interveinal tissues uniformly.

Developmental stages

Stage I: In mild deficiency and during the younger stage of the crop, all leaves on the whole plant become pale green although old leaves remain darker.

Stage II: If the deficiency persists and becomes more severe, the younger leaves turn pale yellow (Plates 167 and 169).

Likely to occur in

1. Soils having low organic matter.
2. Light-textured sandy soils that have been leached by heavy rainfall or excessive irrigation.
3. Soils exhausted by intensive cropping.
4. Soils derived from parent material that is inherently low in sulphur.
5. Acid soils having pH below 6.0.

Integrated nutrient management

1. Get the soil analysed before sowing to measure the amount of 'available' sulphate in the soil and to predict the amount of fertilizer required.

2. Apply analysis-based recommended quantity of sulphur by mixing into the soil, well before sowing, either:
- **a.** elemental sulphur (flowers of sulphur); or
- **b.** gypsum (calcium sulphate).

3. In deficient standing crops apply ammonium sulphate, magnesium sulphate or potassium sulphate with irrigation water.

4. Problematic acid soils should be reclaimed.

Further reading

Blake-Kalff, M.M.A., Hawkesford, M.J., Zhao, F.J. and McGrath, S.P. (2000) Diagnosis of sulfur deficiency in field grown oilseed rape (*Brassica napus* L.) and wheat (*Triticum aestivum* L.). *Plant and Soil* 225, 95–107.

Grundon, N.J. (1987) *Hungry Crops: A Guide to Nutrient Deficiencies in Field Crops*. Queensland Department of Primary Industries, Brisbane, Australia.

Inal, A., Gunes, A., Alpaslan, M., Adak, M.S., Taban, S. and Eraslan, F. (2003) Diagnosis of sulfur deficiency and effects of sulfur on yield and yield components of wheat grown in Central Anatolia, Turkey. *Journal of Plant Nutrition* 26, 1483–1498.

Snowball, K. and Robson, A.D. (1991) *Nutrient Deficiencies and Toxicities in Wheat: A Guide for Field Identification*. International Maize and Wheat Improvement Center (CIMMYT), Mexico City.

Plate 171. Iron-deficient wheat plant. (Photo by Dr Prakash Kumar.)

Plate 172. Fading of interveinal tissues (Photo by Dr Prakash Kumar.)

Plate 173. Leaf showing advanced stage of interveinal chlorosis. (Photo by Dr Prakash Kumar and Dr Manoj Kumar Sharma.)

Plate 174. Close-up of a bleached leaf. (Photo by Dr Prakash Kumar and Dr Manoj Kumar Sharma.)

WHEAT (*Triticum aestivum* Linn.) IRON (Fe) DEFICIENCY

Symptoms

1. Iron deficiency symptoms appear first on the younger leaves. The older leaves remain normal and apparently healthy (Plate 171).
2. In mild deficiency, the younger leaves of the plants develop temporary fading of interveinal tissues with prominent green veins (Plate 172).
3. If the deficiency persists and becomes more severe, a bright pale yellow chlorosis develops in interveinal tissues (tissues between the veins), leaving the veins green and prominent (Plate 171).
4. As the symptoms advance, prominent green veins also fade and become light green to pale yellow (Plate 173).
5. In acute deficiency conditions, the entire leaf bleaches to pale yellow or sometimes white (Plate 174).
6. Iron deficiency sharply reduces tiller production. Plants with a single main shoot without any tillers are common in severe deficiency conditions.

Developmental stages

Stage I: The younger leaves of the plant develop temporary fading of interveinal tissues with prominent green veins (Plate 172).
Stage II: Interveinal chlorosis develops on younger leaves. Interveinal chlorosis extends uniformly along the full length of the leaf (Plate 171).
Stage III: The prominent green veins also fade and become light green to pale yellow (Plate 173).
Stage IV: Affected leaves bleach to pale yellow or sometimes white (Plate 174).

Likely to occur in

1. Sandy soils having low total iron.
2. Calcareous soils where solubility of iron is very low.
3. Peat and muck soils where organic matter ties up iron and reduces its availability in soil solution.
4. Acid soils where, despite high availability of iron in soil solution, excessively high levels of soluble zinc, manganese, copper or nickel hamper the uptake of iron by the plant.
5. Waterlogged soils.
6. Alkaline soils having pH above 7.5.

Integrated nutrient management

1. Get the soil analysed before sowing to measure the amount of 'available' iron in the soil.
2. Problematic alkaline soils should be reclaimed.
3. As inorganic soluble salts of iron such as iron sulphates or chlorides quickly become insoluble in soil, a soil dressing of organic forms of iron such as iron chelates (10 kg/ha) is more effective. These organic iron forms should be used as per soil pH. If the soil is neutral (around pH 7.0), the most suitable chelates are FeHEDTA and FeDTPA. For acid soils FeEDTA is the most effective chelate, while on alkaline calcareous soils FeEDDHA is recommended.
4. If deficiency appears on standing crops apply organic iron chelates or inorganic soluble salts such as iron sulphates or chlorides (0.5 to 1.0% w/v solution) as a foliar spray. Apply two or three sprays at intervals of 10–15 days.

Further reading

Bennett, W.F. (1993) *Nutrient Deficiencies and Toxicities in Crop Plants*. American Phytopathological Society, St Paul, Minnesota.
Grundon, N.J. (1987) *Hungry Crops: A Guide to Nutrient Deficiencies in Field Crops*. Queensland Department of Primary Industries, Brisbane, Australia.
Snowball, K. and Robson, A.D. (1991) *Nutrient Deficiencies and Toxicities in Wheat: A Guide for Field Identification*. International Maize and Wheat Improvement Center (CIMMYT), Mexico City.

Plate 175. Zinc-deficient wheat plant. (Photo by Dr Prakash Kumar and Dr Manoj Kumar Sharma.)

Plate 176. Pale grey spot on the middle of the leaf. (Photo by Dr Prakash Kumar and Dr Manoj Kumar Sharma.)

Plate 177. Pale grey to brown necrotic region. (Photo by Dr Prakash Kumar.)

Plate 178. Collapse of an affected leaf in the middle region. (Photo by Dr Prakash Kumar.)

WHEAT (*Triticum aestivum* Linn.) ZINC (Zn) DEFICIENCY

Symptoms

1. Zinc deficiency causes reduced tillering, ceases plant growth and gives a grassy, tufted appearance resulting from the few heads with few or no grains.
2. Durum wheat is more sensitive to zinc deficiency. Deficiency symptoms of zinc are easily visible on durum varieties, so durum wheat works as an indicator of zinc deficiency in wheat-growing areas. Other wheat types show comparatively strong hidden hunger to zinc deficiency and the recognizable symptoms appear only in severe deficiency conditions.
3. Zinc is partly mobile in wheat; therefore, the deficiency symptoms first appear on the middle leaves. The youngest leaves at the top and the oldest ones at the bottom remain initially unaffected. In some wheat varieties symptoms begin from the older leaves and soon cover middle leaves. Younger leaves usually remain normal.
4. Initial deficiency symptoms develop on the lower half of the leaf, leaving the tip and base green (Plate 175).
5. A muddy grey–green spot appears on the middle of the leaf that soon turns to pale grey or brown necrotic patches.
6. As the deficiency becomes more severe, necrotic spots extend outwards towards the tip and base of the leaf. The necrotic patches become larger and larger.
7. In acute deficiency conditions the affected leaves collapse in the middle region.

Developmental stages

Stage I: A muddy grey–green spot appears on the middle of the leaf that soon turns to pale grey to brown necrotic patches (Plate 176).
Stage II: Necrotic patches extend outwards and become larger and larger (Plate 177).
Stage III: Eventually the affected leaf collapses in the middle region (Plate 178).

Likely to occur in

1. Leached sandy soils where total zinc is low.
2. Just-levelled soils where the subsoil is exposed for cultivation. Available zinc in surface soil is often double that of the subsoil.
3. Soil having heavy and excessive application of phosphate fertilizers, which may hamper use of zinc by the crop.
4. Acid soils having pH below 5.0.
5. Alkaline soils having pH above 7.5.

Integrated nutrient management

1. Get the soil analysed before sowing to estimate the amount of 'available' zinc in the soil.
2. Problematic alkaline or acid soils should be reclaimed.
3. Add organic manures well before sowing.
4. Apply 25–30 kg of zinc sulphate or 10 kg of zinc chelate per hectare every 2 years in zinc-deficient soils.
5. Zinc fertilizers should not be mixed with phosphate fertilizers.
6. In fields with known zinc deficiency, if deficiency appears in the standing crop, apply a foliar application of 0.5% w/v solution of a soluble zinc salt, Zinc sulphate with 0.25% w/v solution of unsoaked lime.

Further reading

Cakmak, I., Ekiz, H., Yilmaz, A., Torun, B., Koreli, N., Gultekin, I., Alkan, A. and Eker, S. (1997) Differential response of rye, triticale, bread and durum wheats to zinc deficiency in calcareous soils. *Plant and Soil* 188, 1–10.
Grundon, N.J. (1987) *Hungry Crops: A Guide to Nutrient Deficiencies in Field Crops*. Queensland Department of Primary Industries, Brisbane, Australia.
Singh, J.P., Karamanos, R.E. and Stewart, J.W.B. (1986) Phosphorus-induced zinc deficiency in wheat on residual phosphorus plots. *Agronomy Journal* 78, 668–675.
Snowball, K. and Robson, A.D. (1991) *Nutrient Deficiencies and Toxicities in Wheat: A Guide for Field Identification*. International Maize and Wheat Improvement Center (CIMMYT), Mexico City.

Plate 179. Copper-deficient wheat plant showing 'wither tip'. (Photo by Dr Prakash Kumar and Dr Manoj Kumar Sharma.)

WHEAT (*Triticum aestivum* Linn.) COPPER (Cu) DEFICIENCY

Plate 180. Crop showing wilting, rolling and death of tips. (Photo by Dr Prakash Kumar and Dr Manoj Kumar Sharma.)

Plate 181. Close-up of a deficient leaf. (Photo by Dr Prakash Kumar and Dr Manoj Kumar Sharma.)

Plate 182. Ear head forming a 'rat tail'-like appearance. (Photo by Dr Prakash Kumar.)

Symptoms

1. Wheat is very sensitive to copper deficiency. Deficiency symptoms appear even in mild deficiency conditions.
2. Copper deficiency affects pollen formation. As a result, copper-deficient plants produce heads with few grains or without grains. This way even mild copper deficiency may cause severe damage to crop yields.
3. The first visual symptoms usually appear during the initial tillering stage. Deficient plants appear limp and wilted even in adequate soil moisture conditions.
4. Copper is immobile in plants and under short supply conditions it is not easily mobilized from older to younger leaves. The deficiency symptoms appear first and become more severe on younger leaves while older leaves usually remain unaffected, dark green and healthy.
5. The tips of young leaves die, turn white to pale brown and then twist into a tight tube. 'Wither tip' is the characteristic or specific symptom of copper deficiency in wheat. Except the tip, the remainder of the affected leaf remains green and healthy (Plates 180 and 181).
6. Deficient plants produce deformed and bleached heads. The deformed heads form a 'rat tail'-like appearance. The heads are broad at the base with full grains, narrow at the middle with shrivelled grains and narrower at the tip with no grains (Plates 182). The tips become necrotic in the later stage.
7. Frost damage may also produce a similar type of necrotic head symptoms, but the presence of wither tip confirms the copper deficiency.

Developmental stages

Stage I: In early stage of deficiency, the plant appears limp or wilted.
Stage II: The tips of young leaves die, turn white to pale brown and then twist into a tight tube. The symptom is described as 'wither tip' and is the characteristic symptom of copper deficiency in wheat (Plate 179).
Stage III: Deficient plants produce deformed and bleached heads that appear like a 'rat tail'.

Likely to occur in

1. Calcareous sandy soils low in copper.
2. Leached acid soils low in copper.
3. Peat and muck soils where organic matter ties up copper and reduces its availability in solution.
4. Soils derived from parent material that is inherently low in copper.
5. Acid soils having pH below 5.0.
6. Alkaline soils having pH above 8.5.

Integrated nutrient management

1. Get the soil analysed before sowing to measure the amount of 'available' copper in the soil.
2. Problematic alkaline soils should be reclaimed.
3. Problematic acid soils should be reclaimed.
4. Add organic manures well before sowing.
5. Apply soluble inorganic copper salts such as copper sulphate or organic copper chelates as basal.
6. In deficient standing crops, apply copper sulphate (0.2 to 0.5% w/v solution) as a foliar spray. Repeated sprays may be required if symptoms reappear.

Further reading

Grundon, N.J. (1980) Effectiveness of soil dressing and foliar sprays of copper sulphate in correcting copper deficiency of wheat (*Triticum aestivum*) in Queensland. *Australian Journal of Experimental Agriculture and Animal Husbandry* 20, 717–723.

Grundon, N.J. (1987) *Hungry Crops: A Guide to Nutrient Deficiencies in Field Crops*. Queensland Department of Primary Industries, Brisbane, Australia.

Snowball, K. and Robson, A.D. (1991) *Nutrient Deficiencies and Toxicities in Wheat: A Guide for Field Identification*. International Maize and Wheat Improvement Center (CIMMYT), Mexico City.

Wlese, M.V. (1993) Wheat and other small grains. In: Bennett, W.F. (ed.) *Nutrient Deficiencies and Toxicities in Crop Plants*. American Phytopathological Society, St. Paul, Minnesota, pp. 27–33.

Plate 183. Nitrogen-deficient barley plant. (Photo by Dr Prakash Kumar and Dr Manoj Kumar Sharma.)

BARLEY (*Hordeum vulgare* (L.) emend. Bowden) NITROGEN (N) DEFICIENCY

Plate 184. Close-up of chlorotic leaves. (Photo by Dr Prakash Kumar and Dr Manoj Kumar Sharma.)

Plate 185. Plants showing a thatch of dry leaves around the base. (Photo by Dr Prakash Kumar and Dr Manoj Kumar Sharma.)

Plate 186. Close-up of a dead, whitish brown leaf. (Photo by Dr Prakash Kumar and Dr Manoj Kumar Sharma.)

Symptoms

1. Being a fast-growing grass species, barley is very sensitive to nitrogen deficiency. Deficiency symptoms appear even in mild deficiency conditions.
2. Nitrogen-deficient plants are stunted with thin stems and pale green foliage. Affected plants lack vigour and produce small heads with few grains. In acute deficiency conditions, many young tillers fail to develop heads and die before maturity.
3. Nitrogen is mobile in plants and under short supply conditions it is easily mobilized from older to younger leaves. The deficiency symptoms appear first and become more severe on older leaves.
4. In mild deficiency conditions or during the young stage of the crop, the entire plant becomes pale green to yellow.
5. If deficiency persists and becomes more severe, a pale yellow chlorosis develops at the tip of older leaves and advances in a broad front towards the leaf base.
6. Pale yellow chlorotic leaves turn whitish to pale brown and die (Plate 186). The dead leaves form a thatch of dry leaves around the base of the plant.

Developmental stages

Stage I: During the young stage of the crop or in mild deficiency conditions, the entire plant appears uniformly pale green in colour.
Stage II: If the deficiency persists and becomes more severe, a pale yellow chlorosis begins at the tip of older leaves and proceeds in a broad front towards the base (Plate 183).
Stage III: As the symptoms advance, the whole leaf becomes chlorotic, changing from yellow to almost white (Plate 184).
Stage IV: In the later stage, the chlorotic leaves wither and die. The dead leaves hang down and make a thatch of dry leaves near the base of the plant (Plate 185).

Likely to occur in

1. Soils having low organic matter.
2. Light-textured sandy soils that have been leached by heavy rainfall or excessive irrigation.
3. Soils exhausted by intensive cropping.
4. Waterlogged conditions.
5. Acid soils having pH below 6.0.
6. Alkaline soils having pH above 8.0.

Integrated nutrient management

1. Get the soil analysed before sowing to measure the amount of 'available' nitrogen in the soil.
2. Apply analysis-based recommended quantity of nitrogen as basal by using:
- **a.** organic manures;
- **b.** bio-fertilizers;
- **c.** nitrogenous fertilizers.

3. Top-dress soluble nitrogenous fertilizers such as urea in two split doses.
4. For quick recovery, apply urea with irrigation water or as a foliar spray in the standing crop. Foliar sprays require to be repeated every 10–15 days.
5. Incorporate legume crops in rotation.

Further reading

Drew, M.C. and Sisworo, E.J. (1977) Early effects of flooding on nitrogen deficiency and leaf chlorosis in barley. *New Phytologist* 79, 567–571.
Grundon, N.J. (1987) *Hungry Crops: A Guide to Nutrient Deficiencies in Field Crops*. Queensland Department of Primary Industries, Brisbane, Australia.
Lee, R.B. and Rudge, K.A. (1986) Effects of nitrogen deficiency on the absorption of nitrate and ammonium by barley plants. *Annals of Botany* 57, 471–486.
Novoa, R. and Loomis, R.S. (1981) Nitrogen and plant production. *Plant and Soil* 58, 177–204.

Plate 187. Phosphorus-deficient barley plant. (Photo by Dr Prakash Kumar and Dr Manoj Kumar Sharma.)

Plate 188. Close-up of purple–red pigmentation on a leaf. (Photo by Dr Prakash Kumar and Dr Manoj Kumar Sharma.)

Plate 189. Orange–purple leaf and purple stem. (Photo by Dr Prakash Kumar and Dr Manoj Kumar Sharma.)

Plate 190. Dark brown older leaf. (Photo by Dr Prakash Kumar and Dr Manoj Kumar Sharma.)

BARLEY (*Hordeum vulgare* (L.) emend. Bowden) PHOSPHORUS (P) DEFICIENCY

Symptoms

1. Barley is sensitive to phosphorus deficiency. Opposite to wheat, recognizable phosphorus deficiency symptoms are easily visible on barley. Phosphorus deficiency in the soil adversely affects the absorption of nitrate by the barley plant. Phosphorus is also important for development of the root system.
2. Deficient plants appear stunted with short, stout stems and dark green foliage (Plate 187). Purple colour often develops on older leaves (Plate 188).
3. Phosphorus deficiency delays crop maturity. The number of tillers, head size and grain yields are reduced severely.
4. Phosphorus is mobile in plants and under short supply conditions it is easily mobilized from older to younger leaves. The deficiency symptoms appear first and become more severe on the older leaves, while the young leaves usually remain unaffected.
5. Old leaves develop a dark purple colour on the leaf tips which proceeds in a broad front towards the base.
6. The stem and leaf sheaths of the lower leaves also become purple–red in colour.
7. In acute deficiency conditions the plant does not produce heads.

Developmental stages

Stage I: Plants develop dark green foliage with stunted growth.
Stage II: When deficiency persists, the old leaves develop dark purple margins that eventually cover the entire leaf.
Stage III: If the deficiency becomes severe, the affected leaves become orange–yellow to orange–purple (Plate 189).
Stage IV: As the symptoms advance, the entire leaf turns dark brown and dies (Plate 190).

Likely to occur in

1. Soils having low organic matter.
2. Alkaline and calcareous soils.
3. Soils exhausted by intensive cropping.
4. Acid soils and highly weathered soils.
5. Soils where the topsoil has been removed by erosion.
6. Acid soils having pH below 6.0.
7. Alkaline soils having pH between 7.5 and 8.5.

Integrated nutrient management

1. Get the soil analysed before sowing to measure the amount of 'available' phosphate in the soil.
2. Apply analysis-based recommended quantity of phosphorus as basal by using:
- **a.** organic manures;
- **b.** phosphate-solubilizing microbial cultures;
- **c.** phosphatic fertilizers.

3. In deficient standing crops apply soluble phosphatic fertilizers such as ammonium phosphate with irrigation water.

Further reading

Clarkson, D.T., Sanderson, J. and Scattergood, C.B. (1978) Influence of phosphate-stress on phosphate absorption and translocation by various parts of the root system of *Hordeum vulgare* L. (barley). *Planta* 139, 47–53.

Grundon, N.J. (1987) *Hungry Crops: A Guide to Nutrient Deficiencies in Field Crops*. Queensland Department of Primary Industries, Brisbane, Australia.

Schjorring, J.K. (1986) Nitrate and ammonium absorption by plants growing at a sufficient or insufficient level of phosphorus in nutrient solution. *Plant and Soil* 91, 313–318.

Wang, C. and Tillberg, J.E. (1997) Effects of short-term phosphorus deficiency on carbohydrate storage in sink and source leaves of barley (*Hordeum vulgare*). *New Phytologist* 136, 131–135.

Plate 191. Potassium-deficient barley plant. (Photo by Dr Prakash Kumar and Dr Manoj Kumar Sharma.)

Plate 192. Leaf showing initial stage of marginal chlorosis. (Photo by Dr Prakash Kumar and Dr Manoj Kumar Sharma.)

Plate 193. Leaf showing marginal chlorosis followed by necrosis. (Photo by Dr Prakash Kumar.)

Plate 194. Burning of leaf during the advanced stage. (Photo by Dr Prakash Kumar and Dr Manoj Kumar Sharma.)

BARLEY (*Hordeum vulgare* (L.) emend. Bowden) POTASSIUM (K) DEFICIENCY

Symptoms

1. Barley is comparatively more sensitive to potassium deficiency than wheat. Recognizable deficiency symptoms may appear even during the initial stages of crop growth.
2. Deficient plants are stunted with short stems. In severe deficiency conditions many young tillers die before producing heads, while mature tillers produce small heads with few grains.
3. Because potassium is readily transferred from older to younger leaves, the deficiency symptoms appear first on old leaves. Younger actively growing leaves draw potassium from old parts of the plant, therefore young leaves usually remain green and apparently healthy.
4. Symptoms begin as marginal chlorosis and necrosis on older leaves. A pale yellow chlorosis followed by necrosis begins on the tip of the leaf and advances along the margins towards the base, leaving the midrib and surrounding tissues green.
5. Marginal chlorosis and necrosis of older leaves is the specific symptom of potassium deficiency.
6. In severe deficiency conditions, affected leaves burn and die (Plate 194) and deficiency symptoms move on to upper leaves.

Developmental stages

Stage I: In mild potassium deficiency, plants appear limp, spindly and wilted with pale yellow–green foliage.
Stage II: When deficiency persists and becomes more severe, marginal chlorosis develops on older leaves starting from the tip of the leaves (Plates 192 and 193).
Stage III: In acute deficiency conditions the entire affected lower leaves die, turning dark brown in colour (Plate 191).

Likely to occur in

1. Soils formed from parent material low in potassium.
2. Light-textured soils where potassium has been leached by heavy rainfall or excessive irrigation.
3. Soils low in organic matter.
4. Soils with wide Na:K, Mg:K or Ca:K ratio.
5. Acute sodic/saline/acid soil conditions.
6. Large bicarbonate concentration in irrigation water.
7. Highly acid soils having pH below 6.0.

Integrated nutrient management

1. Get the soil analysed before sowing to measure the amount of 'available' potassium in the soil.
2. Problematic acid/alkaline/saline soils should be reclaimed.
3. Add organic manures well before sowing.
4. Apply potassium nitrate, potassium sulphate or potassium chloride to the soil at or before sowing as per soil testing recommendations.
5. If potassium deficiency appears on the standing crop, apply soluble potassium salts such as potassium nitrate, potassium sulphate or potassium chloride with irrigation water. Foliar sprays of these salts are usually not recommended because a number of sprays are needed to fulfil crop requirements.

Further reading

Bennett, W.F. (1993) *Nutrient Deficiencies and Toxicities in Crop Plants*. American Phytopathological Society, St Paul, Minnesota.

Grundon, N.J. (1987) *Hungry Crops: A Guide to Nutrient Deficiencies in Field Crops*. Queensland Department of Primary Industries, Brisbane, Australia.

Leigh, R.A. and Johnston, A.E. (1983) The effects of fertilizers and drought on the concentrations of potassium in the dry matter and tissue water of field grown spring barley. *Journal of Agricultural Sciences* 101, 741–748.

Miah, Y.M., Wang, M.K., Hayashi, H. and Chino, M. (1999) Occurrence of potassium deficiency in barley (*Hordeum vulgare* L.) grown on volcanic ash with long term application of sewage sludge. *Food Science and Agricultural Chemistry* 1, 180–185.

Plate 195. Sulphur-deficient barley plant. (Photo by Dr Prakash Kumar and Dr Manoj Kumar Sharma.)

Plate 196. Close-up of an affected leaf. (Photo by Dr Prakash Kumar and Dr Manoj Kumar Sharma.)

Plate 197. Leaf showing uniform yellowing. (Photo by Dr Prakash Kumar and Dr Manoj Kumar Sharma.)

Plate 198. Sulphur-deficient leaf (left), compared with a healthy leaf (right). (Photo by Dr Prakash Kumar and Dr Manoj Kumar Sharma.)

BARLEY (*Hordeum vulgare* (L.) emend. Bowden) SULPHUR (S) DEFICIENCY

Symptoms

1. Similar to wheat, sulphur-deficient barley does not produce any type of necrosis or burning of leaves. The change in leaf colour is the only indicative symptom of sulphur deficiency in barley.
2. Deficient plants appear stunted, thin and spindly with pale yellow foliage. The number of tillers and grains per plant are reduced. Sulphur deficiency also delays crop maturity.
3. Sulphur is immobile in plants and under short supply conditions it is not easily mobilized from older to younger leaves. The deficiency symptoms appear first and become more severe on younger leaves (Plate 195). The youngest leaves are most severely affected.
4. The young leaves become dull or bright yellow while the old leaves usually remain green.
5. The pattern of yellowing on the entire leaf appears uniform, covering both veins and interveinal tissues uniformly.
6. In acute deficiency conditions some barley varieties also develop red to purple–red colour on the margins and sheaths of older leaves.

Developmental stages

Stage I: In mild deficiency or during the initial crop stage, all leaves on the plant become pale green although older leaves remain comparatively darker.
Stage II: If the deficiency becomes severe, the younger leaves turn pale yellow giving the crop a definite yellow appearance (Plates 196 and 197).
Stage III: In the later stage, the pale yellow youngest leaves turn whitish yellow but without necrosis (Plate 198).
Stage IV: In acute deficiency conditions some barley varieties also develop red to purple–red colour on the margins and sheaths of older leaves.

Likely to occur in

1. Soils having low organic matter.
2. Light-textured sandy soils that have been leached by heavy rainfall or excessive irrigation.
3. Soils exhausted by intensive cropping.
4. Soils derived from parent material that is inherently low in sulphur.
5. Acid soils having pH below 6.0.

Integrated nutrient management

1. Get the soil analysed before sowing to measure the amount of 'available' sulphate in the soil and to predict the amount of fertilizer required.
2. Apply analysis-based recommended quantity of sulphur by mixing into the soil, well before sowing, either:
- **a.** elemental sulphur (flowers of sulphur); or
- **b.** gypsum (calcium sulphate).

3. In deficient standing crops apply ammonium sulphate, magnesium sulphate or potassium sulphate with irrigation water.
4. Problematic acid soils should be reclaimed.

Further reading

Bennett, W.F. (1993) *Nutrient Deficiencies and Toxicities in Crop Plants*. American Phytopathological Society, St Paul, Minnesota.

Eriksen, J., Nielsen, M., Mortensen, J.V. and Schjorring, J.K. (2001) Redistribution of sulphur during generative growth of barley plants with different sulphur and nitrogen status. *Plant and Soil* 230, 239–246.

Grundon, N.J. (1987) *Hungry Crops: A Guide to Nutrient Deficiencies in Field Crops*. Queensland Department of Primary Industries, Brisbane, Australia.

Masoni, A., Ercoli, L. and Mariotti, M. (1996) Spectral properties of leaves deficient in iron, sulfur, magnesium and manganese. *Agronomy Journal* 88, 937–943.

Plate 199. Iron-deficient barley plant. (Photo by Dr Prakash Kumar and Dr Manoj Kumar Sharma.)

Plate 200. Temporary fading of interveinal tissues. (Photo by Dr Prakash Kumar and Dr Manoj Kumar Sharma.)

Plate 201. Close-up of a leaf showing interveinal chlorosis. (Photo by Dr Prakash Kumar and Dr Manoj Kumar Sharma.)

Plate 202. Leaves showing various stages. (Photo by Dr Prakash Kumar and Dr Manoj Kumar Sharma.)

BARLEY (*Hordeum vulgare* (L.) emend. Bowden) IRON (Fe) DEFICIENCY

Symptoms

1. Barley is highly sensitive to iron deficiency. Deficiency symptoms appear even in mild deficiency conditions.
2. In mild deficiency, the younger leaves of the plants develop temporary fading of interveinal tissues with prominent green veins (Plate 200).
3. If the deficiency persists and becomes more severe, a bright pale yellow chlorosis develops in interveinal tissues, leaving the veins green and prominent. Interveinal chlorosis extends uniformly along the full length of the leaf (Plate 199).
4. In acute deficiency conditions, the new growth may appear completely devoid of chlorophyll and turn overall dark or pale yellow.

Developmental stages

Stage I: In mild deficiency, younger leaves develop temporary fading of interveinal tissues with prominent green veins (Plate 200).
Stage II: When deficiency persists, interveinal tissues of affected leaves turn bright pale yellow with prominent green veins (Plates 199 and 201).
Stage III: In acute deficiency conditions the entire leaf bleaches to bright yellow or almost white (Plate 202).

Likely to occur in

1. Sandy soils having low total iron.
2. Calcareous soils where solubility of iron is very low.
3. Peat and muck soils where organic matter ties up iron and reduces its availability in soil solution.
4. Acid soils where, despite high availability of iron in soil solution, excessively high levels of soluble zinc, manganese, copper or nickel hamper the uptake of iron by the plant.
5. Waterlogged soils.
6. Alkaline soils having pH above 7.5.

Integrated nutrient management

1. Get the soil analysed before sowing to measure the amount of 'available' iron in the soil.
2. Problematic alkaline soils should be reclaimed.
3. As inorganic soluble salts of iron such as iron sulphates or chlorides quickly become insoluble in soil, a soil dressing of organic forms of iron such as iron chelates (10 kg/ha) is more effective. These organic iron forms should be used as per soil pH. If the soil is neutral (around pH 7.0), the most suitable chelates are FeHEDTA and FeDTPA. For acid soils FeEDTA is the most effective chelate, while on alkaline calcareous soils FeEDDHA is recommended.
4. If deficiency appears on standing crops apply organic iron chelates or inorganic soluble salts such as iron sulphates or chlorides (0.5 to 1.0% w/v solution) as a foliar spray. Apply two or three sprays at intervals of 10–15 days.

Further reading

Grundon, N.J. (1987) *Hungry Crops: A Guide to Nutrient Deficiencies in Field Crops*. Queensland Department of Primary Industries, Brisbane, Australia.

Masoni, A., Ercoli, L. and Mariotti, M. (1996) Spectral properties of leaves deficient in iron, sulfur, magnesium and manganese. *Agronomy Journal* 88, 937–943.

Yousfi, S., Wissal, M., Mahmoudi, H., Abdelly, C. and Gharsalli, M. (2007) Effect of salt on physiological responses of barley to iron deficiency. *Plant Physiology and Biochemistry* 45, 309–314.

Plate 203. Zinc-deficient barley plant. (Photo by Dr Prakash Kumar and Dr Manoj Kumar Sharma.)

Plate 204. Pale yellow linear chlorotic areas on a leaf. (Photo by Dr Prakash Kumar and Dr Manoj Kumar Sharma.)

Plate 205. Grey–white or brown necrotic lesions. (Photo by Dr Prakash Kumar and Dr Manoj Kumar Sharma.)

Plate 206. Affected leaf collapses in the middle region. (Photo by Dr Prakash Kumar and Dr Manoj Kumar Sharma.)

BARLEY (*Hordeum vulgare* (L.) emend. Bowden) ZINC (Zn) DEFICIENCY

Symptoms

1. Barley is very sensitive to zinc deficiency. Barley crops may suffer with severe losses in zinc-deficient conditions. Even in mild deficiency, only a few tillers produce heads but in severe deficiency conditions barley plants hardly develop any head and this way the crop suffers severe yield losses.
2. Zinc is partly mobile in barley; therefore, the deficiency symptoms first appear on the middle leaves. The youngest leaves at the top and the oldest ones at the bottom remain initially unaffected. In some barley varieties symptoms begin from the older leaves and soon cover middle leaves. Younger leaves usually remain normal.
3. Initial deficiency symptoms appear in the mid-section of the leaf, leaving the base and tip green (Plates 203 and 204).
4. Pale yellow linear chlorotic areas develop in the mid-section of the leaf that rapidly convert to grey–white or brown necrotic lesions (Plate 204).
5. In acute deficiency conditions the affected leaves collapse in the middle region (Plate 206).
6. Zinc deficiency prevents internode elongation, resulting in a stunted and grass-tufted appearance of the plants.

Developmental stages

Stage I: Deficiency symptoms first develop on middle-aged or older leaves. A pale yellow linear chlorotic area appears in the mid-section of the leaf (Plate 203).
Stage II: Chlorotic areas quickly convert to grey–white or brown necrotic lesions (Plates 204 and 205).
Stage III: In acute deficiency conditions the affected leaves collapse in the middle region (Plate 206).

Likely to occur in

1. Leached sandy soils where total zinc is low.
2. Just-levelled soils where the subsoil is exposed for cultivation. Available zinc in surface soil is often double that of the subsoil.
3. Soil having heavy and excessive application of phosphate fertilizers, which may hamper use of zinc by the crop.
4. Acid soils having pH below 5.0.
5. Alkaline soils having pH above 7.5.

Integrated nutrient management

1. Get the soil analysed before sowing to estimate the amount of 'available' zinc in the soil.
2. Problematic alkaline or acid soils should be reclaimed.
3. Add organic manures well before sowing.
4. Apply 25–30 kg of zinc sulphate or 10 kg of zinc chelate per hectare every 2 years in zinc-deficient soils.
5. Zinc fertilizers should not be mixed with phosphate fertilizers.
6. In fields with known zinc deficiency, if deficiency appears in the standing crop, apply a foliar application of 0.5% w/v solution of a soluble zinc salt, Zinc sulphate with 0.25% w/v solution of unsoaked lime.

Further reading

Genc, Y., McDonald, G.K. and Graham, R.D. (2000) Effect of seed zinc content on early growth of barley (*Hordeum vulgare* L.) under low and adequate soil zinc supply. *Australian Journal of Agricultural Research* 51, 37–46.

Genc, Y., McDonald, G.K. and Graham, R.D. (2004) Differential expressions of zinc deficiency during the growing season of barley. *Plant and Soil* 263, 273–282.

Grundon, N.J. (1987) *Hungry Crops: A Guide to Nutrient Deficiencies in Field Crops*. Queensland Department of Primary Industries, Brisbane, Australia.

Gupta, U.C. (1989) Effect of zinc fertilization on plant zinc concentration of forages and cereals. *Canadian Journal of Soil Science* 69, 473–479.

PART II

Nutrient Deficiencies in Pulse Crops

Plate 207. Nitrogen-deficient pigeon pea plant. (Photo by Dr Prakash Kumar.)

Plate 208. Entire plant appearing light green. (Photo by Dr Prakash Kumar.)

Plate 209. Interveinal chlorosis on an older leaf. (Photo by Dr Prakash Kumar.)

Plate 210. Older leaves pale yellow. (Photo by Dr Prakash Kumar.)

PIGEON PEA (*Cajanus cajan* (L.) Millsp.) NITROGEN (N) DEFICIENCY

Symptoms

1. Nitrogen deficiency in pigeon pea is usually found during the initial stages of crop growth when root symbiotic nitrogen fixation nodules are yet to develop. The deficiency may also occur during later stages of crop growth when the symbiotic nitrogen-supplying mechanism is disturbed for some reason such as nodule infestation, nodule pathogenic disease or physiological causes.
2. The deficiency symptoms appear first and more severely on the old leaves. The younger leaves usually remain green and apparently healthy (Plate 207).
3. In mild deficiency, the entire plant appears uniformly light green (Plate 208).
4. If deficiency persists and become more severe, the older leaves show chlorosis. Interveinal chlorosis appears on the oldest leaves in the beginning of the deficiency symptom (Plate 209), which soon converts to a uniform pale green, greenish yellow or pale yellow colour. The midrib remains green and turns yellow at last. Interveinal chlorosis stage is mostly missing in severe deficiency conditions.
5. Affected older leaves soon abscise (shed).

Developmental stages

Stage I: The entire plant appears uniformly light green in colour (Plate 208).
Stage II: The older leaves show interveinal chlorosis (Plate 209). This stage is mostly missing in severe deficiency conditions.
Stage III: The affected older leaves become uniform pale green, greenish yellow or pale yellow. The midrib remains green and turns yellow at last (Plate 210).
Stage IV: In the most advanced stage affected leaves abscise.

Likely to occur in

1. Soils having low organic matter.
2. Light-textured sandy soils that have been leached by heavy rainfall or excessive irrigation.
3. Acid soils having pH below 6.0.
4. Alkaline soils having pH above 8.0.
5. Crops with poor *Rhizobium* nodulation or with damaged *Rhizobium* nodules.

Integrated nutrient management

1. Get the soil analysed before sowing to measure the amount of 'available' nitrogen in the soil.
2. Inoculate the crop with an appropriate strain of *Rhizobium* culture through seed treatment.
3. The crop needs nitrogen during the initial stage of growth, when symbiotic nitrogen fixation by the plant is yet to start. Thus, a basal starter application of nitrogen at 20–25 kg/ha is important in nitrogen-deficient soils.
4. Nitrogen deficiency in existing crops can be managed by applying urea with irrigation water or as a foliar spray.

Further reading

Edwards, D.G. (1981) Development of research on pigeonpea nutrition. In: *Proceedings of the International Workshop on Pigeonpea, ICRISAT Center, India, 15–19 December 1980, Vol. 1*. International Crops Research Institute for the Semi-Arid Tropics (ICRISAT), Patancheru, India, pp. 205–211.

Nicholas, R. (1964) Studies on the major element deficiencies of the pigeonpea (*Cajanus cajan* L.) in sand culture – 1. Foliar symptoms of the major-element deficiencies. *Plant and Soil* 21, 377–387.

Plate 211. Phosphorus-deficient pigeon pea plants. (Photo by Dr Prakash Kumar.)

Plate 212. Branch with bluish dark green leaves. (Photo by Dr Prakash Kumar.)

Plate 213. Comparison of a darker leaf (left) with a normal leaf (right). (Photo by Dr Prakash Kumar.)

Plate 214. Orange–yellow older leaf. (Photo by Dr Prakash Kumar.)

PIGEON PEA (*Cajanus cajan* (L.) Millsp.) PHOSPHORUS (P) DEFICIENCY

Symptoms

1. Pigeon pea has very strong hidden hunger to phosphorus deficiency. Recognizable deficiency symptoms usually appear in severe deficiency conditions.
2. Deficient plants appear stunted with dark green foliage. Phosphorus deficiency badly affects elongation of internodes and growth of roots.
3. Phosphorus deficiency delays flowering and maturity. Deficient plants produce fewer flowers and bear only a few pods, resulting in a drastic reduction of crop yield.
4. Phosphorus is mobile in plants and under short supply conditions it is easily mobilized from older to younger leaves. The deficiency symptoms appear first and become more severe on older leaves.
5. The deficient plant develops a dark green to bluish dark green colour of the foliage. The older leaves become darker than the younger. The change in leaf colour is the only recognizable symptom of phosphorus deficiency in pigeon pea.
6. In the most advanced stage, affected older leaves turn orange–yellow in colour and shed.

Developmental stages

Stage I: The deficient plant produces dark foliage and short internodes (Plate 211).
Stage II: If deficiency persists and becomes more severe, the plant develops a dark green to bluish dark green colour of the foliage. The older leaves become darker than the younger (Plates 212 and 213).
Stage III: In acute deficiency conditions, the older leaves turn orange–yellow in colour and shed (Plate 214).

Likely to occur in

1. Soils having low organic matter.
2. Alkaline and calcareous soils.
3. Soils exhausted by intensive cropping.
4. Acid soils and highly weathered soils.
5. Soils where the topsoil has been removed by erosion.
6. Acid soils having pH below 6.0.
7. Alkaline soils having pH between 7.5 and 8.5.

Integrated nutrient management

1. Get the soil analysed before sowing to measure the amount of 'available' phosphate in the soil.
2. Apply analysis-based recommended quantity of phosphorus as basal by using:
- **a.** organic manures;
- **b.** phosphate-solubilizing microbial cultures; or
- **c.** phosphatic fertilizers.

3. In deficient standing crops apply soluble phosphatic fertilizers such as ammonium phosphate with irrigation water.

Further reading

Adu-Gyamfi, J.J., Fujita, K. and Ogata, S. (1990) Phosphorus fractions in relation to growth in pigeonpea (*Cajanus cajan* (L.) Millsp.) at various levels of P supply. *Soil Science and Plant Nutrition* 36, 531–543.

Ascencio, J. (1996) Growth strategies and utilization of phosphorus in *Cajanus cajan* L. Millsp. and *Desmodium tortuosum* (Sw.) DC under phosphorus deficiency. *Communications in Soil Science and Plant Analysis* 27, 1971–1993.

Dalal, R.C. and Quilt, P. (1977) Effects of N, P, liming and Mo on nutrition and grain yield of pigeonpea. *Agronomy Journal* 69, 854–857.

Nicholas, R. (1964) Studies on the major element deficiencies of the pigeonpea (*Cajanus cajan* L.) in sand culture – 1. Foliar symptoms of the major-element deficiencies. *Plant and Soil* 21, 377–387.

Plate 215. Potassium-deficient pigeon pea plant. (Photo by Dr Prakash Kumar.)

Plate 216. Marginal chlorosis starts from the tip of the leaf. (Photo by Dr Prakash Kumar.)

Plate 217. Yellowing advances around the leaf margin. (Photo by Dr Prakash Kumar.)

Plate 218. Scorching spreads around the leaf margin. (Photo by Dr Prakash Kumar.)

PIGEON PEA (*Cajanus cajan* (L.) Millsp.) POTASSIUM (K) DEFICIENCY

Symptoms

1. Pigeon pea, usually grown in the semi-arid tropics, is frequently exposed to water-deficit conditions. As potassium plays a key role in plant–water relationships and provides the plant an inbuilt strength to fight water-deficit situations, potassium is one of the most important limiting mineral nutrient elements for growth and development of pigeon pea. Potassium-deficient pigeon pea plants become stunted.

2. Potassium moves readily from old to young leaves, therefore deficiency symptoms appear first on older leaves (Plate 215).

3. The symptoms begin as marginal yellowing of older leaves. Yellowing starts from leaf tips and advances around the leaf margin (Plates 216 and 217).

4. If deficiency persists, leaf tips become scorched. The scorching spreads around the leaf margin (Plate 218).

5. A yellow band is found between the scorched area and healthy green tissues.

6. In acute deficiency conditions, the symptoms move towards upper leaves.

Developmental stages

Stage I: A pale yellow chlorosis appears on older leaves starting from the tip of the leaf (Plate 216).

Stage II: As the symptoms advance, pale yellow chlorosis covers the margins of the lamina (Plate 217).

Stage III: The leaf tip becomes scorched as the symptoms become severe. Yellow chlorosis is followed by brown necrosis along the margins of the leaf (Plate 218).

Stage IV: In acute deficiency conditions, the symptoms move towards upper leaves and may cover the entire plant.

Likely to occur in

1. Soils formed from parent material low in potassium.

2. Light-textured soils where potassium has been leached by heavy rainfall or excessive irrigation.

3. Soils low in organic matter.

4. Soils with wide Na:K, Mg:K or Ca:K ratio.

5. Large bicarbonate concentration in irrigation water.

6. Highly acid soils having pH below 6.0.

Integrated nutrient management

1. Get the soil analysed before sowing to measure the amount of 'available' potassium in the soil.

2. Problematic acid/alkaline/saline soils should be reclaimed.

3. Add organic manures well before sowing.

4. Apply potassium nitrate, potassium sulphate or potassium chloride to the soil at or before sowing as per soil testing recommendations.

5. If potassium deficiency appears on the standing crop, apply soluble potassium salts such as potassium nitrate, potassium sulphate or potassium chloride with irrigation water. Foliar sprays of these salts are usually not recommended because a number of sprays are needed to fulfil crop requirements.

Further reading

Edwards, D.G. (1981) Development of research on pigeonpea nutrition. In: *Proceedings of the International Workshop on Pigeonpea, ICRISAT Center, India, 15–19 December 1980, Vol. 1*. International Crops Research Institute for the Semi-Arid Tropics (ICRISAT), Patancheru, India, pp. 205–211.

Nicholas, R. (1964) Studies on the major element deficiencies of the pigeonpea (*Cajanus cajan* L.) in sand culture – 1. Foliar symptoms of the major-element deficiencies. *Plant and Soil* 21, 377–387.

Rao, K.V. and Rao, K.V.M. (1983) Influence of potassium nutrition on stomatal behavior, transpiration rate and leaf water potential of pigeonpea (*Cajanus cajan* (L.) Millsp.) in sand culture. *Proceedings of the Indian Academy of Sciences, Plant Science* 92, 323–330.

Singh, F. and Oswalt, D.L. (1992) *Pigeonpea Botany and Production Practices. Skill Development Series No. 9*. International Crops Research Institute for the Semi-Arid Tropics (ICRISAT), Patancheru, India, pp. 14–16.

Plate 219. Magnesium-deficient pigeon pea plant. (Photo by Dr Prakash Kumar.)

Plate 220. Fading of interveinal tissues. (Photo by Dr Prakash Kumar.)

Plate 221. Interveinal chlorosis of lower leaves. (Photo by Dr Prakash Kumar.)

Plate 222. Close-up of affected leaves. (Photo by Dr Prakash Kumar.)

PIGEON PEA (*Cajanus cajan* (L.) Millsp.) MAGNESIUM (Mg) DEFICIENCY

Symptoms

1. Deficient plants appear stunted with short and thin stems.
2. Magnesium is mobile in the plant and under short supply conditions it is transferred from older to younger leaves. The deficiency symptoms appear first and more severely on the older leaves. The youngest leaves remain green and apparently healthy (Plate 219).
3. The symptoms begin with fading of interveinal tissues of older leaves (Plate 220).
4. As the symptoms advance, a greenish yellow to pale yellow interveinal chlorosis develops on older leaves (Plate 221).
5. All veins (primary and secondary) of the leaf remain green (Plate 222).
6. If the deficiency persists and becomes more severe, these symptoms move up the plant towards upper leaves.
7. In acute deficiency conditions, affected interveinal regions turn necrotic.

Developmental stages

Stage I: In mild deficiency the lower leaves develop temporary fading of interveinal tissues with prominent green veins (Plates 219 and 220).
Stage II: A pale yellow interveinal chlorosis develops on older leaves. Veins remain green and prominent (Plate 221).
Stage III: If the deficiency persists and becomes more severe, these symptoms move up the plant towards upper leaves (Plate 221).
Stage IV: In acute deficiency conditions, interveinal regions become pale yellow, leading to eventual necrosis.

Likely to occur in

1. Acid sandy soils from which magnesium has been leached by heavy rainfall.
2. Peat and muck soils that are low in total magnesium.
3. Soils derived from parent material that is inherently low in magnesium.
4. Soils with heavy and excess application of potassium fertilizers.
5. Soils with heavy and excess application of lime (calcium carbonate) or other calcium fertilizers.
6. Acid soils having pH below 6.5.
7. Alkaline soils having pH above 8.5.

Integrated nutrient management

1. Get the soil analysed before sowing to estimate the amount of soluble and exchangeable magnesium in the soil.
2. Apply analysis-based recommended quantity of magnesium before sowing using soluble salts such as magnesium sulphate or magnesium chloride.
3. In deficient standing crops, apply soluble salts such as magnesium sulphate, chloride or nitrate with irrigation water. Foliar sprays of these salts are usually not advised as many sprays at frequent intervals are required to fulfil the crop need.
4. Magnesium deficiency in acid soils may be corrected by applying dolomite (a mixture of calcium carbonate and magnesium carbonate, $CaCO_3 \cdot MgCO_3$) through broadcasting and mixing in the soil a few months before the sowing.
5. Reclamation of problematic acid soils or alkaline soils should be done to regulate the proper supply of magnesium.

Further reading

Edwards, D.G. (1981) Development of research on pigeonpea nutrition. In: *Proceedings of the International Workshop on Pigeonpea, ICRISAT Center, India, 15–19 December 1980, Vol. 1*. International Crops Research Institute for the Semi-Arid Tropics (ICRISAT), Patancheru, India, pp. 205–211.

Nicholas, R. (1964) Studies on the major element deficiencies of the pigeonpea (*Cajanus cajan* L.) in sand culture – 1. Foliar symptoms of the major-element deficiencies. *Plant and Soil* 21, 377–387.

Nicholas, R. (1965) Studies on the major element deficiencies of the pigeonpea (*Cajanus cajan* L.) in sand culture – 2. The effects of major element deficiencies on nodulation, growth, and mineral composition. *Plant and Soil* 22, 112–126.

Singh, F. and Oswalt, D.L. (1992) *Pigeonpea Botany and Production Practices. Skill Development Series No. 9*. International Crops Research Institute for the Semi-Arid Tropics (ICRISAT), Patancheru, India, pp. 14–16.

Plate 223. Sulphur-deficient pigeon pea branch. (Photo by Dr Prakash Kumar.)

Plate 224. Pale green sulphur-deficient upper leaves. (Photo by Dr Prakash Kumar.)

Plate 225. Close-up of uniform pale yellow leaves. (Photo by Dr Prakash Kumar.)

Plate 226. Affected leaf (left) in comparison to a normal leaf (right). (Photo by Dr Prakash Kumar.)

PIGEON PEA (*Cajanus cajan* (L.) Millsp.) SULPHUR (S) DEFICIENCY

Symptoms

1. Sulphur deficiency, even if severe, does not produce any type of necrosis or burning of leaves. The change in leaf colour is the only indicative symptom of sulphur deficiency.
2. At the initial stage, sulphur-deficiency symptoms are often confused with those caused by nitrogen deficiency or iron deficiency.
3. A close observation is required to see whether the older leaves are more dark green and the younger ones are more pale (case of sulphur deficiency) or whether the younger leaves are more dark green and the older ones are more pale (case of nitrogen deficiency).
4. Sulphur is immobile in plants and does not mobilize from older to younger leaves in short supply conditions. Thus, the deficiency symptoms appear first on the younger leaves while the older leaves remain green and healthy (Plate 223).
5. Deficiency symptoms appear as an even and uniform pale green to pale yellow chlorosis across the lamina of young leaves. The colour of the midrib and other veins becomes very similar to the colour of interveinal areas of the leaf (in the case of iron deficiency, these veins remain green and prominent).
6. If deficiency persists and becomes more severe, symptoms eventually move downwards, covering more leaves.
7. Plant vigour, flowering and fruiting reduce drastically, resulting in poor crop yields.

Developmental stages

Stage I: Symptoms begin with uniform fading of young leaves from green to pale green colour.
Stage II: If the deficiency persists and becomes more severe, an even and uniform pale green to pale yellow chlorosis develops across the lamina of young leaves (Plates 223, 225 and 226).
Stage III: In acute conditions the deficiency symptoms move downwards, covering more leaves (Plate 224).

Likely to occur in

1. Soils low in organic matter.
2. Light-textured sandy soils that have been leached by heavy rainfall or excessive irrigation.
3. Soils exhausted by intensive cropping.
4. Soils derived from parent material that is inherently low in sulphur (for example, soils formed from volcanic rocks).
5. Acid soils having pH below 6.0.

Integrated nutrient management

1. Get the soil analysed before sowing to measure the amount of 'available' sulphate in the soil and to predict the amount of fertilizer required.
2. Apply analysis-based recommended quantity of sulphur by mixing into the soil, well before sowing, either:
  a. elemental sulphur (flowers of sulphur); or
  b. gypsum (calcium sulphate).
3. In deficient standing crops apply ammonium sulphate, magnesium sulphate or potassium sulphate with irrigation water.
4. Problematic acid soils should be reclaimed.

Further reading

Edwards, D.G. (1981) Development of research on pigeonpea nutrition. In: *Proceedings of the International Workshop on Pigeonpea, ICRISAT Center, India, 15–19 December 1980*, Vol. 1. International Crops Research Institute for the Semi-Arid Tropics (ICRISAT), Patancheru, India, pp. 205–211.

Johansen, C. (1990) Pigeonpea: mineral nutrition. In: Nene, Y.L., Hall, S.D. and Sheila, V.K. (eds) *The Pigeonpea*. CABI, Wallingford, UK, pp. 209–231.

Oke, O.L. (1969) Sulphur nutrition in legumes. *Experimental Agriculture* 5, 111–116.

Singh, F. and Oswalt, D.L. (1992) *Pigeonpea Botany and Production Practices. Skill Development Series No. 9*. International Crops Research Institute for the Semi-Arid Tropics (ICRISAT), Patancheru, India, pp. 14–16.

Plate 227. Iron-deficient pigeon pea plant. (Photo by Dr Prakash Kumar.)

Plate 228. Plant showing interveinal chlorosis on top leaves. (Photo by Dr Prakash Kumar.)

Plate 229. Prominent green veins fade to light green. (Photo by Dr Prakash Kumar and Dr Manoj Kumar Sharma.)

Plate 230. Leaf showing papery white appearance. (Photo by Dr Prakash Kumar.)

PIGEON PEA (*Cajanus cajan* (L.) Millsp.) IRON (Fe) DEFICIENCY

Symptoms

1. Iron is immobile in plants. So, deficiency symptoms appear first and more severely on the younger leaves. The older leaves remain normal and apparently healthy (Plate 227).
2. In mild deficiency conditions or at the initial stage of deficiency, the topmost younger leaves develop temporary fading of interveinal tissues to a pale green to pale yellow colour.
3. If the deficiency persists and becomes more severe, a bright pale yellow chlorosis develops in interveinal tissues (tissues between the veins), leaving the veins green and prominent. Interveinal chlorosis of top leaves is the specific symptom of iron deficiency (Plate 228). As the symptoms advance, prominent green veins also fade and become light green to pale yellow (Plate 229).
4. In acute deficiency conditions, the entire leaf bleaches to papery white (Plate 230).

Developmental stages

Stage I: The topmost leaves develop temporary fading of interveinal tissues to pale yellow colour with prominent green veins (Plate 227).
Stage II: The interveinal tissues of the affected leaves turn bright yellow with prominent green veins (Plate 228).
Stage III: The prominent green veins also fade and become light green (Plate 229).
Stage IV: The affected top leaves bleach to papery white (Plate 230). The veins may or may not remain green.

Likely to occur in

1. Sandy soils having low total iron.
2. Calcareous soils where solubility of iron is very low.
3. Peat and muck soils where organic matter ties up iron and reduces its availability in soil solution.
4. Acid soils where, despite high availability of iron in soil solution, excessively high levels of soluble zinc, manganese, copper or nickel hamper the uptake of iron by the plant.
5. Waterlogged soils.
6. Alkaline soils having pH above 7.5.

Integrated nutrient management

1. Get the soil analysed before sowing to measure the amount of 'available' iron in the soil.
2. As inorganic soluble salts of iron such as iron sulphates or chlorides quickly become insoluble in soil, soil dressing with organic forms of iron such as iron chelates (10 kg/ha) is more effective. These organic iron forms should be used as per soil pH. If the soil is neutral (around pH 7.0), the most suitable chelates are FeHEDTA and FeDTPA. For acid soils FeEDTA is the most effective chelate, while on alkaline calcareous soils FeEDDHA is recommended.

Further reading

Agrawal, S.C. (2003) *Diseases of Pigeon Pea*. Concept Publishing Company, New Delhi, pp. 314–319.
Katyal, J.C. (1981) Micronutrient research in pigeonpea. In: *Proceedings of the International Workshop on Pigeonpea, ICRISAT Center, India, 15–19 December 1980*, Vol. 1. International Crops Research Institute for the Semi-Arid Tropics (ICRISAT), Patancheru, India, pp. 221–226.
Khan, B., Baloch, M.S. and Hussain, S.M. (1999) Micro-nutritional studies in pigeonpea. *Pakistan Journal of Biological Sciences* 2, 399–401.

Plate 231. Manganese-deficient pigeon pea plant. (Photo by Dr Prakash Kumar.)

Plate 232. Fading of lamina without affecting veins. (Photo by Dr Prakash Kumar.)

Plate 233. Affected leaf showing interveinal chlorosis. (Photo by Dr Prakash Kumar.)

Plate 234. Small brown spots on leaf. (Photo by Dr Prakash Kumar.)

PIGEON PEA (*Cajanus cajan* (L.) Millsp.) MANGANESE (Mn) DEFICIENCY

Symptoms

1. Manganese is not readily transferred from older to younger leaves. Therefore deficiency symptoms appear first on younger parts of the plant. Deficiency symptoms appear first and more severely on developing younger leaves near the growing point. The emerging leaf and the top one or two leaves below the growing point may look normal. This gives the impression that the topmost emerging leaves are normal and the symptoms are appearing on middle leaves.
2. The older leaves remain normal and apparently healthy.
3. The symptoms start with fading of the green lamina without affecting the veins. The veins remain green and prominent (Plates 231 and 232).
4. When deficiency persists a pale yellow interveinal chlorosis develops, leaving the veins green and clearly visible (Plate 232).
5. As symptoms advance, small white and brown spots appear on affected leaves, which eventually convert to brown necrotic lesions.

Developmental stages

Stage I: Fading of the green lamina occurs without affecting veins on the upper developing leaves (Plates 231 and 232).
Stage II: The interveinal tissues of the affected leaves turn pale yellow with clearly visible green veins (Plate 233).
Stage III: White and brown spots appear on affected leaves (Plate 234).
Stage IV: White and brown spots convert to brown necrotic lesions.

Likely to occur in

1. Light-textured sandy soils that have been leached by heavy rainfall or excessive irrigation.
2. Calcareous and alkaline soils where the solubility of manganese is very low.
3. Waterlogged peat soils where organic matter ties up manganese and reduces its availability in solution.
4. Soils derived from parent material that is inherently low in manganese.
5. Acid soils having pH below 5.0.
6. Alkaline soils having pH above 7.5.

Integrated nutrient management

1. Get the soil analysed before sowing to measure the amount of 'available' manganese in the soil.
2. Apply analysis-based recommended quantity of manganese as basal by using manganese salts such as manganese sulphate. Soil dressing, once applied, works for up to 5–6 years.
3. Add organic manures well before sowing.
4. In deficient standing crops, apply manganese salts such as manganese sulphate (0.2 to 0.3% w/v solution) as a foliar spray. Repeated sprays may be required if symptoms reappear.

Further reading

Agrawal, S.C. (2003) *Diseases of Pigeon Pea*. Concept Publishing Company, New Delhi, pp. 314–319.

Katyal, J.C. (1981) Micronutrient research in pigeonpea. In: *Proceedings of the International Workshop on Pigeonpea, ICRISAT Center, India, 15–19 December 1980, Vol. 1*. International Crops Research Institute for the Semi-Arid Tropics (ICRISAT), Patancheru, India, pp. 221–226.

Sharma, C.P., Kurana, N., Chatterjee, C. and Agrawal, S.C. (1988) Response of pigeon pea to variable levels of manganese. *Proceedings of the Indian Academy of Science, Plant Science* 98, 283–290.

Plate 235. Zinc-deficient pigeon pea plant. (Photo by Dr Prakash Kumar.)

Plate 236. Fading of green colour and inward rolling of margins. (Photo by Dr Prakash Kumar.)

Plate 237. Irregular interveinal chlorosis starting from tips. (Photo by Dr Prakash Kumar.)

Plate 238. Scorched leaves showing midrib and adjacent tissues green. (Photo by Dr Prakash Kumar.)

PIGEON PEA (*Cajanus cajan* (L.) Millsp.) ZINC (Zn) DEFICIENCY

Symptoms

1. The zinc-deficient plant shows stunted growth with reduced size of branches and trifoliate leaves, providing a 'little leaf' appearance to the plant. The flowering is delayed. Flowers and pods fall prematurely.

2. Zinc is partially mobile in pigeon pea, and under short supply conditions it is not readily transferred from older to younger leaves. So, the deficiency symptoms appear first and more severely on middle or older leaves.

3. Symptoms begin as a fading of the green colour of leaves and inward rolling of their margins (Plate 236).

4. If deficiency persists and becomes more severe, a pale yellow irregular interveinal chlorosis starts from the tips of affected leaves and spreads to the remaining areas. The midrib and adjacent tissues remain green (Plate 235).

5. In acute deficiency conditions, brown necrotic spots develop on affected buff areas of the leaf. The affected leaves show a severely scorched appearance. The deficient leaves shed prematurely (Plate 238).

Developmental stages

Stage I: The symptoms begin as a fading of the green colour of leaves and inward rolling of their margins (Plate 236).

Stage II: Pale yellow irregular interveinal chlorosis starts from the tips of affected leaves. All three leaflets of the affected trifoliate leaf show chlorosis on leaf tips (Plate 235).

Stage III: As symptoms advance, the interveinal chlorosis spreads to the remaining parts of the leaf. The midrib and adjacent tissues remain green (Plate 237).

Stage IV: In the most advanced stage, the affected leaves show a scorched appearance with brown necrotic spots. The deficient leaves shed prematurely (Plate 238).

Likely to occur in

1. Leached sandy soils where total zinc is low.

2. Just-levelled soils where the subsoil is exposed for cultivation. Available zinc in surface soil is often double that of the subsoil.

3. Soil having heavy and excessive application of phosphate fertilizers, which hampers use of zinc by the crop.

4. Acid soils having pH below 5.0.

5. Alkaline soils having pH above 7.5.

Integrated nutrient management

1. Get the soil analysed before sowing to estimate the amount of 'available' zinc in the soil.

2. Apply 25–30 kg of zinc sulphate or 10 kg of zinc chelate per hectare every 2 years in zinc-deficient soils.

3. Do not mix zinc fertilizers with phosphate fertilizers.

Further reading

Katyal, J.C. (1981) Micronutrient research in pigeonpea. In: *Proceedings of the International Workshop on Pigeonpea, ICRISAT Center, India, 15–19 December 1980, Vol. 1*. International Crops Research Institute for the Semi-Arid Tropics (ICRISAT), Patancheru, India, pp. 221–226.

Nicholas, R. (1964) Studies on the major element deficiencies of the pigeonpea (*Cajanus cajan* L.) in sand culture – 1. Foliar symptoms of the major-element deficiencies. *Plant and Soil* 21, 377–387.

Reddy, P.R., Subrarao, I.V. and Rao, L.M.A. (1978) Nutritional disorders in pigeon pea (*Cajanus cajan* [L.] Millsp.) cv. Hy-3c and remedial measures. *Science and Culture* 44, 112–126.

Plate 239. Copper-deficient pigeon pea plant. (Photo by Dr Prakash Kumar.)

Plate 240. Chlorotic and bleached leaves with 'boat-like' appearance. (Photo by Dr Prakash Kumar.)

Plate 241. Wilting and rolling of leaves showing 'boat-like' structure. (Photo by Dr Prakash Kumar.)

Plate 242. New emerging leaves fail to unfold. (Photo by Dr Prakash Kumar.)

PIGEON PEA (*Cajanus cajan* (L.) Millsp.) COPPER (Cu) DEFICIENCY

Symptoms

1. Copper affects the ability of the plant to take in water from the soil and transfer it in the plant body. As a result, even mildly deficient plants show wilting during daytime when the transpiration rate is high. This happens even when sufficient moisture is available in the soil. The type of wilting is described as 'incipient wilting' or 'mid day depression'. Wilted plants recover overnight.
2. Copper deficiency affects flowering and pollen formation. As a result, copper-deficient plants produce few flowers and set few pods.
3. Copper is immobile in plants, and under short supply conditions it is not easily mobilized from older to younger leaves. The deficiency symptoms appear first and become more severe on the younger leaves, while the older leaves usually remain unaffected, dark green and healthy.
4. First visual symptoms usually appear 1 month after germination. In mild deficiency conditions, the foliage becomes dark green to bluish green and shows temporary wilting.
5. If deficiency persists and becomes more severe, the younger leaves turn chlorotic and bleached (Plate 240). The newly emerging leaves remain small and folded, providing the leaf a 'boat-shaped' appearance. This is the specific symptom of copper deficiency in pigeon pea (Plates 239, 241 and 242).

Developmental stages

Stage I: In mild deficiency, the plant appears dark green and wilted.
Stage II: If deficiency persists and becomes severe, the younger leaves turn chlorotic and bleached (Plate 240).
Stage III: As symptoms advance, the emerging leaves fail to unfold and provide the leaves a typical boat-shaped appearance (Plates 241 and 242).

Likely to occur in

1. Calcareous sandy soils low in copper.
2. Leached acid soils low in copper.
3. Peat and muck soils where organic matter ties up copper and reduces its availability in solution.
4. Soils derived from parent material that is inherently low in copper.
5. Acid soils having pH below 5.0.
6. Alkaline soils having pH above 8.5.

Integrated nutrient management

1. Get the soil analysed before sowing to measure the amount of 'available' copper in the soil.
2. Add organic manures well before sowing.
3. Apply soluble inorganic copper salts such as copper sulphate or organic copper chelates as basal.
4. In deficient standing crops, apply copper sulphate (0.2 to 0.5% w/v solution) as a foliar spray. Repeated sprays may be required if symptoms reappear.

Further reading

Agrawal, S.C. (2003) *Diseases of Pigeon Pea*. Concept Publishing Company, New Delhi, pp. 314–319.

Katyal, J.C. (1981) Micronutrient research in pigeonpea. In: *Proceedings of the International Workshop on Pigeonpea, ICRISAT Center, India, 15–19 December 1980, Vol. 1*. International Crops Research Institute for the Semi-Arid Tropics (ICRISAT), Patancheru, India, pp. 221–226.

Khan, B., Baloch, M.S. and Hussain, S.M. (1999) Micro-nutritional studies in pigeonpea. *Pakistan Journal of Biological Sciences* 2, 399–401.

Plate 243. Stunted nitrogen-deficient plant having light green coloured leaves. (Photo by Dr Prakash Kumar and Dr Manoj Kumar Sharma.)

GREEN GRAM (*Vigna radiata* Linn.) NITROGEN (N) DEFICIENCY

Plate 244. Paling of leaves more prominent on the lower part of the plant. (Photo by Dr Prakash Kumar and Dr Manoj Kumar Sharma.)

Plate 245. Moderately deficient, evenly pale green leaf. (Photo by Dr Prakash Kumar and Dr Manoj Kumar Sharma.)

Plate 246. Severely deficient white–yellow leaf. (Photo by Dr Prakash Kumar and Dr Manoj Kumar Sharma.)

Symptoms

1. Nitrogen deficiency causes retarded growth of plants. Stems become thin and elongated. Branching and flowering are reduced drastically. Reduced pod formation and poor seed set result in poor yields.

2. Nitrogen is fairly mobile within plants and under restricted supply conditions it is rapidly redistributed from older to younger leaves. The deficiency symptoms typically appear in lower leaves first. If deficiency persists, the symptoms move up the plant to the younger leaves.

3. The old leaves become uniformly pale green (Plate 245) and then turn pale yellow to yellow, while the young leaves remain light green.

4. Later, the yellow older leaves turn white and drop early.

Developmental stages

Stage I: The mildly deficient plant becomes uniformly light green in colour (Plate 243).

Stage II: If deficiency continues, the older leaves turn evenly pale yellow to yellow, the upper leaves remain pale green (Plate 244).

Stage III: In severe deficiency, the old leaves turn dark yellow, then white, and fall off prematurely (Plate 246).

Likely to occur in

1. Soils having low organic matter.

2. Light-textured sandy soils that have been leached by heavy rainfall or excessive irrigation.

3. Soils exhausted by intensive cropping.

4. Waterlogged conditions.

5. Acid soils having pH below 6.0.

6. Alkaline soils having pH above 8.0.

Integrated nutrient management

1. Get the soil analysed before sowing to measure the amount of 'available' nitrogen in the soil.

2. Apply analysis-based recommended quantity of nitrogen as basal by using:

- **a.** organic manures;
- **b.** bio-fertilizers;
- **c.** nitrogenous fertilizers.

3. Top-dress soluble nitrogenous fertilizers such as urea.

4. For quick recovery, apply urea with irrigation water or as a foliar spray in the standing crop. Foliar sprays require to be repeated every 10–15 days.

Further reading

Jones, J.B. (2002) Nutrient element deficiency symptoms. Appendix E. Legumes. In: *Agronomic Handbook: Management of Crops, Soils and Their Fertility*. CRC Press, Boca Raton, Florida, p. 409.

Muchow, R.C., Robertson, M.J. and Pengelly, B.C. (1993) Accumulation and partitioning of biomass and nitrogen by soybean, mungbean and cowpea under contrasting environmental conditions. *Field Crops Research* 33, 13–36.

Senaratne, R. and Ratnasinghe, D.S. (1993) Ontogenic variation in nitrogen fixation and accumulation of nitrogen in mungbean, blackgram, cowpea and groundnut. *Biology and Fertility of Soils* 16, 125–130.

Smith, F.W., Imrie, B.C. and Pieters, W.H.J. (1983) *Foliar Symptoms of Nutrient Disorders in Mung Bean (Vigna radiata). Division of Tropical Crops and Pastures Technical Paper No. 24*. CSIRO Publishing, Melbourne, Australia.

Plate 247. Phosphorus-deficient stunted plant having dull green leaves. (Photo by Dr Prakash Kumar.)

GREEN GRAM (*Vigna radiata* Linn.) PHOSPHORUS (P) DEFICIENCY

Plate 248. Normal green leaves with sufficient phosphorus nutrition. (Photo by Dr Prakash Kumar.)

Plate 249. Close-up of a phosphorus-deficient dark green leaf. (Photo by Dr Prakash Kumar.)

Plate 250. Severely deficient dull bluish green trifoliate with distinct purple veins. (Photo by Dr Prakash Kumar.)

Symptoms

1. Phosphorus deficiency results in reduced biomass, photosynthetic activity and nitrogen-fixing ability in green gram.
2. In phosphorus-deficient plants the growth is retarded. Stems become thin and spindly with short internodes. Maturity gets delayed. The size and number of pods are reduced. The decreased number of seeds in pods results in poor yields.
3. Phosphorus is considered a mobile nutrient within plants. Under deficient conditions it is easily moved from older to younger tissues. Thus, the deficiency symptoms occur primarily in lower leaves.
4. Deficient plants appear dark green. Leaves cup upwards and stems may become purple or red.
5. The leaves initially look dark green compared to plants with sufficient phosphorus supply (Plates 248 and 249) and then turn a dull bluish green (Plate 250).
6. Purple pigmentation often develops on the lower leaves, then works up the plant to the upper leaves.

Developmental stages

Stage I: In mild deficiency, the plant becomes stunted and the leaves are small and dark green (Plate 247).
Stage II: As symptoms advance, dark green leaves turn a dull bluish green (Plate 250).
Stage III: If the deficiency is severe, the old leaves turn yellow, then brown and drop prematurely.

Likely to occur in

1. Soils having low organic matter.
2. Alkaline and calcareous soils.
3. Soils exhausted by intensive cropping.
4. Acid soils and highly weathered soils.
5. Soils where the topsoil has been removed by erosion.
6. Acid soils having pH below 6.0.
7. Alkaline soils having pH between 7.5 and 8.5.

Integrated nutrient management

1. Get the soil analysed before sowing to measure the amount of 'available' phosphate in the soil.
2. Apply analysis-based recommended quantity of phosphorus as basal by using:
- **a.** organic manures;
- **b.** phosphate-solubilizing microbial cultures;
- **c.** phosphatic fertilizers.

3. If standing crops show the deficiency, apply soluble phosphatic fertilizers such as ammonium phosphate with irrigation water.

Further reading

Chaudhary, M.I., Adu-Gyamfi, J.J., Saneoka, H., Nguyen, N.T., Suwa, R., Kanai, S., El-Shemy, H.A., Lightfoot, D.A. and Fujita, K. (2008) The effect of phosphorus deficiency on nutrient uptake, nitrogen fixation and photosynthetic rate in mashbean, mungbean and soybean. *Acta Physiologiae Plantarum* 30, 537–544.

Kaur, V., Kaur, N. and Gupta, A.K. (1999) Phosphatase activity and phosphorus partitioning in nodules of developing mungbean. *Acta Physiologiae Plantarum* 21, 215–220.

Khan, M.A., Baloch, M.S., Taj, I. and Gandapur, I. (1999) Effect of phosphorus on the growth and yield of mungbean. *Pakistan Journal of Biological Sciences* 2, 667–669.

Smith, F.W., Imrie, B.C. and Pieters, W.H.J. (1983) *Foliar Symptoms of Nutrient Disorders in Mung Bean (Vigna radiata). Division of Tropical Crops and Pastures Technical Paper No. 24*. CSIRO Publishing, Melbourne, Australia.

Plate 251. Potassium deficient Green Gram plant showing chlorosis on older leaves. (Photo by Dr Prakash Kumar.)

Plate 252. Deficient old leaves develop yellowing of tips and margins that also moves inward. (Photo by Dr Prakash Kumar.)

Plate 253. Chlorosis advancing inward. (Photo by Dr Prakash Kumar.)

Plate 254. Advanced stage of potassium deficiency. (Photo by Dr Prakash Kumar.)

GREEN GRAM (*Vigna radiata* Linn.) POTASSIUM (K) DEFICIENCY

Symptoms

1. The plant–water relationship and photosynthetic rate of green gram are improved to a greater extent by potassium under suboptimal soil moisture. Potassium promotes growth under suboptimal soil moisture conditions.
2. The application of higher levels of potassium improves water relationships as well as growth and yield of green gram under mild saline conditions.
3. Improvement of potassium nutritional status of plants might be of great importance for survival of the crop under environmental stress conditions, such as drought, chilling and high light intensity.
4. Potassium-deficient plants become stunted and internodes are shortened. Plants have reduced branching. Plants show poor growth and tend to become wilted readily.
5. Potassium is highly mobile within plants, so it is readily translocated from older to younger leaves under reduced supply conditions. Thus the visible symptoms tend to occur first in older leaves. The young leaves usually stay dark green.
6. Symptoms start as a chlorosis (yellowing) at the tip and along the edges of leaves and also into interveinal areas of old leaves (Plate 252).
7. The chlorosis is then followed by scorching that spreads inward towards the mid-vein.
8. Eventually, old leaves dry and drop prematurely.

Developmental stages

Stage I: In mild deficiency, the plant shows poor growth and readily wilts in dry conditions.
Stage II: If deficiency becomes severe, chlorosis develops at tips and along the leaf margins, moving inward into interveinal areas of older leaves (Plates 251 and 253).
Stage III: In acute deficiency conditions, the chlorotic leaf tips and margins become scorched. The scorching moves inward towards the mid-vein.
Stage IV: In the later stage, the entire leaf becomes scorched and drops early.

Likely to occur in

1. Soils formed from parent material low in potassium.
2. Light-textured soils where potassium has been leached by heavy rainfall or excessive irrigation.
3. Soils low in organic matter.
4. Soils with high Na:K, Mg:K or Ca:K ratio.
5. Acid soils having pH below 6.0.

Integrated nutrient management

1. Analyse the soil before planting to measure the amount of plant-available potassium.
2. Problematic acid/alkaline/saline soils should be reclaimed.
3. Add organic manures well before sowing.
4. Apply potassium chloride, potassium sulphate or potassium nitrate to the soil at or before sowing as per soil testing recommendations.
5. In standing crops, apply soluble potassium salts with irrigation water.

Further reading

Cakmak, I. (2005) The role of potassium in alleviating detrimental effects of abiotic stresses in plants. *Journal of Plant Nutrition and Soil Science* 168, 521–530.

Fatma, A.A., Fardoas, R.H. and Rizk, W.M. (2001) Effect of potassium fertilization on mungbean (*Vigana radiata* L. Wilczek). *Egyptian Journal of Applied Sciences* 16, 156–167.

Kabir, M.E., Karim, M.A. and Azad, M.A.K. (2004) Effect of potassium on salinity tolerance of mungbean (*Vigana radiata* L. Wilczek). *Journal of Biological Sciences* 4, 103–110.

Sangakkara, U.R., Frehner, M. and Nosberger, J. (2001) Influence of soil moisture and fertilizer potassium on the vegetative growth of mungbean (*Vigna radiata* L. Wilczek) and cowpea (*Vigna unguiculata* L. Walp). *Journal of Agronomy and Crop Science* 186, 73–81.

Plate 255. Mildly deficient plant showing light green young leaves and normal green old leaves. (Photo by Dr Prakash Kumar.)

GREEN GRAM (*Vigna radiata* Linn.) SULPHUR (S) DEFICIENCY

Plate 256. Uniform pale yellowing of a younger trifoliate. (Photo by Dr Prakash Kumar and Dr Manoj Kumar Sharma.)

Plate 257. Top leaves appearing dark yellow as the symptoms advance. (Photo by Dr Prakash Kumar.)

Plate 258. Pale yellow sulphur-deficient leaf (left) compared with a sulphur-sufficient green leaf (right). (Photo by Dr Prakash Kumar.)

Symptoms

1. Sulphur influences the protein content of seeds significantly and increases the size and weight of green gram seeds. Sulphur has a significant influence on the pod length and number of pods per plant. Sulphur is involved in the formation of chlorophyll and thereby encourages vegetative growth.
2. Sulphur has a significant positive effect on the activities of catalase, ascorbate peroxidise and guaiacol peroxidase, the synthesis of chlorophyll and the active iron content of green leaves.
3. Sulphur-deficient plants become stunted with short internodes.
4. Under short supply conditions, sulphur is not readily moved from older to newly growing tissues because it is immobile within the plant.
5. The deficiency symptoms are first observed in younger leaves and then progress to lower leaves if the deficiency continues.
6. The younger leaves turn chlorotic, initially becoming pale green and then pale yellow (Plate 258).
7. The pattern of chlorosis appears uniform on the entire leaf lamina including the veins (Plate 256).
8. In acute deficiency conditions the entire plant appears chlorotic.

Developmental stages

Stage I: In mildly deficient conditions, younger leaves become light green whereas older leaves remain darker (Plate 255).
Stage II: If deficiency continues, the light green younger leaves turn yellow (Plate 257).
Stage III: Severely deficient plants become entirely chlorotic, with the youngest leaves being the most affected.
Stage IV: Later on, leaf margins cup inwards and become necrotic.

Likely to occur in

1. Soils having low organic matter.
2. Light-textured sandy soils that have been leached by heavy rainfall or excessive irrigation.
3. Soils exhausted by intensive cropping.
4. Soils derived from parent material that is inherently low in sulphur.
5. Acid soils having pH below 6.0.

Integrated nutrient management

1. Get the soil analysed before sowing to measure the amount of 'available' sulphate in the soil.
2. Apply analysis-based recommended quantity of sulphur by mixing into the surface soil, well before sowing, either:
- **a.** elemental sulphur; or
- **b.** gypsum.

3. In deficient standing crops apply ammonium sulphate, magnesium sulphate or potassium sulphate with irrigation water.

Further reading

Jones, J.B. (2002) Nutrient element deficiency symptoms. Appendix E. Legumes. In: *Agronomic Handbook: Management of Crops, Soils and Their Fertility*. CRC Press, Boca Raton, Florida, p. 409.

Kaisher, M.S., Rahman, M.A., Amin, M.H.A., Amanullah, A.S.M. and Ahsanullah, A.S.M. (2010) Effect of sulphur and boron on the seed yield and protein content of mungbean. *Bangladesh Research Publications Journal* 3, 1181–1186.

Kumawat, R.N., Rathore, P.S., Nathawat, N.S. and Mahatma, M. (2006) Effect of sulfur and iron on enzymatic activity and chlorophyll content of mungbean. *Journal of Plant Nutrition* 29, 1451–1467.

Macnicol, P.K. and Bergmann, L. (1984) A role for homoglutathione in organic sulphur transport to the developing mung bean seed. *Plant Science Letters* 36, 219–223.

Plate 259. Iron-deficient plant having yellow to white chlorotic younger leaves while the bottom leaves are green. (Photo by Dr Prakash Kumar.)

Plate 260. Plants showing interveinal chlorosis on newly emerged leaves, less severe (right) and comparatively more severe (left). (Photo by Dr Prakash Kumar.)

Plate 261. Severely deficient trifoliate with white interveinal tissues and veins staying green. (Photo by Dr Prakash Kumar and Dr Manoj Kumar Sharma.)

Plate 262. Extremely deficient papery white trifoliate with faded green veins and brown necrotic tissues along the margins. (Photo by Dr Prakash Kumar and Dr Manoj Kumar Sharma.)

GREEN GRAM (*Vigna radiata* Linn.) IRON (Fe) DEFICIENCY

Symptoms

1. The frequent foliar application of iron in green gram has an appreciable effect on chlorophyll synthesis, enzymatic activities and active iron content in green leaves.

2. The green gram cultivars vary greatly in their resistance to iron deficiency in view of chlorosis symptoms, plant growth and seed yield. Some cultivars are highly susceptible to iron deficiency, whereas some are very tolerant to iron deficiency.

3. Deficient plants become thin-stemmed and have smaller leaves. The number and size of pods are reduced. Decreased number and size of seeds per pod result in low yields. Crop maturity gets delayed.

4. Iron is immobile within plants and hence it is not readily redistributed from older to younger plant tissues under reduced supply conditions. Therefore the deficiency symptoms become evident first on younger leaves.

5. Young leaves become yellow (chlorotic) with contrasting narrow, dark green main veins, while older leaves remain green (Plate 260).

6. Chlorotic young leaves then turn yellow to white and symptoms spread down the plant to lower leaves (Plate 259).

7. Dead tissues in the form of spots appear particularly at leaf margins (Plate 262).

Developmental stages

Stage I: In mild deficiency, interveinal tissues become pale green to pale yellow in younger leaves.

Stage II: If deficiency becomes severe, the leaves become yellow and chlorotic while the principal veins remain green (Plate 261).

Stage III: In very severe deficiency conditions, the entire leaf turns white and the veins remain faded green (Plate 260).

Stage IV: In the last stage, the bleached papery white leaves develop brown necrosis particularly at leaf margins (Plate 262).

Likely to occur in

1. Sandy soils having low total iron content.

2. Alkaline and calcareous soils where solubility of iron is very low.

3. Peat and muck soils where organic matter ties up iron and reduces its availability in soil solution.

4. Acid soils with excessively high levels of soluble zinc, manganese, copper or nickel that can hinder the uptake of iron despite high iron availability.

5. Alkaline soils having pH above 7.5.

Integrated nutrient management

1. Analyse the soil before sowing to measure the 'available' iron in the soil.

2. Reclaim problematic alkaline soils.

3. Add organic manure well before sowing.

4. Apply basal dose of soluble iron fertilizers such as $FeSO_4$ (commonly at 25 kg/ha) or Fe chelates (10 kg/ha). Use of organic chelates proves to be more promising as they maintain iron in soil solution.

5. In standing crops, apply $FeSO_4$ or Fe chelates (0.5% w/v solution) and 0.1% w/v citric acid as foliar sprays. Foliar sprays are required to be repeated every 10–15 days.

Further reading

Jones, J.B. (2002) Nutrient element deficiency symptoms. Appendix E. Legumes. In: *Agronomic Handbook: Management of Crops, Soils and Their Fertility*. CRC Press, Boca Raton, Florida, p. 409.

Kumawat, R.N., Rathore, P.S., Nathawat, N.S. and Mahatma, M. (2006) Effect of sulfur and iron on enzymatic activity and chlorophyll content of mungbean. *Journal of Plant Nutrition* 29, 1451–1467.

Ohwaki, Y., Kraokaw, S., Chotechuen, S., Egawa, Y. and Sugahara, K. (1997) Differences in responses to iron deficiency among various cultivars of mungbean (*Vigna radiata* L. Wilczek). *Plant and Soil* 192, 107–114.

Smith, F.W., Imrie, B.C. and Pieters, W.H.J. (1983) *Foliar Symptoms of Nutrient Disorders in Mung Bean (Vigna radiata). Division of Tropical Crops and Pastures Technical Paper No. 24*. CSIRO Publishing, Melbourne, Australia.

Plate 263. Zinc-deficient green gram plant. (Photo by Dr Prakash Kumar.)

Plate 264. Interveinal chlorosis starting from the tip of the leaf. (Photo by Dr Prakash Kumar.)

Plate 265. Chocolate brown necrotic spots on an affected leaf. (Photo by Dr Prakash Kumar.)

Plate 266. Leaf showing chocolate brown spots and holes. (Photo by Dr Prakash Kumar.)

GREEN GRAM (*Vigna radiata* Linn.) ZINC (Zn) DEFICIENCY

Symptoms

1. Green gram is very sensitive to zinc deficiency. Deficiency symptoms appear more prominently during the initial stages of crop growth, usually within 2–3 weeks after sowing.
2. Zinc-deficient green gram plants are stunted with short stems and reduced branches. Leaflets are small and faded with a diseased appearance.
3. Zinc is mobile in green gram plants and under short supply conditions it is transferred from older to younger leaves. So, the deficiency symptoms appear first and more severely on older leaves.
4. Symptoms begin as a faint, pale green interveinal chlorosis of older leaves. Interveinal chlorosis starts from the tip of the leaf and proceeds towards the base. The loss of green colour of interveinal tissues looks like a bleaching effect on affected leaves. The main veins remain green and prominent.
5. If deficiency persists and becomes more severe, the affected leaves develop chocolate brown necrotic spots and lesions on interveinal areas.
6. As symptoms advance, tissues of the necrotic spot areas drop from the leaf lamina, making small holes in the affected leaf (Plate 266).

Developmental stages

Stage I: Faint pale green interveinal chlorosis appears on older leaves. The chlorosis starts from the tips of the leaf and proceeds towards the base (Plate 264).
Stage II: The pale green interveinal chlorosis covers the entire affected leaf. The main veins remain green and prominent (Plate 263).
Stage III: The affected leaves develop chocolate brown necrotic spots and lesions on interveinal areas (Plate 265).
Stage IV: The tissues of necrotic spot areas drop from the leaf lamina, making small holes in the affected leaf.

Likely to occur in

1. Leached sandy soils where total zinc is low.
2. Just-levelled soils where the subsoil is exposed for cultivation. Available zinc in surface soil is often double that of the subsoil.
3. Cool wet weather.
4. Soil having heavy and excessive application of phosphate fertilizers, which hampers use of zinc by the crop.
5. Acid soils having pH below 5.0.
6. Alkaline soils having pH above 7.5.

Integrated nutrient management

1. Get the soil analysed before sowing to estimate the amount of 'available' zinc in the soil.
2. Problematic alkaline soils should be reclaimed.
3. Add organic manures well before sowing.
4. Apply 25–30 kg of zinc sulphate or 10 kg of zinc chelate per hectare every 2 years in zinc-deficient soils.
5. Do not mix zinc fertilizers with phosphate fertilizers.

Further reading

Andrew, C.S., Johnson, A.D. and Haydock, K.P. (1981) The diagnosis of zinc deficiency and effect of zinc on the growth and chemical composition of some tropical and sub-tropical legumes. *Communications in Soil Science and Plant Analysis* 12, 1–18.

DebRoy, P., Narwal, R.P., Malik, R.S. and Gupta, V.K. (2011) Response and enrichment of green gram (*Vigna radiata* L.) genotypes with respect to zinc application. Presented at *3rd International Zinc Symposium – Improving Crop Production and Human Health*, Hyderabad, India, 10–14 October 2011.

Gupta, V.K. (1983) Studies of the comparative response of some kharif crops to zinc application. *Agricultural Science Digest* 3(2), 79–80.

Viets, F.G. Jr., Boawn, L.C. and Crawford, C.L. (1954) Zinc content and deficiency symptoms of 26 crops grown on zinc deficient soil. *Soil Science* 70(4), 305–316.

Plate 267. Nitrogen deficient Black Gram Plant.
(Photo by Dr Prakash Kumar.)

BLACK GRAM (*Phaseolus mungo* var. *radiatus* Linn.) NITROGEN (N) DEFICIENCY

Plate 268. Uniform paleness appearing on the entire plant. (Photo by Dr Prakash Kumar.)

Plate 269. Severely deficient white–yellow leaf. (Photo by Dr Prakash Kumar.)

Plate 270. Very severely deficient white–yellow necrotic leaf. (Photo by Dr Prakash Kumar and Dr Manoj Kumar Sharma.)

Symptoms

1. Plants inoculated with H_2-uptake positive *Rhizobium* strains produce higher nitrogen content and dry matter than plants inoculated with H_2-uptake negative *Rhizobium* strains.
2. There is a rapid increase in nitrogen fixation between flowering and early pod fill. The nodules senesce progressively following the mid-pod filling stage.
3. Insufficient nitrogen supply restricts plant height. The leaf size and number of branches are reduced.
4. Nitrogen deficiency reduces flowering, decreases the number of pods and pod length and reduces the number of seeds and seed size, ultimately resulting in low yields.
5. In short supply conditions, nitrogen is readily transferred from older to younger tissues because it is fairly mobile within plants. Therefore, the deficiency symptoms tend to occur first and become more severe on the lower leaves, then working up the plant to the younger leaves.
6. The old leaves appear pale yellow to yellow, while the young leaves remain pale green.
7. Later, the yellow old leaves turn white and fall off prematurely.

Developmental stages

Stage I: Early deficiency symptoms are expressed as a uniform light green or pale green appearance of leaves in the entire plant (Plate 268).
Stage II: In prolonged deficiency conditions, the lower leaves turn pale yellow while the upper leaves appear light green (Plate 267).
Stage III: In severe deficiency conditions, the older leaves become yellow whereas the younger leaves also turn pale yellow.
Stage IV: In acute deficiency conditions, the leaves turn white (Plate 269), become necrotic and shed prematurely (Plate 270).

Likely to occur in

1. Soils having low organic matter.
2. Light-textured sandy soils that have been leached by heavy rainfall or excessive irrigation.
3. Soils exhausted by intensive cropping.
4. Waterlogged conditions.
5. Acid soils having pH below 6.0.
6. Alkaline soils having pH above 8.0.

Integrated nutrient management

1. Get the soil analysed before sowing to measure the amount of 'available' nitrogen in the soil.
2. Apply analysis-based recommended quantity of nitrogen as basal by using:
  a. organic manures;
  b. bio-fertilizers;
  c. nitrogenous fertilizers.
3. Top-dress soluble nitrogenous fertilizers such as urea.
4. For quick recovery, apply urea (2% w/v solution) as a foliar spray in the standing crop. Foliar sprays must be repeated every 10–15 days.

Further reading

Jones, J.B. (2002) Nutrient element deficiency symptoms. Appendix E. Legumes. In: *Agronomic Handbook: Management of Crops, Soils and Their Fertility*. CRC Press, Boca Raton, Florida, p. 409.

Kulsum, M.U., Baque, M.A. and Karim, M.A. (2007) Effects of different nitrogen levels on the morphology and yield of blackgram. *Journal of Agronomy* 6, 125–130.

Pahwa, K. and Dogra, R.C. (1983) Uptake hydrogenase system in urd bean (*Vigna mungo*) *Rhizobium* in relation to nitrogen fixation. *Journal of Applied Bacteriology* 54, 405–408.

Senaratne, R. and Ratnasinghe, D.S. (1993) Ontogenic variation in nitrogen fixation and accumulation of nitrogen in mungbean, blackgram, cowpea and groundnut. *Biology and Fertility of Soils* 16, 125–130.

Plate 271. Phosphorus-deficient stunted plant with dark green lower leaves. (Photo by Dr Prakash Kumar.)

Plate 272. Entire plant appearing dark green contrasting with dull green bottom leaves. (Photo by Dr Prakash Kumar.)

Plate 273. Dull bluish green leaves and purple pigmentation on stem. (Photo by Dr Prakash Kumar.)

Plate 274. Comparison of phosphorus-deficient trifoliate (left) with normal trifoliate (right). (Photo by Dr Prakash Kumar.)

BLACK GRAM (*Phaseolus mungo* var. *radiatus* Linn.) PHOSPHORUS (P) DEFICIENCY

Symptoms

1. Phosphorus deficiency reduces biomass, photosynthetic activity and nitrogen-fixing ability in black gram.
2. Phosphorus deficiency results in reduced plant growth. Plants appear bluish green and stunted. The root development is restricted and flowering is reduced.
3. Crop maturity is delayed. The number and size of pods are decreased which leads to poor yields.
4. Phosphorus is mobile within plants and is readily translocated from older to younger tissues of the plant under restricted supply conditions. Therefore, older leaves display deficiency symptoms first.
5. Deficient plants become dark green in appearance and the lower stems turn purplish.
6. If deficiency persists for long, the dark green leaves turn bluish green.
7. Phosphorus-deficient plants often develop purple pigmentation on older leaves.

Developmental stages

Stage I: In mild deficiency, plants appear stunted, lower leaves become smaller and dark green (Plates 271, 272 and 274).
Stage II: In prolonged deficient conditions, dark green leaves turn dull bluish green and stems often become purple or red (Plate 273).
Stage III: If deficiency becomes severe, the old leaves turn yellow, then brown and drop prematurely.
Stage IV: In acute deficiency conditions, the leaves turn brown necrotic and shed prematurely.

Likely to occur in

1. Soils having low organic matter.
2. Alkaline and calcareous soils.
3. Soils exhausted by intensive cropping.
4. Acid soils and highly weathered soils.
5. Soils where the topsoil has been removed by erosion.
6. Acid soils having pH below 6.0.
7. Alkaline soils having pH between 7.5 and 8.5.

Integrated nutrient management

1. Get the soil analysed to measure the amount of 'available' phosphate in the soil well before sowing.
2. Apply analysis-based recommended quantity of phosphorus as basal by using:
- **a.** organic manures;
- **b.** phosphate-solubilizing microbial cultures;
- **c.** phosphatic fertilizers.

3. If standing crops show the deficiency, apply soluble phosphatic fertilizers such as ammonium phosphate with irrigation water.

Further reading

Bidwell, R.G.S. (1979) *Plant Physiology*, 2nd edn. Macmillan Publishing Co., Inc., New York.

Chaudhary, M.I., Adu-Gyamfi, J.J., Saneoka, H., Nguyen, N.T., Suwa, R., Kanai, S., El-Shemy, H.A., Lightfoot, D.A. and Fujita, K. (2008) The effect of phosphorus deficiency on nutrient uptake, nitrogen fixation and photosynthetic rate in mashbean, mungbean and soybean. *Acta Physiologiae Plantarum* 30, 537–544.

Jones, J.B. (2002) Nutrient element deficiency symptoms. Appendix E. Legumes. In: *Agronomic Handbook: Management of Crops, Soils and Their Fertility*. CRC Press, Boca Raton, Florida, p. 409.

Romheld, V. and Neumann, G. (1999) Root excretions of carboxylic acids and protons in phosphorus-deficient plants. *Plant and Soil* 211, 121–130.

Plate 275. Potassium-deficient plant expressing marginal yellowing in older leaves that has moved inward, leaving the main veins green. (Photo by Dr Prakash Kumar and Dr Manoj Kumar Sharma.)

Plate 276. Marginal yellowing of a leaf. (Photo by Dr Prakash Kumar and Dr Manoj Kumar Sharma.)

Plate 277. Yellow and necrotic border around the tip end and along the leaf edge. (Photo by Dr Prakash Kumar and Dr Manoj Kumar Sharma.)

Plate 278. Marginal yellowing extending inward with main veins and adjoining tissues left green. (Photo by Dr Prakash Kumar and Dr Manoj Kumar Sharma.)

BLACK GRAM (*Phaseolus mungo* var. *radiatus* Linn.) POTASSIUM (K) DEFICIENCY

Symptoms

1. Potassium-deficient plants appear stunted and fewer branches are produced. Plants exhibit poor growth. Leaves cup inwards and plants are seen as if wilted in hot and dry weather conditions.
2. Potassium has high mobility within plants. If potassium supply to the plant becomes restricted then it is quickly transferred from older to younger plant parts. Therefore, the older leaves display deficiency symptoms first. The young leaves usually remain dark green.
3. The visible symptoms appear as a chlorosis in the form of yellow borders around the tip ends and along the edges of leaves that rapidly moves inward, leaving only the main veins and adjacent tissues green (Plate 275).
4. The yellowing (chlorosis) is followed by scorching that extends rapidly inward between the main veins to the entire leaf.
5. Eventually, the old leaves turn brown and shed prematurely.

Developmental stages

Stage I: As early symptoms, deficient plants exhibit poor growth and wilting of the leaves in dry conditions.
Stage II: If deficiency persists, marginal chlorosis develops in the older leaves (Plates 276 and 277).
Stage III: With advancement of the symptoms the yellowing rapidly extends inward, leaving only the main veins and adjoining tissues green (Plate 278).
Stage IV: Later, the chlorotic leaves become scorched, then brown and fall off early.

Likely to occur in

1. Soils formed from parent material low in potassium.
2. Light-textured soils where potassium has been leached by heavy rainfall or excessive irrigation.
3. Soils low in organic matter.
4. Soils with high Na:K, Mg:K or Ca:K ratio.
5. Acid soils having pH below 6.0.

Integrated nutrient management

1. Analyse the soil before sowing or planting to measure the amount of plant-available potassium.
2. Problematic acid/alkaline soils should be reclaimed.
3. Add organic manures well before planting.
4. Apply soluble potassium salts such as potassium chloride, potassium sulphate or potassium nitrate to the soil at or before planting as per soil testing recommendations.
5. If deficiency appears in standing crops, apply soluble potassium salts with irrigation water.

Further reading

Asghar, A.U., Asghar, M., Abid, R.A., Tahir, M. and Arif, H. (1994) Effect of potassium on yield and yield components of black gram. *Pakistan Journal of Agricultural Sciences* 31, 275–278.

Rajendran, J., Sivaphah, A.N. and Krishnamoorthy, K.K. (1974) Effect of fertilization on yield and nutrient concentration of black gram. *Madras Agricultural Journal* 61, 447–450.

Raval, D.R. and Yadav, G.L. (1986) Fertilizer requirement of urd (*Phaseolus mungo* L.) in dryland conditions on cultivated field in Chittorgarh district. *Legume Research* 9, 111–113.

Subrahmanyam, K. (1987) Effect of N, K and Ca deficiency on growth pattern of blackgram. *Indian Journal of Plant Physiology* 300, 205–207.

Plate 279. Stunted plant showing yellow top leaves, pale yellow middle leaves and light green bottom leaves. (Photo by Dr Prakash Kumar and Dr Manoj Kumar Sharma.)

Plate 280. Yellowing of entire plants with the young leaves being palest. (Photo by Dr Prakash Kumar and Dr Manoj Kumar Sharma.)

Plate 281. Young leaves appearing yellow and lower leaves staying green. (Photo by Dr Prakash Kumar and Dr Manoj Kumar Sharma.)

Plate 282. Uniform yellowing of the entire leaf and veins. (Photo by Dr Prakash Kumar and Dr Manoj Kumar Sharma.)

BLACK GRAM (*Phaseolus mungo* var. *radiatus* Linn.) SULPHUR (S) DEFICIENCY

Symptoms

1. Sulphur deficient-plants appear stunted with short internodes and have reduced branching.
2. The number and size of pods are decreased. Fewer seeds are formed, resulting in poor yields.
3. Sulphur deficiency symptoms are often seen in the early growth stage of the crop.
4. Sulphur is an immobile nutrient within plants and is not readily translocated from older to younger tissues in the plant. Thus, the deficiency symptoms typically appear first on younger leaves (Plate 281).
5. Initially, the entire plant becomes chlorotic. As the deficiency advances, the most yellowing occurs on younger leaves (Plate 280).
6. The yellowing (chlorosis) appears uniformly on the entire leaf (Plate 282).
7. In acute deficiency or prolonged deficiency conditions, the entire plant turns yellow with the youngest leaves being the most affected.

Developmental stages

Stage I: In early deficiency symptoms, plants become stunted with a light green colour of the entire plant.
Stage II: If deficiency persists, the whole plant turns chlorotic and the young leaves appear palest (Plate 280).
Stage III: In severe deficiency conditions, the youngest leaves turn yellow (Plate 279).
Stage IV: In the later stage the entire plant turns yellow.

Likely to occur in

1. Soils having low organic matter.
2. Light-textured sandy soils that have been leached by heavy rainfall or excessive irrigation.
3. Soils exhausted by intensive cropping.
4. Soils derived from parent material that is inherently low in sulphur.
5. Acid soils having pH below 6.0.

Integrated nutrient management

1. Get the soil analysed before sowing to measure the amount of 'available' sulphate in the soil.
2. Apply analysis-based recommended quantity of sulphur by mixing into the surface soil, well before sowing, either:
  a. elemental sulphur; or
  b. gypsum.
3. In deficient standing crops apply ammonium sulphate, magnesium sulphate or potassium sulphate with irrigation water.

Further reading

Khan, T.A. and Majid, M. (2011) Nutritional significance of sulphur in pulse cropping system. Special Issue. *Biology and Medicine* 3(2), 114–133.

Rathi, B.K., Jain, A.K., Kumar, S. and Panwar, J.D.S. (2009) Response of *Rhizobium* inoculation with sulphur and micronutrients on yield and yield attributes of blackgram [*Vigna mungo* (L.) Hepper]. *Legume Research* 32, 62–64.

Singh, Y.P. and Aggarwal, R.L. (1998) Effect of sulphur levels on yields, nutrient uptake and quality of blackgram (*Phaseolus mungo*). *Indian Journal of Agronomy* 43, 448–452.

Singh, Y.P. and Ranbir, S. (2004) Interaction effect of sulphur and phosphorus on growth and nutrient content of black gram (*Phaseolus mungo* L.). *Journal of the Indian Society of Soil Science* 25, 266–269.

Plate 283. Yellow (completely devoid of chlorophyll) younger leaves and normal green old leaves. (Photo by Dr Prakash Kumar and Dr Manoj Kumar Sharma.)

Plate 284. Interveinal chlorotic tissues and easily recognizable veins in top leaves. (Photo by Dr Prakash Kumar and Dr Manoj Kumar Sharma.)

Plate 285. Severely deficient evenly yellow trifoliate. (Photo by Dr Prakash Kumar and Dr Manoj Kumar Sharma.)

Plate 286. Acutely deficient papery white trifoliate with some brown necrotic tissues. (Photo by Dr Prakash Kumar and Dr Manoj Kumar Sharma.)

BLACK GRAM (*Phaseolus mungo* var. *radiatus* Linn.) IRON (Fe) DEFICIENCY

Symptoms

1. Iron deficiency depresses nodule development and nitrogenase activity.
2. Iron deficiency causes reductions in dry matter yield of black gram.
3. Stems become thin and elongated. The leaves become smaller in size. Reduced flowering and poor pod set result in poor yields. Deficient plants get delayed in maturity.
4. Iron is considered an immobile nutrient within plants and is not rapidly transferred from older to younger leaves under restricted supply conditions. Hence, deficiency symptoms are observed primarily on the younger leaves.
5. The symptoms are manifested as interveinal chlorosis occurring on young leaves while the lower leaves remain green (Plate 283).
6. Yellowing appears in interveinal tissues of young leaves while the main veins appear prominently green (Plate 284).
7. The chlorotic leaves then turn almost white and the veins also disappear (Plate 285).
8. In the later stage, a brown necrosis occurs especially in marginal areas of chlorotic (completely devoid of chlorophyll) leaves.

Developmental stages

Stage I: In mild deficiency, pale yellow chlorosis develops in interveinal tissues of young leaves and old leaves stay green (Plate 284).
Stage II: If deficiency continues, the leaves turn yellow and fading of veins occurs.
Stage III: In severe deficiency conditions the entire leaf turns papery white and the veins disappear (Plate 286).
Stage IV: In the advanced stage leaves develop brown necrotic lesions and die.

Likely to occur in

1. Sandy soils having low total iron content.
2. Alkaline and calcareous soils where solubility of iron is reduced.
3. Peat and muck soils where organic matter ties up iron and reduces its availability in soil solution.
4. Acid soils with excessively high levels of soluble zinc, manganese, copper or nickel that can hinder the uptake of iron despite high iron availability.
5. Soils with high levels of phosphorus.
6. Alkaline soils having pH above 7.5.

Integrated nutrient management

1. Analyse the soil before sowing to measure the 'available' iron in the soil.
2. Reclaim problematic alkaline soils.
3. Add organic manure well before sowing.
4. Apply basal dose of soluble iron fertilizers such as $FeSO_4$ (commonly at 25 kg/ha) or Fe chelates (10 kg/ha). Use of organic chelates proves to be more promising as they maintain iron in soil solution.
5. In standing crops, apply $FeSO_4$ or Fe chelates (0.5% w/v solution) as a foliar spray. Foliar sprays must be repeated every 10–15 days.

Further reading

Brown, J.C. (1961) Iron chlorosis in plants. In: Norman, A.G. (ed.) *Advances in Agronomy*. Academic Press, Inc., London, pp. 329–366.

Katyal, J.C. and Vlek, P.L.G. (1984) Micronutrient problems in tropical Asia. *Fertilizer Research* 7, 69–94.

Rao, J.S.P. and Narayanan, A. (1990) Physiological aspects of iron deficiency in groundnut (*Arachis hypogaea* L.) and black gram (*Vigna mungo* L.). *Madras Agricultural Journal* 77, 151–157.

Singh, M.V. (2008) Micronutrient deficiencies in crops and soils in India. In: Alloway, B.J. (ed.) *Micronutrient Deficiencies in Global Crop Production*. Springer, Dordrecht, the Netherlands, pp. 93–126.

Plate 287. Zinc-deficient black gram plant. (Photo by Dr Prakash Kumar.)

Plate 288. Interveinal chlorosis of leaf. (Photo by Dr Prakash Kumar.)

Plate 289. Leaf showing advanced stage of interveinal chlorosis. (Photo by Dr Prakash Kumar.)

Plate 290. Chocolate brown necrotic lesions and holes in an affected leaf. (Photo by Dr Prakash Kumar.

BLACK GRAM (*Phaseolus mungo* var. *radiatus* Linn.) ZINC (Zn) DEFICIENCY

Symptoms

1. The zinc deficiency symptoms of black gram are similar to those of green gram. Both black gram and green gram are very sensitive to zinc deficiency. Deficiency symptoms appear more prominently during the initial stages of crop growth, usually within 2–3 weeks after sowing.
2. Zinc-deficient black gram plants are stunted with short stems and reduced branches. Leaflets are small and faded with a diseased appearance.
3. Zinc is mobile in black gram plants and under short supply conditions it is transferred from older to younger leaves. So, the deficiency symptoms appear first and more severely on the older leaves.
4. Symptoms begin as a faded, pale green interveinal chlorosis of older leaves. Interveinal chlorosis starts from the tip of the leaf and proceeds towards the base. The main veins remain green and prominent.
5. If deficiency persists and becomes more severe, the affected leaves develop chocolate brown necrotic spots and lesions on interveinal areas.
6. As symptoms advance tissues of necrotic spot areas drop from the leaf lamina, making small holes in the affected leaf.

Developmental stages

Stage I: Faded pale green interveinal chlorosis appears on older leaves. The chlorosis starts from the tip of the leaf and proceeds towards the base (Plate 287).
Stage II: The pale green interveinal chlorosis covers the entire affected leaf. The main veins remain green and prominent (Plates 288 and 289).
Stage III: The affected leaves develop chocolate brown necrotic spots and lesions on interveinal areas. The tissues of necrotic spot areas drop from the leaf lamina, making small holes in the affected leaf (Plate 290).

Likely to occur in

1. Leached sandy soils where total zinc is low.
2. Just-levelled soils where the subsoil is exposed for cultivation. Available zinc in surface soil is often double that of the subsoil.
3. Cool, wet weather.
4. Soil having heavy and excessive application of phosphate fertilizers, which hampers use of zinc by the crop.
5. Acid soils having pH below 5.0.
6. Alkaline soils having pH above 7.5.

Integrated nutrient management

1. Get the soil analysed before sowing to estimate the amount of 'available' zinc in the soil.
2. Problematic alkaline soils should be reclaimed.
3. Add organic manures well before sowing.
4. Apply 25–30 kg of zinc sulphate or 10 kg of zinc chelate per hectare every 2 years in zinc-deficient soils.
5. Do not mix zinc fertilizers with phosphate fertilizers.

Further reading

Andrew, C.S., Johnson, A.D. and Haydock, K.P. (1981) The diagnosis of zinc deficiency and effect of zinc on the growth and chemical composition of some tropical and sub-tropical legumes. *Communications in Soil Science and Plant Analysis* 12, 1–18.

Gupta, V.K. (1983) Studies of the comparative response of some kharif crops to zinc application. *Agricultural Science Digest* 3(2), 79–80.

Pandey, N. and Gupta, B. (2012) Improving seed yield of black gram (*Vigna mungo* L. var. DPU-88-31) through foliar fertilization of zinc during the reproductive phase. *Journal of Plant Nutrition* 35, 1683–1692.

Veits, F.G., Brown, L.C. and Crawford, C.L. (1954) Zinc content and deficiency symptoms of 26 crops grown on zinc deficient soil. *Soil Science* 78, 305–316.

Plate 291. Nitrogen-deficient plant showing dark yellow old leaves, pale yellow middle leaves and light green top leaves. (Photo by Dr Prakash Kumar and Dr Manoj Kumar Sharma.)

Plate 292. Stunted plant with entirely pale green leaves. (Photo by Dr Prakash Kumar.)

Plate 293. Yellow chlorotic leaf showing tints of green colour. (Photo by Dr Prakash Kumar.)

Plate 294. Severely nitrogen-deficient dark yellow old leaf. (Photo by Dr Prakash Kumar and Dr Manoj Kumar Sharma.)

COWPEA (*Vigna sinensis* Linn.) NITROGEN (N) DEFICIENCY

Symptoms

1. Rapid increase in nitrogen fixation occurs between flowering and early pod fill stage.
2. Salinity decreases the number of nodules and nitrogen content of the plant.
3. The growth of deficient plants is retarded. Stems become thin and elongated.
4. Initially, the entire plant appears light green to pale yellow (Plate 292).
5. Since nitrogen is mobile within plants, it is rapidly translocated from older to younger leaves if the plant suffers from deficiency.
6. The deficiency symptoms tend to occur first in older leaves and then progress up the plant to new leaves.
7. Older leaves turn yellow while young leaves may remain pale green (Plate 291).
8. The new shoots may turn red to reddish-brown.
9. The old leaves become brown and necrotic. Eventually leaves drop off early.

Developmental stages

Stage I: Early or mild deficiency symptoms are expressed as pale green leaves of the entire plant (Plate 292).
Stage II: In prolonged deficiency, the older leaves turn evenly yellow (Plate 293) and the upper leaves appear pale green (Plate 291).
Stage III: In severe deficiency, the old leaves turn dark yellow and drop early (Plate 294).

Likely to occur in

1. Soils having low organic matter.
2. Light-textured sandy soils that have been leached by heavy rainfall or excessive irrigation.
3. Soils exhausted by intensive cropping.
4. Waterlogged conditions.
5. Acid soils having pH below 6.0.
6. Alkaline soils having pH above 8.0.

Integrated nutrient management

1. Get the soil analysed before sowing to measure the amount of 'available' nitrogen in the soil.
2. Apply analysis-based recommended quantity of nitrogen as basal by using:
  a. organic manures;
  b. bio-fertilizers;
  c. nitrogenous fertilizers.
3. Top-dress soluble nitrogenous fertilizers such as urea.
4. For quick recovery, apply urea (2% w/v solution) as a foliar spray in the standing crop. Foliar sprays must be repeated every 10–15 days.

Further reading

Balasubramanian, V. and Sinha, S.K. (1976) Effects of salt stress on growth, nodulation and nitrogen fixation in cowpea and mung beans. *Physiologia Plantarum* 36, 197–200.

Muchow, R.C., Robertson, M.J. and Pengelly, B.C. (1993) Accumulation and partitioning of biomass and nitrogen by soybean, mungbean and cowpea under contrasting environmental conditions. *Field Crops Research* 33, 13–36.

Senaratne, R. and Ratnasinghe, D.S. (1993) Ontogenic variation in nitrogen fixation and accumulation of nitrogen in mungbean, blackgram, cowpea and groundnut. *Biology and Fertility of Soils* 16, 125–130.

Plate 295. Stunted plant showing severely deficient bottom leaves, moderately deficient middle leaves and normal top leaves. (Photo by Dr Prakash Kumar and Dr Manoj Kumar Sharma.)

COWPEA (*Vigna sinensis* Linn.) MAGNESIUM (Mg) DEFICIENCY

Plate 296. Moderately deficient plant with pale yellow chlorosis on older leaves while the top leaves remain green. (Photo by Dr Prakash Kumar.)

Plate 297. Pale yellow chlorotic trifoliate with clearly visible veins and sub-veins. (Photo by Dr Prakash Kumar.)

Plate 298. Whitish chlorotic interveinal tissues with necrotic areas in a severely deficient trifoliate. (Photo by Dr Prakash Kumar and Dr Manoj Kumar Sharma.)

Symptoms

1. Magnesium deficiency results in the retarded growth of the plant. Leaves become thin and brittle. The number of pods is reduced and fewer seeds are formed per pod.
2. If magnesium supply is reduced, then it is readily transferred from older to the upper leaves of the plant because magnesium is mobile within plants.
3. The older leaves display visible symptoms first, which then work up the plant to younger leaves (Plate 295).
4. The older leaves display marginal and interveinal chlorosis, leaving the veins prominently green (Plate 296).
5. In severe symptoms, necrosis occurs in interveinal areas (Plate 298).
6. Eventually, leaves die and fall off.

Developmental stages

Stage I: In early deficiency, interveinal yellowing appears in lower leaves (Plates 296 and 297).
Stage II: In prolonged deficient conditions, bottom leaves become yellow to white chlorotic, leaving the veins dark green, and symptoms also proceed to younger leaves (Plate 295).
Stage III: In severe deficiency, necrotic lesions develop on the margins and interveinal tissues of most affected bottom leaves (Plate 295).
Stage IV: In the later stage, the affected leaves shed prematurely.

Likely to occur in

1. Acid sandy soils that have been leached by heavy rainfall.
2. Peat and muck soils that are low in total magnesium.
3. Soils having higher quantity of calcium or potassium.
4. Soils derived from parent material that is inherently low in magnesium.
5. Acid soils having pH below 6.5.
6. Alkaline soils having pH above 8.5.

Integrated nutrient management

1. Get the soil analysed before sowing to measure the amount of 'available' magnesium in the soil.
2. Apply analysis-based recommended quantity of magnesium before sowing using soluble salts such as magnesium sulphate or magnesium chloride.
3. In deficient standing crops, apply soluble salts such as magnesium sulphate, chloride or nitrate with irrigation water.

Further reading

Shengguang, X., Xinrong, L., Peiling, L., Shuyi, L., Huidian, Z., Yucan, Z. and Riqiang, L. (2005) Effects of magnesium and micro-elements fertilizers on the quality and yields of *Vigna sinensis* (L.) Savi in two different soil types. *Journal of Nanjing Agricultural University* 28, 59–63.

Walsh, T. and O'Donohoe, T.F. (1945) Magnesium deficiency in some crop plants in relation to the level of potassium nutrition. *Journal of Agricultural Science* 35, 254–263.

Wilkinson, S.R., Welch, R.M., Mayland, H.F. and Grunes, D.L. (1990) Magnesium in plants: uptake, distribution, function, and utilization by man and animals. In: Sigel, H. (ed.) *Compendium on Magnesium and Its Role in Biology, Nutrition, and Physiology, Vol. 26*. Institute of Inorganic Chemistry, University of Basel, Basel, Switzerland, pp. 34–56.

Woodruff, J.R. (1972) Plant chemistry of magnesium. In: Jones, J.B., Blount, M.C. and Wilkinson, S.R. (eds) *Magnesium in the Environment, Soils, Crops, Animals, and Man. Proceedings of a Symposium held at Fort Valley State College, Fort Valley, Georgia, 5–6 October 1972*. Taylor County Printing Company, Reynolds, Georgia, pp. 41–60.

Plate 299. Plant in early developmental stage showing pale green young leaves and normal green old leaves. (Photo by Dr Prakash Kumar.)

Plate 300. Uniformly pale yellow young trifoliate. (Photo by Dr Prakash Kumar and Dr Manoj Kumar Sharma.)

Plate 301. Top leaves appearing dark yellow, middle leaves pale yellow and bottom leaves light green. (Photo by Dr Prakash Kumar and by Dr Manoj Kumar Sharma.)

Plate 302. Severe deficiency revealed by uniform yellowing in a young trifoliate. (Photo by Dr Prakash Kumar and Dr Manoj Kumar Sharma.)

COWPEA (*Vigna sinensis* Linn.) SULPHUR (S) DEFICIENCY

Symptoms

1. Sulphur-deficient plants become stunted. Yields are reduced markedly. Sulphur deficiency results in delayed maturity.
2. Sulphur deficiency can occur at any developmental stage of the crop, whenever the sulphur supply is reduced.
3. Initially, the whole plant may appear light green.
4. Since sulphur is considered an immobile nutrient within plants, under deficient conditions it is not easily mobilized from older to younger tissues. Thus, the deficiency symptoms typically appear first on younger leaves.
5. The younger leaves become pale green to pale yellow (Plate 299).
6. The chlorosis develops evenly in the leaves, covering entire tissues including the veins (Plates 300 and 302).
7. In severe deficiency conditions or in prolonged deficiency, the young leaves turn yellow whereas the old leaves may appear pale green.

Developmental stages

Stage I: In mild deficiencies, fading of green colour occurs on the entire plant; however, old leaves look darker (Plate 299).
Stage II: If the deficiency exists for long, the younger leaves turn yellow and the lower leaves may appear pale green (Plate 301).
Stage III: In acute deficiency conditions, the whole plant turns yellow.

Likely to occur in

1. Soils having low organic matter.
2. Light-textured sandy soils that have been leached by heavy rainfall or excessive irrigation.
3. Soils exhausted by intensive cropping.
4. Soils derived from parent material that is inherently low in sulphur.
5. Acid soils having pH below 6.0.

Integrated nutrient management

1. Get the soil analysed before sowing to measure the amount of 'available' sulphate in the soil.
2. Apply analysis-based recommended quantity of sulphur by mixing into the surface soil, well before sowing, either:
  a. elemental sulphur; or
  b. gypsum.
3. In deficient standing crops apply ammonium sulphate, magnesium sulphate or potassium sulphate with irrigation water.

Further reading

Aulakh, M.S. (2003) Sulphur nutrition to crops. In: Abrol, Y.P. and Ahmed, A. (eds) *Sulphur in Plants*. Kluwer Academic Publishers, Dordrecht, the Netherlands, pp. 341–358.

Aulakh, M.S. and Chibba, I.M. (1992) Sulphur in soils and responses of crops to its application in Punjab. *Fertilizer News* 37(9), 33–45.

Aulakh, M.S. and Pasricha, N.S. (1986) Role of sulphur in the production of the grain legumes. *Fertilizer News* 31(9), 31–35.

Aulakh, M.S., Sidhu, B.S., Arora, B.R. and Singh, B. (1985) Contents and uptake of nutrients by pulses and oilseed crops. *Indian Journal of Ecology* 12, 238–242.

Plate 303. Severely deficient plant having yellow to white younger leaves with sharp green veins and green bottom leaves. (Photo by Dr Prakash Kumar.)

COWPEA (*Vigna sinensis* Linn.) IRON (Fe) DEFICIENCY

Plate 304. Yellow chlorotic trifoliate with prominent green veins. (Photo by Dr Prakash Kumar and Dr Manoj Kumar Sharma.)

Plate 305. As deficiency advances, yellow trifoliate turns to white and veins stay green. (Photo by Dr Prakash Kumar and Dr Manoj Kumar Sharma.)

Plate 306. Very severely deficient youngest trifoliate turns papery white and has faded green veins. (Photo by Dr Prakash Kumar.)

Symptoms

1. Iron is essential for chlorophyll synthesis in plants. In deficient iron supply, plants become unable to form chlorophyll.
2. Cowpea is sensitive to iron deficiency. Deficient plants become stunted and thin stemmed. Fewer pods are formed and the number of seeds per pod is reduced, consequently producing poor yields. The crop takes more time to mature.
3. Under reduced supply conditions, iron is not readily moved from older to younger tissues because it is immobile within plants. The younger leaves display deficiency symptoms first.
4. The younger leaves become yellow with recognizable dark green veins; the yellowing extends the full length of the leaves (Plates 304 and 305).
5. The yellow leaves then turn white. The chlorosis also spreads to lower leaves of the plant (Plate 303).
6. Eventually, the leaf margins become scorched and the leaf tips show browning.

Developmental stages

Stage I: In mild deficiency, pale green to pale yellow interveinal chlorosis develops in younger leaves.
Stage II: When deficiency persists, pale green young leaves turn yellow with green veins (Plate 304).
Stage III: In severe deficiency conditions, the entire leaf blade becomes white and the veins remain faded green (Plate 306).
Stage IV: In the later stage, leaves may develop marginal scorching and brown necrosis on leaf tips.

Likely to occur in

1. Sandy soils having low total iron content.
2. Alkaline and calcareous soils where solubility of iron is very low.
3. Peat and muck soils where organic matter ties up iron and reduces its availability in soil solution.
4. Acid soils with excessively high levels of soluble zinc, manganese, copper or nickel that can hinder the uptake of iron despite high iron availability.
5. Alkaline soils having pH above 7.5.

Integrated nutrient management

1. Analyse the soil before sowing to measure the 'available' iron in the soil.
2. Reclaim problematic alkaline soils.
3. Add organic manure well before sowing.
4. Apply basal dose of soluble iron fertilizers such as $FeSO_4$ (commonly at 25 kg/ha) or Fe chelates (10 kg/ha). Use of organic chelates proves to be more promising as they maintain iron in soil solution.
5. In standing crops, apply $FeSO_4$ or Fe chelates (0.5% w/v solution) as a foliar spray. Foliar sprays must be repeated every 10–15 days.

Further reading

Ajakaiya, C.O. (1986) Effect of P and Fe applications to cowpea in three soil types. *Plant and Soil* 94, 235–245.

Gupta, V.K. and Kala, R. (1980) Zinc, iron, and manganese nutrition in cowpeas (*Vigna sinensis* L.) as influenced by molybdenum and copper. *Haryana Agricultural University Journal of Research* 10, 380–385.

Jacobs, J.W. and Walker, R.B. (1977) Localization of iron in *Vigna sinensis* (L.) and *Zea mays* (L.). *Journal of Agricultural and Food Chemistry* 25, 803–806.

Marsh, H.V., Evans, H.J. Jr and Matrone, G. (1963) Investigations of the role of iron in chlorophyll metabolism. II. Effect of iron deficiency on chlorophyll synthesis. *Plant Physiology* 38, 638–642.

Plate 307. Nitrogen-deficient cluster bean plant. (Photo by Dr Prakash Kumar.)

CLUSTER BEAN (*Cyamopsis tetragonoloba* (L.) Taub) NITROGEN (N) DEFICIENCY

Plate 308. Entire plant appearing light green. (Photo by Dr Prakash Kumar and Dr Manoj Kumar Sharma.)

Plate 309. Comparison of a deficient plant (right) with a healthy plant (left). (Photo by Dr Prakash Kumar and Dr Manoj Kumar Sharma.)

Plate 310. Close-up of pale yellow older leaves. (Photo by Dr Prakash Kumar.)

Symptoms

1. Nitrogen deficiency is usually found during the initial stages of crop growth when root symbiotic nitrogen fixation nodules are yet to develop.
2. Nitrogen deficiency may also occur during later stages of crop growth when the symbiotic nitrogen-supplying mechanism is disturbed for some reason such as nodule infestation, nodule pathogenic disease or physiological causes.
3. Nitrogen is mobile in plants and under short supply conditions it is easily mobilized from older to younger leaves. The deficiency symptoms appear first and more severely on old leaves (Plate 307).
4. In mild deficiency conditions or when deficiency occurs in the young stage, the entire plant appears uniformly light green in colour (Plates 308 and 309).
5. If deficiency persists and becomes more severe, the older leaves show uniform pale green to pale yellow chlorosis (Plate 307).
6. Affected older leaves abscise (shed).

Developmental stages

Stage I: The entire plant appears uniformly light green in colour (Plates 308 and 309).
Stage II: If the deficiency persists, the older leaves show chlorosis (Plate 307).
Stage III: As the symptoms advance, affected older leaves become uniform pale yellow and middle leaves pale green, while top leaves remain green and healthy (Plates 307 and 310).
Stage IV: In the most advanced stage the affected leaves abscise.

Likely to occur in

1. Soils having low organic matter.
2. Light-textured sandy soils that have been leached by heavy rainfall or excessive irrigation.
3. Soils exhausted by intensive cropping.
4. Waterlogged conditions.
5. Acid soils having pH below 6.0.
6. Alkaline soils having pH above 8.0.
7. Crops with poor *Rhizobium* nodulation or with damaged *Rhizobium* nodules.

Integrated nutrient management

1. Get the soil analysed before sowing to measure the amount of 'available' nitrogen in the soil.
2. Inoculate the crop with an appropriate strain of *Rhizobium* culture through seed treatment.
3. The crop needs nitrogen during the initial stage of growth when symbiotic nitrogen fixation by the plant is yet to start. Thus, basal starter application of nitrogen at 20–25 kg/ha is important in nitrogen-deficient soils.

Further reading

Haag, H.P., Campora, P. and Forti, L.H.S.P. (1990) Nutrient deficiency symptoms and foliar nutrient levels in cluster bean plants. *Anais da Escola Superior de Agricultura 'Luiz de Queiroz' (Brazil)* 47, 251–260.

Keating, B.A. (1984) *Foliar Symptoms of Nutrient Disorders in Guar (Cyamopsis tetragonoloba). Division of Tropical Crops and Pastures Technical Paper No. 27*. CSIRO Publishing, Melbourne, Australia.

Plate 311. Potassium-deficient cluster bean plant. (Photo by Dr Prakash Kumar.)

Plate 312. Marginal chlorosis starts from the tip of the leaf. (Photo by Dr Prakash Kumar.)

Plate 313. Yellow chlorosis covering leaf margins. (Photo by Dr Prakash Kumar.)

Plate 314. Chlorosis followed by scorching around the margin. (Photo by Dr Prakash Kumar.)

CLUSTER BEAN (*Cyamopsis tetragonoloba* (L.) Taub) POTASSIUM (K) DEFICIENCY

Symptoms

1. Cluster bean is very sensitive to potassium deficiency. Deficiency symptoms appear even in mild deficiency conditions.
2. The potassium-deficient crop lacks vigour, matures slowly and yields less grains.
3. Potassium moves readily from old to young leaves; therefore deficiency symptoms appear first and more severely on older leaves (Plate 311).
4. The symptoms begin as marginal yellowing of older leaves. Yellowing starts from the leaf tip and advances around the leaf margin.
5. If deficiency persists, the leaf tip becomes scorched. The scorching spreads around the leaf margin (Plate 314).
6. A yellow band is found between the scorched area and healthy green tissues.
7. In acute deficiency conditions, the symptoms move towards upper leaves and may cover the entire plant.

Developmental stages

Stage I: A pale yellow chlorosis appears on older leaves starting from the tips of the leaf (Plate 312).
Stage II: As the symptoms advance, pale yellow chlorosis covers the margins of the lamina (Plate 313).
Stage III: The leaf tip becomes scorched as the symptoms become severe. Yellow chlorosis is followed by brown necrosis along the margins of the leaf (Plate 314).
Stage IV: In acute deficiency conditions, the symptoms move towards upper leaves and may cover the entire plant.

Likely to occur in

1. Soils formed from parent material low in potassium.
2. Light-textured soils where potassium has been leached by heavy rainfall or excessive irrigation.
3. Soils low in organic matter.
4. Soils with wide Na:K, Mg:K or Ca:K ratio.
5. Large bicarbonate concentration in irrigation water.
6. Highly acid soils having pH below 6.0.

Integrated nutrient management

1. Get the soil analysed before sowing to measure the amount of 'available' potassium in the soil.
2. Problematic acid soils should be reclaimed.
3. Add organic manures well before sowing, if organic matter is low.
4. Apply potassium nitrate, potassium sulphate or potassium chloride to the soil at or before sowing as per soil testing recommendations.
5. If potassium deficiency appears on the standing crop, apply soluble potassium salts such as potassium nitrate, potassium sulphate or potassium chloride with irrigation water. Foliar application of these salts is usually not recommended because a number of sprays are needed to fulfil crop requirements.

Further reading

Haag, H.P., Campora, P. and Forti, L.H.S.P. (1990) Nutrient deficiency symptoms and foliar nutrient levels in cluster bean plants. *Anais da Escola Superior de Agricultura 'Luiz de Queiroz' (Brazil)* 47, 251–260.

Keating, B.A. (1984) *Foliar Symptoms of Nutrient Disorders in Guar (Cyamopsis tetragonoloba). Division of Tropical Crops and Pastures Technical Paper No. 27.* CSIRO Publishing, Melbourne, Australia.

Rao, C.S. (2012) Impacts of participatory site specific potassium management in several crops in rainfed regions of Andhra Pradesh, India. *E-ifc* no. 30, March 2012; available at http://www.ipipotash.org/en/eifc/2012/30/3 (accessed 3 April 2013).

Smartt, J. (1976) *Tropical Pulses*. CAB International, Wallingford, UK.

Plate 315. Magnesium-deficient cluster bean plant. (Photo by Dr Prakash Kumar.)

CLUSTER BEAN (*Cyamopsis tetragonoloba* (L.) Taub) MAGNESIUM (Mg) DEFICIENCY

Plate 316. Interveinal chlorosis of lower leaves. (Photo by Dr Prakash Kumar.)

Symptoms

1. Magnesium deficiency in cluster bean is often confused with the symptoms of white fly infestation on the crop. White fly infestation also creates a resembling interveinal chlorosis on lower leaves. Confusion may be removed by observing the lower surface of affected leaves. In the case of infestation, white fly eggs, nymphs and flies, which are easily visible to the naked eye, are found on the lower leaf surfaces.
2. Magnesium is mobile in plants. The deficiency symptoms appear first and more severely on the older leaves. The youngest leaves remain green and apparently healthy (Plate 315).
3. A whitish pale yellow interveinal chlorosis develops on older leaves. All veins (primary and secondary) of the affected leaves remain green (Plates 316 and 317).
4. If the deficiency persists and becomes more severe, a bright yellow interveinal chlorosis develops on lower leaves (Plate 318).

Plate 317. Close-up of an affected leaf. (Photo by Dr Prakash Kumar.)

Developmental stages

Stage I: A whitish pale yellow interveinal chlorosis develops on older leaves. Veins remain green and prominent (Plates 315 and 316).
Stage II: If the deficiency persists and becomes more severe, interveinal chlorosis turns bright yellow (Plate 318).
Stage III: As symptoms advance, interveinal regions turn necrotic.

Likely to occur in

1. Acid sandy soils from which magnesium has been leached by heavy rainfall.
2. Peat and muck soils that are low in total magnesium.
3. Soils derived from parent material that is inherently low in magnesium.
4. Soils with heavy and excess application of potassium fertilizers.
5. Soils with heavy and excess application of lime (calcium carbonate) or other calcium fertilizers.
6. Acid soils having pH below 6.5.
7. Alkaline soils having pH above 8.5.

Integrated nutrient management

1. Get the soil analysed before sowing to estimate the amount of soluble and exchangeable magnesium in the soil.
2. Apply analysis-based recommended quantity of magnesium before sowing using soluble salts such as magnesium sulphate or magnesium chloride.
3. In deficient standing crops, apply soluble salts such as magnesium sulphate, chloride or nitrate with irrigation water. Foliar sprays of these salts are usually not advised as many sprays at frequent intervals are required to fulfil the crop need.
4. Magnesium deficiency in acid soils may be corrected by applying dolomite (a mixture of calcium carbonate and magnesium carbonate, $CaCO_3 \cdot MgCO_3$) through broadcasting and mixing into the soil a few months before the sowing.

Plate 318. Interveinal chlorosis in advanced stage. (Photo by Dr Prakash Kumar.)

Further reading

Haag, H.P., Campora, P. and Forti, L.H.S.P. (1990) Nutrient deficiency symptoms and foliar nutrient levels in cluster bean plants. *Anais da Escola Superior de Agricultura 'Luiz de Queiroz' (Brazil)* 47, 251–260.

Keating, B.A. (1984) *Foliar Symptoms of Nutrient Disorders in Guar (Cyamopsis tetragonoloba). Division of Tropical Crops and Pastures Technical Paper No. 27*. CSIRO Publishing, Melbourne, Australia.

Smartt, J. (1976) *Tropical Pulses*. CAB International, Wallingford, UK.

Plate 319. Sulphur-deficient cluster bean. (Photo by Dr Prakash Kumar.)

Plate 320. Pale green sulphur-deficient upper leaves. (Photo by Dr Prakash Kumar.)

Plate 321. Close-up of uniform pale yellow leaves. (Photo by Dr Prakash Kumar and Dr Manoj Kumar Sharma.)

Plate 322. Close-up of affected leaves. (Photo by Dr Prakash Kumar.)

CLUSTER BEAN (*Cyamopsis tetragonoloba* (L.) Taub) SULPHUR (S) DEFICIENCY

Symptoms

1. The sulphur-deficient cluster bean crop lacks vigour and yields poorly. Affected plants are stunted with thin stems and pale green to pale yellow foliage.
2. Sulphur deficiency, even if severe, does not produce any type of necrosis or burning of leaves. The change in leaf colour is the only indicative symptom of sulphur deficiency.
3. At the initial stage, sulphur deficiency symptoms are often confused with those caused by nitrogen deficiency and iron deficiency.
4. Sulphur is immobile in plants and does not mobilize from older to younger leaves in short supply conditions. Thus, just the reverse to nitrogen deficiency, sulphur-deficiency symptoms appear first and more severely on younger leaves while older leaves remain green and healthy (Plates 319 and 320).
5. Deficiency symptoms appear as an even and uniform pale green to pale yellow chlorosis across the lamina of young leaves. The colour of the midrib and other veins becomes very similar to that of interveinal areas of the leaf (in the case of iron deficiency, these veins remain green and prominent).
6. If deficiency persists and becomes more severe, symptoms eventually move downwards covering more leaves.

Developmental stages

Stage I: Symptoms begin with a uniform fading of young leaves from green to pale green colour. Older leaves remain green and healthy (Plates 319 and 320).
Stage II: If the deficiency persists and becomes more severe, an even and uniform pale green to pale yellow chlorosis develops across the lamina of young leaves (Plates 321 and 322).
Stage III: In acute conditions, the deficiency symptoms move downwards covering more leaves.

Likely to occur in

1. Soils low in organic matter.
2. Light-textured sandy soils that have been leached by heavy rainfall or excessive irrigation.
3. Soils exhausted by intensive cropping.
4. Soils derived from parent material that is inherently low in sulphur (for example, soils formed from volcanic rocks).
5. Acid soils having pH below 6.0.

Integrated nutrient management

1. Get the soil analysed before sowing to measure the amount of 'available' sulphate in the soil and to predict the amount of fertilizer required.
2. Apply analysis-based recommended quantity of sulphur by mixing into the surface soil, well before sowing, either:
  a. elemental sulphur; or
  b. gypsum.
3. In deficient standing crops apply ammonium sulphate, magnesium sulphate or potassium sulphate with irrigation water.
4. Problematic acid soils should be reclaimed.

Further reading

Haag, H.P., Campora, P. and Forti, L.H.S.P. (1990) Nutrient deficiency symptoms and foliar nutrient levels in cluster bean plants. *Anais da Escola Superior de Agricultura 'Luiz de Queiroz' (Brazil)* 47, 251–260.

Keating, B.A. (1984) *Foliar Symptoms of Nutrient Disorders in Guar (Cyamopsis tetragonoloba). Division of Tropical Crops and Pastures Technical Paper No. 27*. CSIRO Publishing, Melbourne, Australia.

Oke, O.L. (1969) Sulphur nutrition in legumes. *Experimental Agriculture* 5, 111–116.

Smartt, J. (1976) *Tropical Pulses*. CAB International, Wallingford, UK.

Plate 323. Iron-deficient cluster bean plant. (Photo by Dr Prakash Kumar and Dr Manoj Kumar Sharma.)

Plate 324. Plant showing interveinal chlorosis on top leaves. (Photo by Dr Prakash Kumar and Dr Manoj Kumar Sharma.)

Plate 325. Close-up of affected leaves. (Photo by Dr Prakash Kumar and Dr Manoj Kumar Sharma.)

Plate 326. Leaf bleached to papery white. (Photo by Dr Prakash Kumar.)

CLUSTER BEAN (*Cyamopsis tetragonoloba* (L.) Taub) IRON (Fe) DEFICIENCY

Symptoms

1. Cluster bean is very sensitive to iron deficiency.
2. Iron is immobile in plants. Deficiency symptoms appear first and more severely on the younger leaves. The older leaves remain normal and apparently healthy (Plate 324).
3. In mild deficiency conditions or at the initial stage of deficiency, the topmost younger leaves develop temporary fading of interveinal tissues to pale yellow in colour.
4. If the deficiency persists and becomes more severe, a bright pale yellow chlorosis develops in interveinal tissues (tissues between the veins) leaving the veins green and prominent. Interveinal chlorosis of top leaves is the specific symptom of iron deficiency (Plates 324 and 325).
5. As symptoms advance, the entire leaf bleaches to papery white (Plate 326).
6. In acute deficiency conditions, the affected leaf eventually burns and dies.

Developmental stages

Stage I: The topmost leaves develop fading of interveinal tissues to pale yellow colour while veins remain prominent and green (Plate 323).
Stage II: Interveinal tissues of the affected leaves turn bright yellow with prominent green veins (Plates 324 and 325).
Stage III: The affected top leaves bleach to papery white (Plate 326).
Stage IV: The affected leaves eventually burn and die.

Likely to occur in

1. Sandy soils having low total iron.
2. Calcareous soils where solubility of iron is very low.
3. Peat and muck soils where organic matter ties up iron and reduces its availability in soil solution.
4. Acid soils where, despite high availability of iron in soil solution, excessively high levels of soluble zinc, manganese, copper or nickel hamper the uptake of iron by the plant.
5. Waterlogged soils.
6. Alkaline soils having pH above than 7.5.

Integrated nutrient management

1. Get the soil analysed before sowing to measure the amount of 'available' iron in the soil.
2. As inorganic soluble salts of iron such as iron sulphates or chlorides quickly become insoluble in soil, a soil dressing of organic forms of iron such as iron chelates (10 kg/ha) is more effective than inorganic forms of iron. These organic iron forms should be used as per soil pH. If the soil is neutral (around pH 7.0), the most suitable chelates are FeHEDTA and FeDTPA. For acid soils FeEDTA is the most effective chelate, while on alkaline calcareous soils FeEDDHA is recommended.

Further reading

Keating, B.A. (1984) *Foliar Symptoms of Nutrient Disorders in Guar (Cyamopsis tetragonoloba). Division of Tropical Crops and Pastures Technical Paper No. 27*. CSIRO Publishing, Melbourne, Australia.

Rathore, V.S., Singh, J.P., Soni, M.L., Yadav, N.D. and Beniwal, R.K. (2007) Productivity, quality and resource utilization by cluster bean (*Cyamopsis tetragonoloba*) as influenced by nutrient management. *Indian Journal of Agronomy* 52, 243–246.

Smartt, J. (1976) *Tropical Pulses*. CAB International, Wallingford, UK.

Plate 327. Zinc-deficient cluster bean plant. (Photo by Dr Prakash Kumar.)

Plate 328. Chlorosis of trifoliate leaf starting from the tips. (Photo by Dr Prakash Kumar.)

Plate 329. Close-up of an affected leaf. (Photo by Dr Prakash Kumar.)

Plate 330. Severely affected zinc-deficient leaf. (Photo by Dr Prakash Kumar.)

CLUSTER BEAN (*Cyamopsis tetragonoloba* (L.) Taub) ZINC (Zn) DEFICIENCY

Symptoms

1. Cluster bean has strong hidden hunger to zinc deficiency. Recognizable deficiency symptoms appear only in severe deficiency conditions.
2. The zinc-deficient cluster bean plant shows stunted growth with reduced size of branches and trifoliate leaves, providing a 'little leaf' appearance to the plant. Leaflets are very small, usually less than half the normal size. Flowering is delayed.
3. Zinc is partially mobile in cluster bean plants and, under short supply conditions, is not readily transferred from older to younger leaves. So, the deficiency symptoms appear first and more severely on middle leaves.
4. Symptoms begin as a fading of the green colour of leaves (Plate 329). If deficiency persists and becomes more severe, interveinal chlorosis starts from the tips of affected leaves and spreads to the remaining areas (Plate 330). The midrib and adjacent tissues remain green.
5. In acute deficiency conditions necrotic spots develop on affected areas of the leaf. In the most advanced stage, the faded leaf symptoms cover even the younger leaves.

Developmental stages

Stage I: The deficient plant produces small leaves, usually less than half of normal size.
Stage II: The recognizable visual symptoms begin as fading of the green colour of leaves (Plate 327).
Stage III: Interveinal chlorosis starts from the tips of affected leaves. All three leaflets of the affected trifoliate show chlorosis on leaf tips (Plate 328).
Stage IV: Necrotic spots develop on affected leaves in acute deficiency conditions.

Likely to occur in

1. Leached sandy soils where total zinc is low.
2. Just-levelled soils where the subsoil is exposed for cultivation. Available zinc in surface soil is often double that of the subsoil.
3. Cool wet weather.
4. Soils having heavy and excessive application of phosphate fertilizers, which hampers use of zinc by the crop.
5. Acid soils having pH below 5.0.
6. Alkaline soils having pH above 7.5.

Integrated nutrient management

1. Get the soil analysed before sowing to estimate the amount of 'available' zinc in the soil.
2. Add organic manures well before sowing.
3. Apply 25–30 kg of zinc sulphate or 10 kg of zinc chelate per hectare every 2 years in zinc-deficient soils.
4. Do not mix zinc fertilizers with phosphate fertilizers.
5. In fields with known zinc deficiency, if deficiency appears in the standing crop, apply a foliar application of 0.5% w/v solution of a soluble zinc salt (e.g. 5 g of $ZnSO_4 \cdot 7H_2O$ per litre of water) at 15–20 days after seedling emergence or as soon as symptoms appear.

Further reading

Haag, H.P., Campora, P. and Forti, L.H.S.P. (1990) Nutrient deficiency symptoms and foliar nutrient levels in cluster bean plants. *Anais da Escola Superior de Agricultura 'Luiz de Queiroz' (Brazil)* 47, 251–260.

Keating, B.A. (1984) *Foliar Symptoms of Nutrient Disorders in Guar (Cyamopsis tetragonoloba). Division of Tropical Crops and Pastures Technical Paper No. 27.* CSIRO Publishing, Melbourne, Australia.

Singh, K. (1986) The critical level of zinc in soil and plant for predicting response of cluster bean to zinc fertilization. *Plant and Soil* 94, 285–288.

Plate 331. Plant showing bottom leaves white, middle leaves yellow and top leaves green. (Photo by Dr Prakash Kumar.)

Plate 332. Nitrogen-deficient crop in foreground compared with nitrogen-fertilized crop behind. (Photo by Dr Prakash Kumar.)

Plate 333. Pink pigmentation on lower stem and pale yellow to white chlorotic older leaves. (Photo by Dr Prakash Kumar.)

Plate 334. Severely deficient whitish yellow leaflets with reddish pink coloration on the edges. (Photo by Dr Prakash Kumar.)

CHICKPEA (*Cicer arietinum* Linn.) NITROGEN (N) DEFICIENCY

Symptoms

1. The formation and development of nodules are restricted under soil salinity and sodicity conditions.
2. Nitrogen fixation reaches the maximum level at flowering stage and then declines sharply during pod filling.
3. Nitrogen deficiency restricts plant growth and reduces branching. Plants have fewer flowers. Fewer pods are formed resulting in poor yields.
4. When nitrogen supply becomes restricted the older leaves display deficiency symptoms first.
5. The entire plant appears chlorotic, while older leaves turn more yellow than upper leaves (Plate 332).
6. Pink pigmentation develops on the lower part of the stem (Plate 333).
7. In the later stage, the yellow older leaves turn white and drop prematurely.

Developmental stages

Stage I: Early deficiency symptoms are expressed as stunted growth and a uniform pale green appearance of the entire plant.
Stage II: In prolonged deficiency conditions, the lower leaves turn yellow with reddish pink margins and a pink coloration develops on the lower stem (Plate 331).
Stage III: In severely deficient conditions, the yellow older leaves turn white (Plate 334) whereas the younger leaves also become more chlorotic.
Stage IV: In acute deficiency conditions, the older leaves become necrotic and drop early.

Likely to occur in

1. Soils having low organic matter.
2. Light-textured sandy soils that have been leached by heavy rainfall or excessive irrigation.
3. Soils exhausted by intensive cropping.
4. Waterlogged conditions.
5. Acid soils having pH below 6.0.
6. Alkaline soils having pH above 8.0.

Integrated nutrient management

1. Get the soil analysed before sowing to measure the amount of 'available' nitrogen in the soil.
2. Apply analysis-based recommended quantity of nitrogen as basal by using:
- **a.** organic manures;
- **b.** bio-fertilizers;
- **c.** nitrogenous fertilizers.

3. Top-dress soluble nitrogenous fertilizers such as urea.
4. For quick recovery, apply urea (2% w/v solution) as a foliar spray in the standing crop. Foliar sprays require to be repeated every 10–15 days.

Further reading

Ahlawat, I.P.S. (1990) Diagnosis and alleviation of mineral nutrient constraints in chickpea. In: *Chickpea in the Nineties: Proceedings of the Second International Workshop on Chickpea Improvement, ICRISAT Center, India, 4–8 December 1989*. International Crop Research Institute for the Semi-Arid Tropics (ICRISAT), Patancheru, India, pp. 93–100.

Jones, J.B. (2002) Nutrient element deficiency symptoms. Appendix E. Legumes. In: *Agronomic Handbook: Management of Crops, Soils and Their Fertility*. CRC Press, Boca Raton, Florida, p. 409.

Kurdali, F. (1996) Nitrogen and phosphorus assimilation, mobilization and partitioning in rainfed chickpea (*Cicer arietinum* L.). *Field Crops Research* 47, 81–92.

Rao, D.L.N., Giller, K.E., Yeo, A.R. and Flowers, T.J. (2002) The effects of salinity and sodicity upon nodulation and nitrogen fixation in chickpea (*Cicer arietinum*). *Annals of Botany* 89, 563–570.

Plate 335. Plant showing dark green leaves with reddish purple discoloration in older leaves. (Photo by Dr Prakash Kumar and Dr Manoj Kumar Sharma.)

Plate 336. Purpling appearing on the edges of leaflets. (Photo by Dr Prakash Kumar.)

Plate 337. Purple pigmentation spreading inward from the edges. (Photo by Dr Prakash Kumar.)

Plate 338. Purple discoloration spreads over the upper surface of leaflets. (Photo by Dr Prakash Kumar.)

CHICKPEA (*Cicer arietinum* Linn.) PHOSPHORUS (P) DEFICIENCY

Symptoms

1. Phosphorus deficiency strongly increases the net release of protons from the roots. Phosphorus deficiency induces exudation of carboxylic acids in the roots of chickpea.
2. Balanced phosphorus and zinc nutrition is essential for the maximum activity of *Rhizobium* for nitrogen fixation.
3. Phosphorus-deficient plants become short and dark green in appearance. Root development is restricted. Maturity gets delayed and yield is reduced.
4. Phosphorus is a mobile nutrient within plants and under short supply conditions phosphorus is moved from older to younger tissues rapidly. The visible deficiency symptoms are first observed on older leaves.
5. The stem develops a reddish purple pigmentation that intensifies and becomes darker in prolonged deficiency conditions (Plate 335).
6. In phosphorus-deficient plants the top edges and upper surface of the leaflets of older leaves exhibit reddish purple discoloration (Plate 336).
7. The discoloration begins from the bottom leaflets of older leaves and proceeds to upper leaflets, then working up the plant to younger leaves.
8. In some cases, older leaves develop a reddish brown discoloration on the upper surfaces that tends to be more pronounced at their tips and margins.
9. Later, the leaflets lose their colour and become bronze.

Developmental stages

Stage I: In the early stage of deficiency, plants appear small and dark green.
Stage II: As the symptoms advance, a reddish purple discoloration begins at the top edges and upper surface of bottom leaflets in older leaves (Plates 337 and 338).
Stage III: If deficiency becomes severe, the old leaves turn yellow–green.
Stage IV: In the later stage, the leaves turn bronze and fall off early.

Likely to occur in

1. Soils having low organic matter.
2. Alkaline and calcareous soils.
3. Soils exhausted by intensive cropping.
4. Acid soils and highly weathered soils.
5. Soils where topsoil has been removed by erosion.
6. Acid soils having pH below 6.0.
7. Alkaline soils having pH between 7.5 and 8.5.

Integrated nutrient management

1. Get the soil analysed to measure the amount of 'available' phosphate in the soil well before sowing.
2. Apply analysis-based recommended quantity of phosphorus as basal by using:
- **a.** organic manures;
- **b.** phosphate-solubilizing microbial cultures;
- **c.** phosphatic fertilizers.

3. If standing crops show the deficiency symptoms, apply soluble phosphatic fertilizers such as ammonium phosphate with irrigation water.

Further reading

Idri, M., Mahmood, T. and Malik, K.A. (1989) Response of field grown chickpea (*Cicer arietinum* L.) to phosphorus fertilization for yield and nitrogen fixation. *Plant and Soil* 114, 135–138.
Romheld, V. and Neumann, G. (1999) Root excretions of carboxylic acids and protons in phosphorus-deficient plants. *Plant and Soil* 211, 121–130.
Shukla, U.C. and Yadav, O.P. (1982) Effect of phosphorus and zinc on nodulation and nitrogen fixation in chickpea (*Cicer arietinum* L.). *Plant and Soil* 65, 239–248.
Singh, O.N. and Ram, H. (1990) Effect of phosphorus and sulphur application on protein and amino acid contents in chickpea. *Indian Journal of Pulses Research* 3, 36–39.

Plate 339. Yellowing followed by necrosis at tips and margins in older leaflets, contrasted with dark green younger leaflets. (Photo by Dr Prakash Kumar.)

Plate 340. Potassium-deficient leaflets showing marginal yellowing at tips and margins. (Photo by Dr Prakash Kumar.)

Plate 341. Chlorosis of margins and brown necrosis at the tips of affected leaflets. (Photo by Dr Prakash Kumar.)

Plate 342. Yellowing extending inwards and necrotic patches occurring at tips. (Photo by Dr Manoj Kumar Sharma.)

CHICKPEA (*Cicer arietinum* Linn.) POTASSIUM (K) DEFICIENCY

Symptoms

1. Potassium-deficient plants appear stunted and produce less branching. They have reduced flowering and poor pod development.
2. Potassium-deficient plants easily show the sign of early wilting in dry conditions.
3. Potassium is regarded as a highly mobile nutrient within plants and has a tendency to move rapidly from older to younger tissues when its supply becomes restricted to the plant.
4. The visible deficiency symptoms tend to occur first in the lower leaves with severity and then progress to the upper leaves.
5. Lower leaves develop yellowing (chlorosis) at the tips and along the margins (Plate 340).
6. The stem turns reddish brown on the side that is exposed to direct sunlight at the base of the plant (Plate 339).
7. Brown necrosis begins in the form of patches at the tips of older leaves; ultimately the tips die and become light brown.
8. Eventually the affected leaflets shed prematurely.

Developmental stages

Stage I: In mild deficiency symptoms, plants lack vigour and show the sign of early wilting.
Stage II: If deficiency persists, a pale yellow chlorosis develops at the tips and along the margins of lower leaves (Plate 340).
Stage III: In severe deficiency conditions, the yellowing proceeds inwards rapidly, followed by necrosis and dying of tissues at leaf tips (Plates 341 and 342).
Stage IV: In the later stage, the older leaflets dry completely and drop early (Plate 339).

Likely to occur in

1. Soils formed from parent material low in potassium.
2. Light-textured soils where potassium has been leached by heavy rainfall or excessive watering.
3. Soils low in organic matter.
4. Soils having wide Na:K, Mg:K or Ca:K ratio.
5. Acid soils having pH below 6.0.

Integrated nutrient management

1. Analyse the soil before sowing to measure the amount of plant-available potassium.
2. Problematic acid/alkaline soils should be reclaimed.
3. Add organic manures well before planting.
4. Apply soluble potassium salts such as potassium chloride, sulphate or nitrate to the soil at or before planting.
5. In standing crops, apply soluble potassium salts with irrigation water.

Further reading

Ahlawat, I.P.S. (1990) Diagnosis and alleviation of mineral nutrient constraints in chickpea. In: *Chickpea in the Nineties: Proceedings of the Second International Workshop on Chickpea Improvement, ICRISAT Center, India, 4–8 December 1989*. International Crop Research Institute for the Semi-Arid Tropics (ICRISAT), Patancheru, India, pp. 93–100.

Jones, J.B. (2002) Nutrient element deficiency symptoms. Appendix E. Legumes. In: *Agronomic Handbook: Management of Crops, Soils and Their Fertility*. CRC Press, Boca Raton, Florida, p. 409.

Kurdali, F., Al-Ain, F. and Al-Shamma, M. (2002) Nodulation, dry matter production and N_2 fixation by fababean and chickpea as affected by soil moisture and potassium fertilizer. *Journal of Plant Nutrition* 25, 355–368.

Smith, F.W. and Pieters, W.H.J. (1983) *Foliar Symptoms of Nutrient Disorders in Chickpea (Cicer arietinum). Division of Tropical Crops and Pasture Technical Paper No. 23*. CSIRO Publishing, Melbourne, Australia.

Plate 343. Sulphur-deficient plant showing pale green younger leaves contrasted with dark green older leaves. (Photo by Dr Prakash Kumar.)

Plate 344. Yellowing intensified on younger leaflets. (Photo by Dr Prakash Kumar.)

Plate 345. Chlorosis starting from the younger leaves. (Photo by Dr Prakash Kumar.)

Plate 346. Leaflets showing uniform yellowing. (Photo by Dr Prakash Kumar.)

CHICKPEA (*Cicer arietinum* Linn.) SULPHUR (S) DEFICIENCY

Symptoms

1. Sulphur nutrition imparts cold tolerance in chickpea under low-temperature stress conditions.
2. Sulphur-deficient plants become smaller and slender.
3. The yield is severely reduced as the deficient plants produce fewer pods and smaller seeds.
4. Sulphur deficiency symptoms are often seen in the early growth stage of the crop.
5. Sulphur is fairly immobile within the plant body and it is not easily mobilized from older to younger leaves under reduced supply of sulphur to the plant.
6. The deficiency symptoms of sulphur first appear and become more severe in younger leaves (Plates 344 and 345).
7. The younger leaves turn pale green to pale yellow while the lower leaves remain dark green (Plate 343).
8. In severe deficiency conditions the entire plant turns chlorotic. However, the younger leaves appear more chlorotic.

Developmental stages

Stage I: In early symptoms, the younger leaves turn pale green while the older leaves remain dark green (Plate 343).
Stage II: As the symptoms advance, the chlorosis advances down the plant to lower leaves while the younger leaves turn palest.
Stage III: In severe deficiency conditions, the youngest leaflets turn completely yellow (Plate 346).

Likely to occur in

1. Soils having low organic matter.
2. Light-textured sandy soils that have been leached by heavy rainfall or excessive irrigation.
3. Soils exhausted by intensive cropping.
4. Soils derived from a parent material that is inherently low in sulphur.
5. Acid soils having pH below 6.0.

Integrated nutrient management

1. Get the soil analysed before sowing to measure the amount of 'available' sulphate in the soil.
2. Apply analysis-based recommended quantity of sulphur by mixing into the surface soil, well before sowing, either:
  a. elemental sulphur; or
  b. gypsum.
3. In deficient standing crops apply ammonium sulphate, magnesium sulphate or potassium sulphate with irrigation water.

Further reading

Gupta, N. and Singh, R.S. (1983) Response of Bengal gram (*Cicer arietinum* L.) to nitrogen, phosphorus and sulphur. *Journal of the Indian Society of Soil Sciences* 31, 156–159.
McLachlan, K.D. (1978) *An Atlas of Sulphur Deficiency in Commercial Plants*. CSIRO Publishing, Melbourne, Australia.
Rao, S.S. and Sahu, M.P. (1991) Effect of sulphur and foliar applied chemicals on cold tolerance in chickpea (*Cicer arietinum* L.). *Journal of Agronomy and Crop Science* 167, 3206–325.
Smith, F.W. and Pieters, W.H.J. (1983) *Foliar Symptoms of Nutrient Disorders in Chickpea (Cicer arietinum). Division of Tropical Crops and Pasture Technical Paper No. 23*. CSIRO Publishing, Melbourne, Australia.

Plate 347. Yellowish green younger leaves and dark green older leaves. (Photo by Dr Prakash Kumar.)

Plate 348. Leaflets of younger leaves are uniformly bright yellow to white. (Photo by Dr Prakash Kumar and Dr Manoj Kumar Sharma.)

Plate 349. Chlorosis advancing down the plant to lower leaves. (Photo by Dr Prakash Kumar.)

Plate 350. Newly emerged leaves are bright yellow. (Photo by Dr Prakash Kumar.)

CHICKPEA (*Cicer arietinum* Linn.) IRON (Fe) DEFICIENCY

Symptoms

1. Chickpea genotypes vary in their sensitivity to iron deficiency. Considerable yield losses due to iron-deficiency chlorosis may occur when susceptible genotypes are grown on calcareous soils with high pH.
2. Iron deficiency generally results in stunted growth. Deficient plants show poor nodulation, low leghaemoglobin production and reduced nitrogenase activity.
3. Iron is immobile within plants and is not readily transferred from lower leaves to upper leaves if its supply becomes restricted.
4. Plants display deficiency symptoms first on younger leaves (Plate 347).
5. Younger leaves turn bright yellow then white, while older leaves remain dark green and healthy (Plate 348).
6. As symptoms advance, white necrotic areas develop on the distal half of the leaflets in young leaves.
7. In the later stage, the white necrotic areas get enlarged and the leaves wither and die.
8. Eventually, the uppermost leaves die, become white and drop off.

Developmental stages

Stage I: In mild deficiency, the newly formed young leaves turn uniformly yellowish green (Plate 349).
Stage II: When deficiency advances, the young leaves turn bright yellow (Plate 350).
Stage III: In severe deficiency conditions, the entire leaflets of young leaves turn white (Plate 348).
Stage IV: Ultimately, the leaflets dry up and drop off.

Likely to occur in

1. Sandy soils having low total iron content.
2. Alkaline and calcareous soils where solubility of iron is very low.
3. Peat and muck soils where organic matter ties up iron and reduces its availability in soil solution.
4. Acid soils with excessively high levels of soluble zinc, manganese, copper or nickel that can hinder the uptake of iron despite high iron availability.
5. Alkaline soils having pH above 7.5.

Integrated nutrient management

1. Get the soil analysed before sowing to measure the 'available' iron in the soil.
2. Reclaim problematic alkaline soils.
3. Add organic manure well before sowing.
4. Apply basal dose of soluble iron fertilizers such as $FeSO_4$ (commonly at 25 kg/ha) or Fe chelates (10 kg/ha). Use of organic chelates proves to be more promising as they maintain iron in soil solution.
5. For standing crops, apply $FeSO_4$ or Fe chelates (0.5% w/v solution) as a foliar spray. Foliar sprays have to be repeated every 10–15 days.

Further reading

Ahlawat, I.P.S. (1990) Diagnosis and alleviation of mineral nutrient constraints in chickpea. In: *Chickpea in the Nineties: Proceedings of the Second International Workshop on Chickpea Improvement, ICRISAT Center, India, 4–8 December 1989*. International Crop Research Institute for the Semi-Arid Tropics (ICRISAT), Patancheru, India, pp. 93–100.

Briat, J.F., Fobis-Loisy, I., Grignon, N., Lobréaux, S., Pascal, N., Savino, G., Thoiron, S., Von Wirén, N. and Van Wuytswinkel, O. (1995) Cellular and molecular aspects of iron metabolism in plants. *Biology of the Cell* 84, 69–81.

Tang, C., Robson, A.D. and Dilworth, M.J. (1992) The role of iron in the (brady) *Rhizobium* legume symbiosis. *Journal of Plant Nutrition* 15, 2235–2252.

Toker, C., Canci, H. and Inci, N.E. (2012) Pyramiding of the resistance to Fe-deficiency chlorosis and leaf miner (*Liriomyza cicerina* Rond.) in chickpea (*Cicer arietinum* L.) by mutation breeding. *Turkish Journal of Field Crops* 17, 41–45.

Plate 351. Deficient leaflets showing reddish pigmentation and necrosis on the margins. (Photo by Dr Prakash Kumar.)

Plate 352. Zinc-deficient leaflet becomes pale green. (Photo by Dr Prakash Kumar.)

Plate 353. Reddish brown pigmentation spreading on the entire upper surface of the leaflets. (Photo by Dr Prakash Kumar.)

Plate 354. The upper portion of leaflets becomes necrotic. (Photo by Dr Prakash Kumar.)

CHICKPEA (*Cicer arietinum* Linn.) ZINC (Zn) DEFICIENCY

Symptoms

1. Chickpea is generally considered a sensitive crop to zinc deficiency. Genotypic variations exist among the cultivars of chickpea to zinc deficiency.
2. Zinc efficiency in some genotypes is probably related to an efficient zinc absorption coupled with a better transport from root to shoot.
3. Zinc deficiency affects plant–water relationships, induces stomatal closure and decreases transpiration in plants.
4. Zinc-deficient plants appear stunted and have fewer branches. The size of leaflets is reduced. Crop maturity gets delayed.
5. Zinc has only limited mobility within plants and is partly mobile from older to younger plant parts.
6. The younger leaves become pale green first, then a reddish brown discoloration appears on margins of leaflets and on the lower parts of the stem.
7. In the later stage, the upper portion of the leaflets turns bronzed and necrotic.

Developmental stages

Stage I: In the early stage of the deficiency, plants show stunted growth and smaller leaves.
Stage II: The leaflets of younger leaves turn pale green and develop marginal reddish brown pigmentation (Plates 351 and 352).
Stage III: In severe deficiency, bronzing and necrosis occurs on the leaflets (Plates 353 and 354).
Stage IV: Eventually, the leaflets die and shed prematurely.

Likely to occur in

1. Leached, light sandy soils where zinc content is low.
2. Alkaline and calcareous soils, where zinc availability is depressed.
3. Recently levelled soils where the subsoil is exposed for cultivation. Plant-available zinc in surface soil is often double that of the subsoil.
4. Soils to which high rates of phosphatic fertilizers have been applied, which can hamper zinc uptake by crops.
5. Acid soils having pH below 5.0.
6. Alkaline soils having pH above 7.5.

Integrated nutrient management

1. Analyse the soil before sowing to estimate the amount of plant-available zinc in the soil.
2. Reclaim problematic alkaline soils.
3. Incorporate any organic manure well before sowing.
4. Apply zinc sulphate (commonly at 25–30 kg/ha) or zinc chelates (10 kg/ha) once every 2 years in zinc-deficient soils.
5. If deficiency symptoms appear in standing crops, spray 5 kg of zinc sulphate and 2.5 kg of unslaked lime in 500 l of water.

Further reading

Chen, W., Sharma, H.L. and Muehlbauer, F.J. (2011) *Compendium of Chickpea and Lentil Diseases and Pests*. American Phytopathological Society, St Paul, Minnesota.

Khan, H.R., McDonald, G.K. and Rengel, Z. (1998) Chickpea genotypes differ in their sensitivity to Zn deficiency. *Plant and Soil* 198, 11–18.

Khan, H.R., McDonald, G.K. and Rengel, Z. (2004) Zinc fertilization and water stress affects plant water relations, stomatal conductance and osmotic adjustment in chickpea (*Cicer arientinum* L.). *Plant and Soil* 267, 271–284.

Valenciano, J.B., Miguelez-Frade, M.M. and Marcelo, V. (2009) Response of chickpea (*Cicer arietinum* L.) to soil zinc application. *Spanish Journal of Agricultural Research* 7, 952–956.

Plate 355. Nitrogen-deficient kidney bean plant. (Photo by Dr Prakash Kumar.)

Plate 356. Older leaves with uniform pale green colour. (Photo by Dr Prakash Kumar.)

Plate 357. Normal green leaf (left) with nitrogen-deficient pale green leaf (right). (Photo by Dr Prakash Kumar.)

Plate 358. Close-up of a leaf showing yellow chlorosis. (Photo by Dr Prakash Kumar.)

KIDNEY BEAN (*Phaseolus vulgaris* Linn.) NITROGEN (N) DEFICIENCY

Symptoms

1. Nitrogen deficiency in kidney bean is mostly found during the initial stages of crop growth (15–20 days after emergence), when root symbiotic nitrogen-fixation nodules are yet to begin a sufficient nitrogen supply to the plant.

2. Nitrogen deficiency may occur during later stages of crop growth when the symbiotic nitrogen-supplying mechanism is disturbed for some reason such as nodule infestation, nodule pathogenic disease or physiological causes. Poor nodulation because of improper *Rhizobium* strain or unfavourable environmental conditions may also cause nitrogen deficiency.

3. Nitrogen is mobile in plants and under short supply conditions it is easily mobilized from older to younger leaves. The deficiency symptoms appear first and more severely on the old leaves. The younger leaves usually remain green and apparently healthy (Plate 355).

4. Deficiency appears as a uniform pale green to pale yellow discoloration of older leaves (Plate 356).

5. If deficiency persists and becomes more severe, the older leaves show yellow chlorosis (Plate 358). Affected older leaves soon abscise (shed).

Developmental stages

Stage I: When deficiency occurs during the initial stage of crop growth, the entire plant appears uniformly light green in colour.

Stage II: As the symptoms advance, affected older leaves become uniform pale green to pale yellow with the younger leaves healthy and green (Plates 355 and 357).

Stage III: If deficiency persists and becomes more severe, the older leaves show yellow chlorosis (Plate 358).

Likely to occur in

1. Soils having low organic matter.

2. Light-textured sandy soils that have been leached by heavy rainfall or excessive irrigation.

3. Soils exhausted by intensive cropping.

4. Waterlogged conditions.

5. Acid soils having pH below 6.0.

6. Alkaline soils having pH above 8.0.

7. Crops with poor *Rhizobium* nodulation or with damaged *Rhizobium* nodules.

Integrated nutrient management

1. Get the soil analysed before sowing to measure the amount of 'available' nitrogen in the soil.

2. Apply analysis-based recommended quantity of nitrogen as a basal starter dose through organic manures or nitrogenous fertilizers.

3. Inoculate the crop with an appropriate strain of *Rhizobium* culture through seed treatment.

4. The crop needs nitrogen during the initial stage of growth, when symbiotic nitrogen fixation by the plant is yet to start. Thus, a basal starter application of nitrogen is important in nitrogen-deficient soils.

5. Nitrogen deficiency in existing crops can be managed by applying urea with irrigation water or as a foliar spray.

Further reading

Allen, D.J., Ampofo, J.K.O. and Wortmann, C.S. (1996) *Pests, Diseases and Nutritional Disorders of the Common Bean in Africa: A Field Guide*. International Center for Tropical Agriculture (CIAT), Cali, Colombia.

Grundon, N.J. (1987) *Hungry Crops: A Guide to Nutrient Deficiencies in Field Crops*. Queensland Department of Primary Industries, Brisbane, Australia.

Hungria, M., Barradas, C.A.A. and Wallsgrove, R.M. (1991) Nitrogen fixation, assimilation and transport during the initial growth stage of *Phaseolus vulgaris* L. *Journal of Experimental Botany* 42, 839–844.

Piha, M.I and Munns, D.N. (1987) Nitrogen fixation potential of beans (*Phaseolus vulgaris* L.) compared with other grain legumes under controlled conditions. *Plant and Soil* 98, 169–182.

Plate 359. Magnesium-deficient kidney bean plant. (Photo by Dr Prakash Kumar.)

Plate 360. Lower leaves showing interveinal chlorosis. (Photo by Dr Prakash Kumar.)

Plate 361. Close-up of a leaf showing interveinal chlorosis. (Photo by Dr Prakash Kumar.)

Plate 362. Brown necrotic lesions on a leaf. (Photo by Dr Prakash Kumar.)

KIDNEY BEAN (*Phaseolus vulgaris* Linn.) MAGNESIUM (Mg) DEFICIENCY

Symptoms

1. Magnesium is mobile in plants. The deficiency symptoms appear first and more severely on the older leaves. The youngest leaves remain green and apparently healthy (Plate 359).
2. The symptoms begin with fading of the interveinal tissues of older leaves (Plate 359).
3. As symptoms advance, a pale yellow interveinal chlorosis develops on older leaves; veins of the leaf remain green and prominent (Plates 360 and 361).
4. If deficiency persists, brown necrotic lesions develop on the interveinal tissues of affected leaves (Plate 362).
5. In acute deficiency conditions, the necrotic leaves eventually burn and die.

Developmental stages

Stage I: The lower leaves develop temporary fading of interveinal tissues with prominent green veins (Plates 359 and 360).
Stage II: A pale yellow interveinal chlorosis develops on older leaves. Veins remain green and prominent (Plate 361).
Stage III: Brown necrotic lesions appear on leaf interveinal tissues and on the undersides of main veins.
Stage IV: Necrotic leaves eventually burn and die.

Likely to occur in

1. Acid sandy soils from which magnesium has been leached by heavy rainfall.
2. Peat and muck soils that are low in total magnesium.
3. Soils derived from parent material that is inherently low in magnesium.
4. Soils with heavy and excess application of potassium fertilizers.
5. Soils with heavy and excess application of lime (calcium carbonate) or other calcium fertilizers.
6. Acid soils having pH below 6.5.
7. Alkaline soils having pH above 8.5.

Integrated nutrient management

1. Get the soil analysed before sowing to estimate the amount of soluble and exchangeable magnesium in the soil.
2. Apply analysis-based recommended quantity of magnesium before sowing using soluble salts such as magnesium sulphate or magnesium chloride.
3. In deficient standing crops, apply soluble salts such as magnesium sulphate, chloride or nitrate with irrigation water. Foliar sprays of these salts are usually not advised as many sprays at frequent intervals are required to fulfil the crop need.
4. Magnesium deficiency in acid soils may be corrected by applying dolomite (a mixture of calcium carbonate and magnesium carbonate, $CaCO_3 \cdot MgCO_3$) through broadcasting and mixing into the soil a few months before the sowing.
5. Reclamation of problematic acid soils or alkaline soils should be done to regulate a proper supply of magnesium.

Further reading

Allen, D.J., Ampofo, J.K.O. and Wortmann, C.S. (1996) *Pests, Diseases and Nutritional Disorders of the Common Bean in Africa: A Field Guide*. International Center for Tropical Agriculture (CIAT), Cali, Colombia.

Grundon, N.J. (1987) *Hungry Crops: A Guide to Nutrient Deficiencies in Field Crops*. Queensland Department of Primary Industries, Brisbane, Australia.

Wallace, T. (1961) *The Diagnosis of Mineral Deficiencies in Plants by Visual Symptoms*, 2nd edn. Chemical Publishing Co., New York.

Plate 363. Sulphur-deficient kidney bean plant. (Photo by Dr Prakash Kumar.)

Plate 364. Plant with small and pale yellow young leaves. (Photo by Dr Prakash Kumar.)

Plate 365. Deficient leaves, uniformly pale green. (Photo by Dr Prakash Kumar.)

Plate 366. Close-up of a leaf showing pale yellow colour. (Photo by Dr Prakash Kumar.)

KIDNEY BEAN (*Phaseolus vulgaris* Linn.) SULPHUR (S) DEFICIENCY

Symptoms

1. Sulphur deficiency symptoms are often observed in kidney bean at early stages of growth. The deficient plant lacks vigour, appears pale green to pale yellow, produces few branches, bears few pods and yields poorly.

2. Sulphur is immobile in plants and does not mobilize from older to younger leaves in short supply conditions. Thus, the deficiency symptoms appear first on younger leaves while older leaves remain green and healthy (Plate 363).

3. Sulphur deficiency, even if severe, does not produce any type of necrosis or burning of leaves. The change in leaf colour is the only indicative symptom of sulphur deficiency.

4. At the initial stage sulphur deficiency symptoms are often confused with those caused by nitrogen deficiency. A close observation is required to see whether the older leaves are more dark green and the younger ones are more pale (case of sulphur deficiency) or whether the younger leaves are more dark green and the older ones are more pale (case of nitrogen deficiency).

5. Deficiency symptoms appear as an even and uniform pale green to pale yellow chlorosis across the lamina of young leaves. The midrib and other veins become very similar in colour to interveinal areas of the leaf (Plate 366).

6. If deficiency persists and becomes more severe, symptoms eventually move downwards covering more leaves.

Developmental stages

Stage I: Symptoms begin with a uniform fading of young leaves from green to pale green colour (Plate 363).

Stage II: If the deficiency persists and becomes more severe, an even and uniform pale green to pale yellow chlorosis develops across the lamina of young leaves. The youngest leaves become smallest in size and palest in colour (Plates 364 and 365).

Stage III: In acute deficiency conditions, the symptoms move downwards and may cover entire leaves.

Likely to occur in

1. Soils low in organic matter.

2. Light-textured sandy soils that have been leached by heavy rainfall or excessive irrigation.

3. Soils exhausted by intensive cropping.

4. Soils derived from parent material that is inherently low in sulphur (for example, soils formed from volcanic rocks).

5. Acid soils having pH below 6.0.

Integrated nutrient management

1. Get the soil analysed before sowing to measure the amount of 'available' sulphate in the soil and to predict the amount of fertilizer required.

2. Apply analysis-based recommended quantity of sulphur by mixing into the soil, well before sowing, either:

- **a.** elemental sulphur; or
- **b.** gypsum (calcium sulphate).

3. In deficient standing crops apply ammonium sulphate, magnesium sulphate or potassium sulphate with irrigation water.

4. Problematic acid soils should be reclaimed.

Further reading

Allen, D.J., Ampofo, J.K.O. and Wortmann, C.S. (1996) *Pests, Diseases and Nutritional Disorders of the Common Bean in Africa: A Field Guide*. International Center for Tropical Agriculture (CIAT), Cali, Colombia.

Bennett, W.F. (1993) *Nutrient Deficiencies and Toxicities in Crop Plants*. American Phytopathological Society, St Paul, Minnesota.

Grundon, N.J. (1987) *Hungry Crops: A Guide to Nutrient Deficiencies in Field Crops*. Queensland Department of Primary Industries, Brisbane, Australia.

Hitsuda, K., Yamada, M. and Klepker, D. (2005) Sulphur requirement of eight crops at early stages of growth. *Agronomy Journal* 97, 155–159.

Plate 367. Iron-deficient kidney bean plant. (Photo by Dr Prakash Kumar.)

Plate 368. Plant showing interveinal chlorosis on young leaves. (Photo by Dr Prakash Kumar.)

Plate 369. Prominent green veins fade to light green. (Photo by Dr Prakash Kumar.)

Plate 370. Leaf showing papery white appearance. (Photo by Dr Prakash Kumar.)

KIDNEY BEAN (*Phaseolus vulgaris* Linn.) IRON (Fe) DEFICIENCY

Symptoms

1. Iron is immobile in plants. So, deficiency symptoms appear first and more severely on the younger leaves (Plate 367).
2. In mild deficiency conditions or at the initial stage of deficiency, the topmost younger leaves develop temporary fading of interveinal tissues to pale green to pale yellow in colour.
3. If the deficiency persists and becomes more severe, a bright pale yellow chlorosis develops in interveinal tissues, leaving the veins green and prominent. Interveinal chlorosis of younger leaves is the specific symptom of iron deficiency (Plate 368). As the symptoms advance, the prominent green veins also fade and become light green to pale yellow (Plate 369).
4. In acute deficiency conditions the entire leaf bleaches to papery white (Plate 370).

Developmental stages

Stage I: The topmost leaves develop temporary fading of interveinal tissues to pale yellow colour, leaving prominent green veins.
Stage II: Interveinal tissues of the affected leaves turn bright yellow with prominent green veins (Plate 368).
Stage III: The prominent green veins also fade and become light green (Plate 369).
Stage IV: The affected young leaves bleach to papery white (Plate 370). The veins may or may not remain green.

Likely to occur in

1. Sandy soils having low total iron.
2. Calcareous soils where solubility of iron is very low.
3. Peat and muck soils where organic matter ties up iron and reduces its availability in soil solution.
4. Acid soils where, despite high availability of iron in soil solution, excessively high levels of soluble zinc, manganese, copper or nickel hamper the uptake of iron by the plant.
5. Waterlogged soils.
6. Alkaline soils having pH above 7.5.

Integrated nutrient management

1. Get the soil analysed before sowing to measure the amount of 'available' iron in the soil.
2. As inorganic soluble salts of iron such as iron sulphates or chlorides quickly become insoluble in soil, a soil dressing of organic forms of iron such as iron chelates (10 kg/ha) is more effective than inorganic forms of iron. These organic iron forms should be used as per soil pH. If the soil is neutral (around pH 7.0), the most suitable chelates are FeHEDTA and FeDTPA. For acid soils FeEDTA is the most effective chelate, while on alkaline calcareous soils FeEDDHA is recommended.

Further reading

Bennett, W.F. (1993) *Nutrient Deficiencies and Toxicities in Crop Plants*. American Phytopathological Society, St Paul, Minnesota.
Grundon, N.J. (1987) *Hungry Crops: A Guide to Nutrient Deficiencies in Field Crops*. Queensland Department of Primary Industries, Brisbane, Australia.
Wallace, T. (1961) *The Diagnosis of Mineral Deficiencies in Plants by Visual Symptoms*, 2nd edn. Chemical Publishing Co., New York.

Plate 371. Zinc-deficient kidney bean plant. (Photo by Dr Prakash Kumar.)

Plate 372. Interveinal chlorosis on old leaves and faded young leaves. (Photo by Dr Prakash Kumar.)

Plate 373. Brown necrotic lesions on an affected leaf. (Photo by Dr Prakash Kumar.)

Plate 374. Burning of leaves. (Photo by Dr Prakash Kumar.)

KIDNEY BEAN (*Phaseolus vulgaris* Linn.) ZINC (Zn) DEFICIENCY

Symptoms

1. Zinc deficiency symptoms appear first and more severely on older leaves (Plate 371).
2. Symptoms begin as a pale green to pale yellow interveinal chlorosis of older leaves (Plate 372). The main veins and the tissues adjacent to main veins remain green and prominent.
3. If deficiency persists and becomes more severe, the affected leaves develop a puckered appearance with brown necrotic lesions on interveinal areas (Plate 373).
4. In acute deficiency conditions, the deficiency symptoms quickly cover the entire plant including younger leaves. Older leaves burn like blight symptoms and younger leaves show faded signs of interveinal chlorosis.
5. The affected field looks patchy and diseased in appearance.

Developmental stages

Stage I: Pale green to pale yellow interveinal chlorosis develops on older leaves, leaving the veins and adjacent tissues green and prominent.
Stage II: Brown necrotic lesions develop on interveinal chlorotic areas. The leaf also shows a puckered appearance.
Stage III: The new emerging leaflets become brittle and small in size with faded signs of interveinal chlorosis.
Stage IV: The older affected leaves burn and die (Plate 374).

Likely to occur in

1. Leached sandy soils where total zinc is low.
2. Just-levelled soils where subsoil is exposed for cultivation. Available zinc in surface soil is often double that of the subsoil.
3. Cool wet weather.
4. Soils having heavy and excessive application of phosphate fertilizers, which hampers use of zinc by the crop.
5. Acid soils having pH below 5.0.
6. Alkaline soils having pH above 7.5.

Integrated nutrient management

1. Get the soil analysed before sowing to estimate the amount of 'available' zinc in the soil.
2. Apply 25–30 kg of zinc sulphate or 10 kg of zinc chelate per hectare once every 2 years in zinc-deficient soils.
3. Do not mix zinc fertilizers with phosphate fertilizers.
4. In fields with known zinc deficiency, if deficiency appears in the standing crop, apply a foliar application of 0.5% w/v solution of a soluble zinc salt, Zinc sulphate with 0.25% w/v solution of unsoaked lime.

Further reading

Grundon, N.J. (1987) *Hungry Crops: A Guide to Nutrient Deficiencies in Field Crops*. Queensland Department of Primary Industries, Brisbane, Australia.

Marschner, H. and Cakmak, I. (1989) High light intensity enhances chlorosis and necrosis in leaves of zinc, potassium and magnesium deficient bean (*Phaseolus vulgaris*) plants. *Journal of Plant Physiology* 134, 308–315.

Melton, J.R., Ellis, B.G. and Doll, E.C. (1970) Zinc, phosphorus and lime interactions with yield and zinc uptake by *Phaseolus vulgaris*. *Soil Science Society of America Journal* 34, 91–93.

Plate 375. Nitrogen-deficient lentil plant. (Photo by Dr Prakash Kumar.)

Plate 376. Light green foliage of lentil crop. (Photo by Dr Manoj Kumar Sharma.)

Plate 377. Nitrogen-deficient pale green leaflet. (Photo by Dr Prakash Kumar.)

Plate 378. Nitrogen-deficient white leaflet. (Photo by Dr Prakash Kumar.)

LENTIL (*Lens culinaris* Medik.) NITROGEN (N) DEFICIENCY

Symptoms

1. Nitrogen deficiency in lentil is usually found during the initial stages of crop growth when root symbiotic nitrogen-fixation nodules are yet to develop.
2. Nitrogen deficiency may occur during later stages of crop growth when the symbiotic nitrogen-supplying mechanism is disturbed for some reason such as nodule infestation, nodule pathogenic disease or physiological causes. Poor nodulation because of improper *Rhizobium* strain or unfavourable environmental conditions may also cause nitrogen deficiency.
3. Nitrogen-deficient lentil plants are stunted with thin, spindly stems and pale green to pale yellow foliage.
4. Nitrogen is mobile in plants and under short supply conditions it is easily mobilized from older to younger leaves. The deficiency symptoms appear first and more severely on old leaves. The younger leaves usually remain green and apparently healthy (Plate 375).
5. In mild deficiencies or when deficiency occurs in the young plant stage, the entire plant appears uniformly light green in colour (Plate 376).
6. If deficiency persists and becomes more severe, the older leaves show uniform pale green to greenish yellow chlorosis.
7. In acute deficiency conditions, older leaves turn to white and abscise (shed).

Developmental stages

Stage I: In mild deficiency conditions or when deficiency occurs in the young stage, the entire plant appears uniformly light green in colour (Plate 376).
Stage II: As symptoms advance, affected older leaves become uniform pale green to greenish yellow (Plate 377).
Stage III: In acute deficiency conditions leaves turn to white (Plate 378).
Stage IV: In the most advanced stage the affected leaves abscise.

Likely to occur in

1. Soils having low organic matter.
2. Light-textured sandy soils that have been leached by heavy rainfall or excessive irrigation.
3. Soils exhausted by intensive cropping.
4. Waterlogged conditions.
5. Acid soils having pH below 6.0.
6. Alkaline soils having pH above 8.0.
7. Crops with poor *Rhizobium* nodulation or with damaged *Rhizobium* nodules.

Integrated nutrient management

1. Get the soil analysed before sowing to measure the amount of 'available' nitrogen in the soil.
2. Apply analysis-based recommended quantity of nitrogen as a basal starter dose through organic manures or nitrogenous fertilizers.
3. Inoculate the crop with an appropriate strain of *Rhizobium* culture through seed treatment.
4. A basal starter application of nitrogen at 20 kg/ha is important in nitrogen-deficient soils.
5. Nitrogen deficiency in existing crops can be managed by applying urea with irrigation water or as a foliar spray.

Further reading

Ashraf, M. and Zafar, Z.U. (2008) Effect of nitrogen deficiency on growth and some biochemical characteristics in salt tolerant and salt sensitive lines of lentil (*Lens culinaris* Medic.). *Archives of Agronomy and Soil Science* 40, 231–239.

Jones, J.B. (2002) Nutrient element deficiency symptoms. Appendix E. Legumes. In: *Agronomic Handbook: Management of Crops, Soils and Their Fertility*. CRC Press, Boca Raton, Florida, p. 409.

Niri, H.H., Tobeh, A., Gholipouri, A., Zakaria, R.A., Mostafaei, H. and Jamati-e-Somarin, S. (2010) Effect of nitrogen and phosphorus on yield and protein content of lentil in dry land condition. *American-Eurasian Journal of Agricultural & Environmental Science* 8920, 185–188.

Verma, V.S. and Kalara, G.S. (1983) Effect of different levels of irrigation, N and P on growth and yield of lentil. *Indian Journal of Agricultural Science* 17, 124–128.

Plate 379. Phosphorus-deficient lentil plant. (Photo by Dr Prakash Kumar.)

Plate 380. Purple coloration of older leaves. (Photo by Dr Prakash Kumar.)

Plate 381. Purpling starts from leaf margins. (Photo by Dr Prakash Kumar.)

Plate 382. Advanced stage, showing purpling on entire leaf. (Photo by Dr Prakash Kumar.)

LENTIL (*Lens culinaris* Medik.) PHOSPHORUS (P) DEFICIENCY

Symptoms

1. Phosphorus nutrition is important for proper growth and optimum yields in lentil. Affected crops lack vigour, mature slowly and produce very poor yields.
2. Deficient plants are stunted with short, stout stems and dull blue–green foliage.
3. Phosphorus is mobile in plants and under short supply conditions it is easily mobilized from older to younger leaves.
4. The deficiency symptoms appear first and become more severe on older leaves.
5. The phosphorus-deficient plant increases anthocyanin pigmentation in its leaves.
6. The plant develops a characteristic dark green to bluish green coloration that is more prominent on older leaves.
7. Under severe deficiency conditions, some lentil varieties may develop reddish brown or purple tints on older leaves. The symptom of 'purpling of older leaves' is not necessarily found in all lentil varieties or in all environments. So, dark green to bluish green coloration of the foliage is considered the more reliable symptom for visual identification of phosphorus deficiency in all lentil varieties.

Developmental stages

Stage I: The older leaves develop dark green to bluish green coloration. Younger leaves remain green and normal.
Stage II: If the deficiency becomes more severe, some lentil varieties develop reddish brown or purple tints on older leaves (Plates 379 and 380). Pigmentation usually starts from margins of older leaves (Plate 381).
Stage III: As the symptoms advance, purple pigmentation covers the entire leaf (Plate 382).
Stage IV: In acute deficiency conditions, brown necrosis develops on the tip and proceeds along the margins towards the base of the leaves, ultimately burning the entire leaf.

Likely to occur in

1. Soils having low organic matter.
2. Alkaline and calcareous soils.
3. Soils exhausted by intensive cropping.
4. Acid soils and highly weathered soils.
5. Soils where the topsoil has been removed by erosion.
6. Acid soils having pH below 6.0.
7. Alkaline soils having pH between 7.5 and 8.5.

Integrated nutrient management

1. Get the soil analysed before sowing to measure the amount of 'available' phosphate in the soil.
2. Apply analysis-based recommended quantity of phosphorus as basal by using:
  a. organic manures;
  b. phosphate-solubilizing microbial cultures;
  c. phosphatic fertilizers.
3. If standing crops show the deficiency symptoms, apply soluble phosphatic fertilizers such as ammonium phosphate with irrigation water.

Further reading

Ali, A. and Johnson, D.L. (1999) Association of growth habit and anthocyanin pigmentation with winter hardiness in lentil. *Pakistan Journal of Biological Sciences* 2, 1292–1295.

Niri, H.H., Tobeh, A., Gholipouri, A., Zakaria, R.A., Mostafaei, H. and Jamati-e-Somarin, S. (2010) Effect of nitrogen and phosphorus on yield and protein content of lentil in dry land condition. *American-Eurasian Journal of Agricultural & Environmental Science* 8920, 185–188.

Sarkar, B.C. and Karmoker, J.L. (2011) Effects of phosphorus deficiency on accumulation of biochemical compounds in lentil (*Lens culinaris* Medik.). *Bangladesh Journal of Botany* 40, 23–27.

Zafar, M., Maqsood, M., Anser, M.R. and Zahad, A. (2003) Growth and yield of lentil as affected by phosphorus. *International Journal of Agriculture and Biology* 5, 98–100.

Plate 383. Potassium-deficient lentil plant. (Photo by Dr Prakash Kumar.)

LENTIL (*Lens culinaris* Medik.) POTASSIUM (K) DEFICIENCY

Plate 384. Marginal chlorosis starts from the leaf tip. (Photo by Dr Prakash Kumar.)

Plate 385. Chlorotic areas around the edges of leaves. (Photo by Dr Prakash Kumar.)

Plate 386. Scorching starting from leaf tips. (Photo by Dr Prakash Kumar.)

Symptoms

1. The potassium-deficient lentil crop lacks vigour, grows slowly, matures late and yields poorly. Plants are stunted with short stems.
2. Potassium moves readily from old to young leaves, therefore deficiency symptoms appear first on older leaves (Plate 383).
3. The symptoms begin as marginal yellowing of older leaves. Yellowing starts from the leaf tip and advances around the leaf margin (Plates 384 and 385).
4. If deficiency persists, the leaf tip becomes scorched. The scorching spreads around the leaf margin (Plate 386).
5. A yellow band is found between the scorched area and healthy green tissues.
6. In acute deficiency conditions, the symptoms move towards upper leaves.

Developmental stages

Stage I: A pale yellow chlorosis appears on older leaves starting from the tip of the leaf (Plate 384).
Stage II: As the symptoms advance, pale yellow chlorosis covers the margins of the lamina (Plate 385).
Stage III: The leaf tip becomes scorched as the symptoms become severe. Yellow chlorosis is followed by brown necrosis along the leaf margins (Plate 386).
Stage IV: In acute deficiency conditions, the symptoms move towards upper leaves and may cover the entire plant.

Likely to occur in

1. Soils formed from parent material low in potassium.
2. Light-textured soils where potassium has been leached by heavy rainfall or excessive irrigation.
3. Soils low in organic matter.
4. Soils with wide Na:K, Mg:K or Ca:K ratio.
5. Large bicarbonate concentration in irrigation water.
6. Highly acid soils having pH below 6.0.

Integrated nutrient management

1. Get the soil analysed before sowing to measure the amount of 'available' potassium in the soil.
2. Problematic acid soils should be reclaimed.
3. Add organic manures well before sowing, if organic matter is low.
4. Apply potassium nitrate, potassium sulphate or potassium chloride to the soil at or before sowing as per soil testing recommendations.
5. If potassium deficiency appears on the standing crop, apply soluble potassium salts such as potassium nitrate, potassium sulphate or potassium chloride with irrigation water. Foliar application of these salts is usually not recommended because a number of sprays are needed to fulfil crop requirements.

Further reading

Ashraf, M. and Zafar, Z.U. (1997) Effect of potassium deficiency on growth and some biochemical characteristics in two lines of lentil (*Lens culinaris* Medic.). *Acta Physiologiae Plantarum* 19, 9–15.

Jahan, S.A., Alim, M.A., Hasan, M.M., Kabiraj, U.K. and Hossain, M.B. (2009) Effect of potassium levels on the growth, yield and yield attributes of lentil. *International Journal of Sustainable Crop Production* 4(6), 1–6.

Jones, J.B. (2002) Nutrient element deficiency symptoms. Appendix E. Legumes. In: *Agronomic Handbook: Management of Crops, Soils and Their Fertility*. CRC Press, Boca Raton, Florida p. 409.

Tiwari, K.N. and Nigam, V. (1985) Crop responses to potassium fertilization in soils of Utter Pradesh. *Journal of Potassium Research* 1, 62–71.

Plate 387. Iron-deficient lentil plant. (Photo by Dr Prakash Kumar.)

LENTIL (*Lens culinaris* Medik.) IRON (Fe) DEFICIENCY

Plate 388. Interveinal chlorosis of leaves. (Photo by Dr Prakash Kumar.)

Plate 389. Plant showing iron chlorosis on youngest leaves. (Photo by Dr Prakash Kumar.)

Plate 390. Burning of leaves after papery white appearance. (Photo by Dr Prakash Kumar.)

Symptoms

1. Lentil is one of the most sensitive crops to iron deficiency. A large area of lentil grown in calcareous soils, around the world, suffers with iron deficiency.
2. Iron is immobile in plants. So, the deficiency symptoms appear first and more severely on the younger leaves. The older leaves remain normal and apparently healthy (Plate 387).
3. In mild deficiency conditions or at the initial stage of deficiency, the topmost younger leaves develop temporary fading of interveinal tissues to a pale green to pale yellow colour.
4. As the symptoms advance, a bright pale yellow interveinal chlorosis develops on younger leaves leaving the main vein green (Plate 388).
5. If the deficiency persists and becomes more severe, the entire leaf bleaches to papery white (Plate 389).
6. In acute deficiency conditions, affected youngest leaves burn and die (Plate 390).

Developmental stages

Stage I: The youngest leaves develop temporary fading of interveinal tissues.
Stage II: Interveinal tissues in affected leaves turn pale green to yellow (Plates 387 and 388).
Stage III: The entire leaf bleaches to papery white (Plate 389).
Stage IV: The affected top leaves burn and die (Plate 390).

Likely to occur in

1. Sandy soils having low total iron.
2. Calcareous soils where solubility of iron is very low.
3. Peat and muck soils where organic matter ties up iron and reduces its availability in soil solution.
4. Acid soils where, despite high availability of iron in soil solution, excessively high levels of soluble zinc, manganese, copper or nickel hamper the uptake of iron by the plant.
5. Waterlogged soils.
6. Alkaline soils having pH above 7.5.

Integrated nutrient management

1. Get the soil analysed before sowing to measure the amount of 'available' iron in the soil.
2. Problematic alkaline soils should be reclaimed.
3. As inorganic soluble salts of iron such as iron sulphates or chlorides quickly become insoluble in soil, soil dressing of organic forms of iron such as iron chelates (10 kg/ha) is more effective than inorganic forms of iron. These organic iron forms should be used as per soil pH. If the soil is neutral (around pH 7.0), the most suitable chelates are FeHEDTA and FeDTPA. For acid soils FeEDTA is the most effective chelate, while on alkaline calcareous soils FeEDDHA is recommended.

Further reading

Erskine, W., Saxena, N.P. and Saxena, M.C. (1993) Iron deficiency in lentil: yield loss and geographic distribution in a germplasm collection. *Plant and Soil* 151, 249–254.

Jones, J.B. (2002) Nutrient element deficiency symptoms. Appendix E. Legumes. In: *Agronomic Handbook: Management of Crops, Soils and Their Fertility*. CRC Press, Boca Raton, Florida, p. 409.

Rai, R., Prasad, V., Choudhury, S.K. and Sinha, N.P. (2008) Iron nutrition and symbiotic N_2- fixation of lentil (*Lens culinaris*) genotypes in calcareous soil. *Journal of Plant Nutrition* 7, 399–405.

Plate 391. Nitrogen-deficient plant showing pale yellow old leaves and green young leaves. (Photo by Dr Prakash Kumar.)

PEA (*Pisum sativum* var. *arvense* Linn.) NITROGEN (N) DEFICIENCY

Plate 392. The entire plant stunted, light green to pale yellow. (Photo by Dr Manoj Kumar Sharma.)

Plate 393. Chlorotic lower leaves. (Photo by Dr Prakash Kumar.)

Plate 394. Severely deficient, brown necrotic old leaves. (Photo by Dr Prakash Kumar.)

Symptoms

1. Low and high temperature, continuous light and high light intensity conditions can induce nitrosative stress in pea plants.
2. Photosynthetic parameters in plants are affected by the nitrogen-fixation characteristics of the rhizobial symbiont.
3. In the case of deficiency, plants appear stunted and have small leaves.
4. The common symptom of nitrogen deficiency is the general yellowing of the entire plant.
5. When plants are not supplied with sufficient nitrogen, then it is rapidly translocated from older tissues to the younger parts (as nitrogen is mobile within plants).
6. The deficiency symptoms typically appear first on lower leaves (Plate 391).
7. Initially, the whole plant appears light green.
8. The lower leaves gradually turn pale yellow to yellow while the young leaves appear light green.
9. Subsequently the lower leaves become white then brown necrotic.

Developmental stages

Stage I: In the early stage or in mild deficiency, the entire plant appears uniformly pale green (Plate 392).
Stage II: In advanced deficiency, the lower leaves turn uniformly yellow (Plates 391 and 393).
Stage III: In severe deficiency, the entire plant turns yellow.
Stage IV: In acute deficiency, the old leaves turn white, then brown necrotic and fall off prematurely (Plate 394).

Likely to occur in

1. Soils having low organic matter.
2. Light-textured sandy soils that have been leached by heavy rainfall or excessive irrigation.
3. Soils exhausted by intensive cropping.
4. Waterlogged conditions.
5. Acid soils having pH below 6.0.
6. Alkaline soils having pH above 8.0.

Integrated nutrient management

1. Get the soil analysed before sowing to measure the amount of 'available' nitrogen in the soil.
2. Apply analysis-based recommended quantity of nitrogen as basal by using:
  a. organic manures;
  b. bio-fertilizers;
  c. nitrogenous fertilizers.
3. Top-dress soluble nitrogenous fertilizers such as urea.
4. For quick recovery in the standing crop, apply urea (2% w/v solution) as a foliar spray. Foliar sprays require to be repeated every 10–15 days.

Further reading

Bethlenfalvay, G.J., Abu-Shakra, S.S. and Phillips, D.A. (1978) Interdependence of nitrogen nutrition and photosynthesis in *Pisum sativum* L. *Plant Physiology* 62, 131–133.

Corpas, F.J., Chaki, M., Fernandez-Ocana, A., Valderrama, R., Palma, J.M., Carreras, A., Begara-Morales, J.C., Airaki, M., Rio, L.A. and Barroso, J.B. (2008) Metabolism of reactive nitrogen species in pea plants under abiotic stress conditions. *Plant and Cell Physiology* 49, 1711–1722.

Mulder, E.G. (1948) Investigations on the nitrogen nutrition of pea plants. *Plant and Soil* 1, 179–212.

Plate 395. Yellowing develops on the margins on old leaves. (Photo by Dr Prakash Kumar.)

Plate 396. Dark green leaves showing marginal yellowing. (Photo by Dr Prakash Kumar.)

Plate 397. Yellowing advancing into interveinal tissues. (Photo by Dr Prakash Kumar.)

Plate 398. Broad yellow band covering leaf margins and yellowing advancing towards the midrib. (Photo by Dr Prakash Kumar.)

PEA (*Pisum sativum* var. *arvense* Linn.) POTASSIUM (K) DEFICIENCY

Symptoms

1. Potassium deficiency causes putrescine accumulation in the leaves and it may be detected before the deficiency symptoms appear.
2. Potassium-deficient plants become small and thick, having short young internodes.
3. Plant growth is retarded and pods are poorly filled.
4. Potassium is considered highly mobile within plants and has a tendency to move rapidly from older to younger tissues when the supply is reduced. Therefore, symptoms become evident first on older leaves and then progress up the plant to younger leaves.
5. Leaves become dark green with yellowing occurring at the tips and margins of old leaves (Plates 395, 397 and 398).
6. The margins of old leaves become scorched and show inward cupping.

Developmental stages

Stage I: In mild deficiency, plant growth becomes retarded.
Stage II: If deficiency persists, yellowing develops at tips and along the margins of lower leaves (Plates 395 and 396).
Stage III: When deficiency becomes severe, the tips and margins of old leaves become scorched and curled inward.
Stage IV: In the later stage, the old leaves dry completely and shed prematurely.

Likely to occur in

1. Soils formed from parent material low in potassium.
2. Light-textured soils where potassium has been leached by heavy rainfall or excessive watering.
3. Soils low in organic matter.
4. Soils having wide Na:K, Mg:K or Ca:K ratio.
5. Acid soils having pH below 6.0.

Integrated nutrient management

1. Analyse the soil before sowing to measure the amount of plant-available potassium.
2. Problematic acid/alkaline/saline soils should be reclaimed.
3. Add organic manures well before planting.
4. Apply soluble potassium salts such as potassium chloride, potassium sulphate or potassium nitrate to the soil at or before planting.
5. In standing crops, apply soluble potassium salts with irrigation water.

Further reading

Basso, L.C. and Smith, T.A. (1974) Effect of mineral deficiency on amine formation in higher plants. *Phytochemistry* 13, 875–883.

Klein, H., Priebe, A. and Jager, H.J. (1979) Putrescine and spermidine in peas: effects on nitrogen source and potassium supply. *Physiologia Plantarum* 45, 497–499.

Sodek, L., Lee, P.J. and Miflin, B.J. (1980) Distribution and properties of a potassium-dependent asparaginase isolated from developing seeds of *Pisum sativum* and other plants. *Plant Physiology* 65, 22–26.

Wallace, T. (1943) *The Diagnosis of Mineral Deficiencies in Plants by Visual Symptoms. A Colour Atlas and Guide.* His Majesty's Stationery Office, London.

Plate 399. Plant showing interveinal chlorosis on older leaves. (Photo by Dr Manoj Kumar Sharma.)

Plate 400. Chlorosis intensified into interveinal areas. (Photo by Dr Manoj Kumar Sharma.)

Plate 401. Entire interveinal portion turns white chlorotic, leaving green areas only near to veins. (Photo by Dr Manoj Kumar Sharma.)

Plate 402. Severely deficient leaves showing some necrosis. (Photo by Dr Prakash Kumar and Dr Manoj Kumar Sharma.)

PEA (*Pisum sativum* var. *arvense* Linn.) MAGNESIUM (Mg) DEFICIENCY

Symptoms

1. Magnesium deficiency causes putrescine accumulation in pea.
2. Magnesium deficiency has little effect on the proportions of storage proteins in pea seeds.
3. Magnesium is considered a mobile nutrient within plants, and in reduced supply conditions it is readily transferred from older to younger parts. Therefore, the older leaves typically show visual symptoms first.
4. If deficiency persists for long, the symptoms become more severe on older leaves and then progress to the upper leaves.
5. Initially, the symptoms start as a pale green interveinal chlorosis with distinct green veins on older leaves (Plate 400).
6. In severe deficiency conditions, the interveinal tissues become dark yellow and then turn white and necrotic.

Developmental stages

Stage I: In mild/early stage of deficiency, interveinal tissues become pale green in older leaves (Plate 399).
Stage II: In prolonged deficiency conditions, interveinal areas turn dark yellow and then white, leaving the veins of old leaves dark green (Plate 401).
Stage III: In severe deficiency, the interveinal tissues become white with necrotic regions (Plate 402).
Stage IV: In the later stage, affected leaves can die and shed prematurely.

Likely to occur in

1. Acid sandy soils that have been leached by heavy rainfall.
2. Peat and muck soils that are low in total magnesium.
3. Soils having higher quantity of calcium or potassium.
4. Soils derived from parent material inherently low in magnesium.
5. Acid soils having pH below 6.5.
6. Alkaline soils having pH above 8.5.

Integrated nutrient management

1. Get the soil analysed before sowing to estimate the amount of available magnesium in the soil.
2. Apply analysis-based recommended quantity of magnesium before sowing using soluble salts such as magnesium sulphate or magnesium chloride.
3. In deficient standing crops, apply soluble salts such as magnesium sulphate, chloride or nitrate with irrigation water.

Further reading

Basso, L.C. and Smith, T.A. (1974) Effect of mineral deficiency on amine formation in higher plants. *Phytochemistry* 13, 875–883.

Randall, P.J., Thomson, J.A. and Schroeder, H.E. (1979) Cotyledonary storage proteins in *Pisum sativum*. IV. Effects of sulfur, phosphorus, potassium and magnesium deficiencies. *Australian Journal of Plant Physiology* 6, 11–24.

Wallace, T. (1943) *The Diagnosis of Mineral Deficiencies in Plants by Visual Symptoms. A Colour Atlas and Guide*. His Majesty's Stationery Office, London.

Plate 403. Symptoms appearing as chlorosis in young leaves while old leaves are green. (Photo by Dr Manoj Kumar Sharma.)

Plate 404. Distinct uniform paleness appearing in young leaves. (Photo by Dr Manoj Kumar Sharma.)

Plate 405. Early symptoms showing as pale green young leaves. (Photo by Dr Manoj Kumar Sharma.)

Plate 406. As deficiency advances, pale green tissues turn light yellow. (Photo by Dr Prakash Kumar.)

PEA (*Pisum sativum* var. *arvense* Linn.) SULPHUR (S) DEFICIENCY

Symptoms

1. Roots/nodules have a high demand for sulphur and the nitrogen fixation is very sensitive to sulphur deficiency.
2. The reduced legumin accumulation in pea seeds under sulphur deficiency is primarily a consequence of decreased levels of legumin mRNA.
3. Sulphur deficiency results in a relative decrease in legumin and in vicilin peak 3, accompanied by a relative increase in the predominant vicilin, peak 4.
4. Sulphur-deficient plants become stunted and their yields are markedly reduced.
5. Deficient plants are also delayed in maturity.
6. Because sulphur is immobile in plants it is not easily mobilized from older to younger leaves (under poor supply conditions). The visible symptoms appear primarily on younger leaves.
7. The younger leaves become pale green or pale yellow (Plate 404).
8. The entire leaf becomes evenly chlorotic, including the veins and interveinal tissues (Plates 405 and 406).
9. In prolonged deficiency conditions, the whole plant turns yellow.

Developmental stages

Stage I: In mild deficiencies, the entire plant appears light green.
Stage II: As the symptoms progress, younger leaves become pale yellow while old leaves stay green (Plate 403).
Stage III: If the deficiency becomes severe, the younger leaves turn yellow.

Likely to occur in

1. Soils having low organic matter.
2. Light-textured sandy soils that have been leached by heavy rainfall or excessive irrigation.
3. Soils exhausted by intensive cropping.
4. Soils derived from parent material that is inherently low in sulphur.
5. Acid soils having pH below 6.0.

Integrated nutrient management

1. Get the soil analysed before sowing to measure the amount of 'available' sulphate in the soil.
2. Apply analysis-based recommended quantity of sulphur by mixing into the soil, well before sowing, either:
  a. elemental sulphur; or
  b. gypsum (calcium sulphate).
3. In deficient standing crops apply ammonium sulphate, magnesium sulphate or potassium sulphate with irrigation water.

Further reading

Chandler, P.M., Higgins, T.J.V., Randall, P.J. and Spencer, D. (1983) Regulation of legumin levels in developing pea seeds under conditions of sulfur deficiency. *Plant Physiology* 71, 47–54.

Randall, P.J., Thomson, J.A. and Schroeder, H.E. (1979) Cotyledonary storage proteins in *Pisum sativum*. IV. Effects of sulfur, phosphorus, potassium and magnesium deficiencies. *Australian Journal of Plant Physiology* 6, 11–24.

Scherer, H.W., Pacyna, S., Manthey, N. and Schulz, M. (2006) Sulphur supply to peas (*Pisum sativum* L.) influences symbiotic N_2 fixation. *Plant, Soil and Environment* 52, 72–77.

Zhao, F.J., Wood, A.P. and McGrath, S.P. (1999) Effects of sulphur nutrition on growth and nitrogen fixation of pea (*Pisum sativum* L.). *Plant and Soil* 212, 207–217.

Plate 407. Iron-deficient plant showing yellow younger leaves and green old leaves. (Photo by Dr Prakash Kumar.)

Plate 408. Uppermost leaves are pale yellow, while those just below have interveinal chlorosis with green veins. (Photo by Dr Manoj Kumar Sharma.)

Plate 409. Severely deficient chlorotic younger leaves completely devoid of chlorophyll. (Photo by Dr Prakash Kumar.)

Plate 410. Eventually, deficient leaves turn almost white. (Photo by Dr Prakash Kumar.)

PEA (*Pisum sativum* var. *arvense* Linn.) IRON (Fe) DEFICIENCY

Symptoms

1. Iron is responsible for various morphological and biological changes in plants. The yellowing of leaves is a common visible symptom of iron deficiency.
2. Measurement of catalase activity and peroxidase/catalase ratio appears helpful in identifying iron deficiencies in peas.
3. Iron-deficient plant status may promote heavy metal uptake via increased expression of RIT1 (metal ion) transporter.
4. Iron deficiency generally decreases nodule formation, leghaemoglobin production and nitrogenase activity.
5. Since iron is immobile within the plant body, it is not easily translocated from lower to upper parts of the plant when its supply to the plant becomes poor.
6. Visible symptoms become evident first on younger leaves while older leaves remain green and healthy (Plate 407).
7. The symptoms show up as chlorosis on young leaves and tendrils (Plate 408).
8. In the later stage the leaves turn almost white and necrotic spots develop around the margins.

Developmental stages

Stage I: In mild deficiency conditions, yellowing occurs in interveinal areas and veins remain green.
Stage II: When deficiency becomes severe, young leaves evenly turn dark yellow (Plate 409).
Stage III: In acutely deficient conditions, the entire leaf blade becomes bleached and papery white (Plate 410).
Stage IV: In the later stage, leaves may develop brown or necrotic spots around the margins.

Likely to occur in

1. Sandy soils having low total iron content.
2. Alkaline and calcareous soils where solubility of iron is very low.
3. Peat and muck soils where organic matter ties up iron and reduces its availability in soil solution.
4. Acid soils with excessively high levels of soluble zinc, manganese, copper or nickel that can hinder the uptake of iron despite high iron availability.
5. Alkaline soils having pH above 7.5.

Integrated nutrient management

1. Analyse the soil before sowing to measure the 'available' iron in the soil.
2. Reclaim problematic alkaline soils.
3. Add organic manure well before sowing.
4. Apply basal dose of soluble iron fertilizers such as $FeSO_4$ (commonly at 25 kg/ha) or Fe chelates (10 kg/ha). Use of organic chelates proves to be more promising as they maintain iron in soil solution.
5. In standing crops, apply $FeSO_4$ or Fe chelates (0.5% w/v solution) as a foliar spray. Foliar sprays must be repeated every 10–15 days.

Further reading

Briat, J.F., Fobis-Loisy, I., Grignon, N., Lobreaux, S., Pascal, N., Savino, G., Thoiron, S., Von Wirén, N. and Van Wuytswinkel, O. (1995) Cellular and molecular aspects of iron metabolism in plants. *Biology of the Cell* 84, 69–81.

Cohen, C.K., Garvin, D.F. and Kochian, L.V. (2004) Kinetic properties of a micronutrient transporter from *Pisum sativum* indicate a primary function in Fe uptake from the soil. *Planta* 218, 784–792.

Rio, L.A., Gomez, M., Yanez, J., Leal, A. and Gorge, J.L. (1978) Iron deficiency in pea plants effect on catalase, peroxidase, chlorophyll and proteins of leaves. *Plant and Soil* 49, 343–353.

Tang, C., Robson, A.D. and Dilworth, M.J. (1992) The role of iron in the (brady) *Rhizobium* legume symbiosis. *Journal of Plant Nutrition* 15, 2235–2252.

PART III

Nutrient Deficiencies in Oilseed Crops

Plate 411. Plant showing yellowing of older leaf and normal green upper leaves. (Photo by Dr Prakash Kumar.)

Plate 412. Field view of a nitrogen-deficient castor crop. (Photo by Dr Prakash Kumar.)

Plate 413. Uniformly pale green to pale yellow lower leaves. (Photo by Dr Prakash Kumar.)

Plate 414. Dark yellow and brown necrotic severely deficient leaf. (Photo by Dr Prakash Kumar and Dr Manoj Kumar Sharma.)

CASTOR (*Ricinus communis* Linn.) NITROGEN (N) DEFICIENCY

Symptoms

1. Nitrogen deficiency decreases leaf area and photosynthesis of castor plants, leading to lower biomass accumulation.
2. High NH_4^+ concentration in the plant greatly restricts growth.
3. Under deficient conditions, plant growth is retarded remarkably. The leaf dry weight is reduced greatly. The root/shoot ratio is increased.
4. When nitrogen supply is reduced, the deficiency symptoms tend to occur first on lower leaves (Plate 413).
5. The old leaves become pale green to pale yellow while the younger leaves remain normal green (Plate 412).

Developmental stages

Stage I: In mild deficiency conditions, the entire plant may appear light green, having a more pronounced effect on older leaves.
Stage II: Under prolonged deficiency conditions, the lower leaves turn uniformly light yellow (Plate 413).
Stage III: As the symptoms advance, the lower leaves become dark yellow (Plate 411).
Stage IV: Under severe deficiency, the leaves become brown, necrotic and shed prematurely (Plate 414).

Likely to occur in

1. Soils having low organic matter.
2. Light-textured sandy soils that have been leached by heavy rainfall or excessive irrigation.
3. Soils exhausted by intensive cropping.
4. Waterlogged conditions.
5. Acid soils having pH below 6.0.
6. Alkaline soils having pH above 8.0.

Integrated nutrient management

1. Get the soil analysed before sowing to measure the amount of 'available' nitrogen in the soil.
2. Apply analysis-based recommended quantity of nitrogen as basal by using:
- **a.** organic manures;
- **b.** bio-fertilizers;
- **c.** nitrogenous fertilizers.

3. Top-dress soluble nitrogenous fertilizers such as urea in split doses.
4. For quick recovery, apply urea (2% w/v solution) as a foliar spray in the standing crop. Foliar sprays must be repeated every 10–15 days.
5. Incorporate legume crops in rotation.

Further reading

Allen, S. and Smith, J.A.C. (1986) Ammonium nutrition in *Ricinus communis*: its effect on plant growth and the chemical composition of the whole plant, xylem and phloem saps. *Journal of Experimental Botany* 37, 1599–1610.

Kirkby, E.A. and Armstrong, M.J. (1980) Nitrate uptake by roots as regulated by nitrate assimilation in the shoot of castor oil plants. *Plant Physiology* 65, 286–290.

Peuke, A.D., Jeschke, W.D. and Hartung, W. (1998) Foliar application of nitrate or ammonium as sole nitrogen supply in *Ricinus communis*. II. The flows of cations, chloride and abscisic acid. *New Phytologist* 140, 625–636.

Plate 415. Plant showing restricted growth and dark blue/green pigmentation on old leaves. (Photo by Dr Prakash Kumar.)

Plate 416. Normal green leaves of a plant with sufficient phosphorus supply. (Photo by Dr Prakash Kumar.)

Plate 417. Blue/green discoloration appearing in a deficient leaf. (Photo by Dr Prakash Kumar.)

Plate 418. Blue/green pigmentation intensified in a severely deficient leaf. (Photo by Dr Prakash Kumar.)

CASTOR (*Ricinus communis* Linn.) PHOSPHORUS (P) DEFICIENCY

Symptoms

1. The xylem transport of abscisic acid in phosphorus-deficient plants is stimulated largely, whereas the phloem transport is stimulated only very slightly. The increased import of abscisic acid shows a good correlation with the inhibition of leaf conductance under phosphorus deficiency.
2. Phosphorus deficiency leads to severe inhibition of NO_3^- uptake and transport in xylem and an even greater depression of NO_3^- reduction in the root but not in the shoot. In phosphorus-deficient plants the uptake of $H_2PO_4^-$ is severely decreased.
3. Plant growth is inhibited and dry matter production decreases drastically. Shoot/root ratio is decreased under phosphorus deficiency.
4. In phosphorus-deficient plants, leaf growth is particularly depressed and leaves turn dark green to bluish green (Plate 417) as compared with plants having sufficient phosphorus supply (Plate 416).
5. Nitrogen intake and xylem transport to the shoot decrease in phosphorus-deficient plants.
6. Flowering and maturity are delayed. Fewer capsules are formed containing fewer seeds, leading to low yields.
7. Because phosphorus is a fairly mobile nutrient in plants, under reduced supply conditions it is rapidly transferred from older tissues to younger growth. The deficiency symptoms are observed primarily in older leaves (Plate 417).
8. Blue/green pigmentation often develops on the lower leaves (Plates 415 and 418).

Developmental stages

Stage I: In mildly deficient conditions, plant growth is restricted and leaves become small and dark green.
Stage II: Blue/green pigmentation often develops on the lower leaves beginning from margins and proceeding inwards rapidly.
Stage III: In severe deficiency, the old leaves become dark bluish green (Plate 418).
Stage IV: Eventually the leaves turn dark brown and fall off.

Likely to occur in

1. Soils having low organic matter.
2. Alkaline and calcareous soils.
3. Soils exhausted by intensive cropping.
4. Acid soils and highly weathered soils.
5. Soils where the topsoil has been removed by erosion.
6. Acid soils having pH below 6.0.
7. Alkaline soils having pH between 7.5 and 8.5.
8. Cool and wet conditions where uptake of phosphorus is reduced.

Integrated nutrient management

1. Get the soil analysed to measure the amount of 'available' phosphate in the soil.
2. Apply analysis-based recommended quantity of phosphorus as basal by using:
- **a.** organic manures;
- **b.** phosphate-solubilizing microbial cultures;
- **c.** phosphatic fertilizers.

3. If standing crops show the deficiency, apply soluble phosphatic fertilizers such as ammonium phosphate with irrigation water.

Further reading

Jeschke, W.D., Peuke, A.D., Kirkby, E.A., Pate, J.S. and Hartung, W. (1996) Effects of P deficiency on the uptake, flows and utilization of C, N and H_2O within intact plants of *Ricinus communis* L. *Journal of Experimental Botany* 47, 1737–1754.

Jeschke, W.D., Kirkby, E.A., Peuke, A.D., Pate, J.S. and Hartung, W. (1997) Effects of P deficiency on assimilation and transport of nitrate and phosphate in intact plants of castor bean (*Ricinus communis* L.). *Journal of Experimental Botany* 48, 75–91.

Jeschke, W.D., Peuke, A.D., Pate, J.S. and Hartung, W. (1997) Transport, synthesis and catabolism of abscisic acid (ABA) in intact plants of castor bean (*Ricinus communis* L.) under phosphate deficiency and moderate salinity. *Journal of Experimental Botany* 48, 1737–1747.

Rojas, A.I. and Neptune, A.M.L. (1971) Effect of macronutrients and iron upon the growth and mineral composition of castor bean (*Ricinus communis* L.) cultivated in nutrient solution. *Anais da Escola Superior de Agricultura 'Luiz de Queiroz'* 28, 31–67.

Plate 419. Deficient plant displaying marginal yellowing on older leaves. (Photo by Dr Prakash Kumar.)

CASTOR (*Ricinus communis* Linn.) POTASSIUM (K) DEFICIENCY

Plate 420. Leaf showing yellow chlorosis along the edges and necrosis at the leaf tips. (Photo by Dr Prakash Kumar.)

Plate 421. Older leaves are chlorotic while upper leaves are normally green. (Photo by Dr Prakash Kumar.)

Plate 422. Severely deficient leaf showing yellowing and necrosis. (Photo by Dr Prakash Kumar.)

Symptoms

1. Potassium has a favourable effect on the assimilation of carbon dioxide and in particular promotes the export of photosynthates from the leaf.

2. Owing to potassium limitation, photosynthesis decreases slightly while dark respiration of the shoot decreases markedly and root respiration becomes nearly doubled. Low potassium supply also results in an increase in biosynthesis of abscisic acid in the roots.

3. Low potassium content in plants results in premature senescence in the crop. Maturity is also delayed. Plant growth is restricted and the number of flowers is reduced drastically.

4. Because potassium is regarded as a mobile nutrient in plants, under restricted supply conditions it is readily transferred from lower to younger tissues. Thus, the older leaves display deficiency symptoms first (Plate 419).

5. The symptoms appear as pale yellow chlorosis at the tips and along the edges of older leaves, while the upper leaves stay normally green (Plate 421).

6. As the symptoms advance, pale yellow chlorosis progresses inwards towards the main veins.

7. The chlorosis then turns into brown necrosis starting at the tips and along the leaf margins.

Developmental stages

Stage I: In mild deficiency the plants appear stunted, having small green foliage.

Stage II: In prolonged deficient conditions, marginal chlorosis develops on older leaves (Plate 421).

Stage III: Yellow chlorosis then turns into brown necrosis at the leaf tips (Plate 420).

Stage IV: In severe deficiency, the yellow chlorosis and necrosis advance towards the main veins (Plate 422).

Likely to occur in

1. Soils formed from parent material low in potassium.

2. Light-textured soils where potassium has been leached by heavy rainfall or excessive irrigation.

3. Soils low in organic matter.

4. Soils with wide Na:K, Mg:K or Ca:K ratio.

5. Acid soils having pH below 6.0.

Integrated nutrient management

1. Analyse the soil before planting to measure the amount of plant-available potassium.

2. Problematic acid/alkaline soils should be reclaimed.

3. Add organic manures well before planting.

4. Apply potassium chloride, potassium sulphate or potassium nitrate to the soil at or before planting as per soil testing recommendations.

5. In standing crops, apply soluble potassium salts with irrigation water.

Further reading

Cho, B.-H. and Komor, E. (1980) The role of potassium in charge compensation for sucrose-proton-symport by cotyledons of *Ricinus communis*. *Plant Science Letters* 17, 425–435.

Jeschke, W.D. and Wolf, O. (1988) External potassium supply is not required for root growth in saline conditions: experiments with *Ricinus communis* L. grown in a reciprocal split-root system. *Journal of Experimental Botany* 39, 1149–1167.

Mengel, K. and Haeder, H.-E. (1977) Effect of potassium supply on the rate of phloem sap exudation and the composition of phloem sap of *Ricinus communis*. *Plant Physiology* 59, 282–284.

Peuke, A.D., Jeschke, W.D. and Hartung, W. (2001) Flows of elements, ions and abscisic acid in *Ricinus communis* and site of nitrate reduction under potassium limitation. *Journal of Experimental Botany* 53, 241–250.

Plate 423. Interveinal chlorosis (mottling) appearing on lower leaves while younger leaves are normal. (Photo by Dr Prakash Kumar and Dr Manoj Kumar Sharma.)

Plate 424. Stunted plant having pale green leaves and pale yellow interveinal mottling on old leaves. (Photo by Dr Prakash Kumar.)

Plate 425. Leaf showing interveinal mottling beginning from the margins. (Photo by Dr Prakash Kumar.)

Plate 426. Yellow mottled leaf with prominent dark green broad veins. (Photo by Dr Prakash Kumar.)

CASTOR (*Ricinus communis* Linn.) MAGNESIUM (Mg) DEFICIENCY

Symptoms

1. Phosphate availability is considered a decisive factor for the mobilization and translocation of magnesium ions within the plant.
2. Magnesium deficiency limits stem diameter, plant height and dry matter yield of castor.
3. Deficient plants appear pale green. The stems become thin and spindly.
4. A lower number of capsules is produced containing fewer small seeds, contributing to poor yields.
5. Under restricted supply conditions, magnesium is readily mobilized from older to younger tissues (since magnesium is a mobile nutrient within plants).
6. The visual deficiency symptoms typically appear first on the older leaves. The younger leaves remain normal green (Plate 423).
7. Interveinal mottling develops on the older leaves, leaving veins dark green and prominent (Plates 425 and 426).
8. Eventually, the interveinal tissues may develop brown necrotic lesions.

Developmental stages

Stage I: In mild deficiency, the plants appear small and pale green.
Stage II: If deficiency persists, pale yellow interveinal mottling becomes evident in older leaves (Plate 424).
Stage III: As the symptoms advance, pale yellow mottling intensifies into interveinal regions, leaving distinct broad veins.
Stage IV: In severe deficiency, the interveinal tissues turn brown and necrotic.

Likely to occur in

1. Acid sandy soils that have been leached by heavy rainfall.
2. Peat and muck soils that are low in total magnesium.
3. Soils having higher amounts of ammonium, calcium or potassium.
4. Soils derived from parent material that is inherently low in magnesium.
5. Acid soils having pH below 6.5.
6. Alkaline soils having pH above 8.5.
7. Soils highly fertilized with potassium and manganese that hinder root uptake of magnesium.

Integrated nutrient management

1. Get the soil analysed before sowing to measure the amount of 'available' magnesium in the soil.
2. Apply analysis-based recommended quantity of magnesium before sowing using soluble salts such as magnesium sulphate or magnesium chloride.
3. In deficient standing crops apply soluble magnesium salts such as the sulphate, chloride or nitrate with irrigation water.

Further reading

Lavres, J. Jr, Nogueira, T.A.R., Cabral, C.P. and Malavolta, E. (2009) Deficiencies of macronutrient on the growth and biomass yield of castor bean cultivar Iris. *Revista Brasileira de Ciencias Agrarias* 4, 405–413.

Rojas, A.I. and Neptune, A.M.L. (1971) Effect of macronutrients and iron upon the growth and mineral composition of castor bean (*Ricinus communis* L.) cultivated in nutrient solution. *Anais da Escola Superior de Agricultura 'Luiz de Queiroz'* 28, 31–67.

Walsh, T. and O'Donohoe, T.F. (1945) Magnesium deficiency in some crop plants in relation to the level of potassium nutrition. *Journal of Agricultural Science* 35, 254–263.

Zhong, W., Schobert, C. and Komor, E. (1993) Transport of magnesium ions in the phloem of *Ricinus communis* L. seedlings. *Planta* 190, 114–119.

Plate 427. Yellow topmost leaf and dark green older leaves of a sulphur-deficient plant. (Photo by Dr Prakash Kumar.)

Plate 428. Uppermost leaves appearing pale yellow while those just below are light green. (Photo by Dr Prakash Kumar.)

Plate 429. Mildly sulphur-deficient pale green young leaf. (Photo by Dr Prakash Kumar and Dr Manoj Kumar Sharma.)

Plate 430. Severely sulphur-deficient pale yellow young leaf. (Photo by Dr Prakash Kumar and Dr Manoj Kumar Sharma.)

CASTOR (*Ricinus communis* Linn.) SULPHUR (S) DEFICIENCY

Symptoms

1. Sulphur deficiency limits stem diameter, plant height and dry matter yield of castor.
2. The number of branches and flowers is reduced. Fewer capsules are produced, resulting in poor yields.
3. Since sulphur is an immobile nutrient in plants, it is not readily moved from older to younger growth under poor supply conditions. Hence the deficiency symptoms are typically observed first on younger leaves, while older leaves remain green (Plate 427).
4. Initially, the entire plant may appear light green in colour.
5. The light green youngest leaves then turn pale yellow to yellow (Plates 427 and 430).
6. The topmost leaves turn yellow, middle leaves appear pale green while the bottom leaves remain green (Plate 428).
7. The paleness appears uniformly on the entire leaf, covering both the veins and the interveinal tissues evenly (Plate 429).

Developmental stages

Stage I: In mild deficiency conditions, all leaves on the plant become light green.
Stage II: If the deficiency persists, the younger leaves turn pale yellow while the old leaves remain pale green (Plate 428).
Stage III: In severe deficiency the youngest leaves become pale yellow.

Likely to occur in

1. Soils having low organic matter.
2. Light-textured sandy soils that have been leached by heavy rainfall or excessive irrigation.
3. Soils exhausted by intensive cropping.
4. Soils derived from parent material that is inherently low in sulphur.
5. Acid soils having pH below 6.0.

Integrated nutrient management

1. Get the soil analysed before sowing to measure the amount of 'available' sulphate in the soil.
2. Apply analysis-based recommended quantity of sulphur by mixing into the surface soil, well before sowing, either:
- **a.** elemental sulphur; or
- **b.** gypsum.

3. In deficient standing crops apply ammonium sulphate, magnesium sulphate or potassium sulphate with irrigation water.

Further reading

Lavres, J. Jr, Nogueira, T.A.R., Cabral, C.P. and Malavolta, E. (2009) Deficiencies of macronutrient on the growth and biomass yield of castor bean cultivar Iris. *Revista Brasileira de Ciencias Agrarias* 4, 405–413.

Patel, M.K., Klyansundaram, N.K., Pathak, A.R. and Fatteh, U.G. (1995) Sulphur application for castor production. *Indian Farming* 45(6), 19–21.

Rojas, A.I. and Neptune, A.M.L. (1971) Effect of macronutrients and iron upon the growth and mineral composition of castor bean (*Ricinus communis* L.) cultivated in nutrient solution. *Anais da Escola Superior de Agricultura 'Luiz de Queiroz'* 28, 31–67.

Plate 431. Iron deficiency castor plant showing interveinal chlorosis of upper leaves. (Photo by Dr Prakash Kumar.)

CASTOR (*Ricinus communis* Linn.)
IRON (Fe) DEFICIENCY

Plate 432. Yellow chlorotic youngest leaves showing only veins green. (Photo by Dr Prakash Kumar.)

Plate 433. Leaf showing yellow interveinal chlorosis and prominent dark green veins. (Photo by Dr Prakash Kumar and Dr Manoj Kumar Sharma.)

Plate 434. Dark yellow to whitish yellow young leaves with faded veins. (Photo by Dr Prakash Kumar.)

Symptoms

1. Iron is bound to the iron transport protein (ITP) for phloem-mediated long-distance transport in castor plants. It is presumed that Fe is transported as an Fe(III)–ITP complex in the phloem.
2. Deficient plants have stunted growth. The number and size of capsules are reduced, resulting in poor yields.
3. Since iron is considered an immobile nutrient within plants, under reduced supply conditions it is not readily translocated from older to younger leaves. Thus, the younger leaves display deficiency symptoms first.
4. Interveinal chlorosis develops on the younger leaves while the older leaves remain normally green (Plate 431).
5. A pale yellow chlorosis develops in interveinal tissues, leaving the veins prominently dark green (Plate 433).
6. The pale yellow interveinal tissues then turn dark yellow and the veins also become faded (Plate 432).
7. Severely deficient leaves then turn white and the veins almost disappear.

Developmental stages

Stage I: In mild deficiency, the uppermost young leaves develop pale yellow interveinal chlorosis. The older leaves remain green and healthy (Plate 431).
Stage II: If deficiency persists, pale green interveinal tissues turn yellow with prominent green veins (Plate 432).
Stage III: In the advanced stage of deficiency, leaves become yellow and veins are hardly visible (Plate 434).
Stage IV: In severe deficiency, leaves turn almost white and develop brown necrotic spots.

Likely to occur in

1. Sandy soils having low total iron content.
2. Alkaline and calcareous soils where solubility of iron is very low.
3. Peat and muck soils where organic matter ties up iron and reduces its availability in soil solution.
4. Acid soils with excessively high levels of soluble zinc, manganese, copper or nickel that can hinder the uptake of iron despite high iron availability.
5. Alkaline soils having pH above 7.5.

Integrated nutrient management

1. Analyse the soil before sowing to measure the 'available' iron in the soil.
2. Reclaim problematic alkaline soils.
3. Add organic manure well before sowing.
4. Apply a basal dose of soluble iron fertilizers such as $FeSO_4$ (commonly at 25 kg/ha) or Fe chelates (10 kg/ha). Use of organic chelates proves to be more promising as they maintain iron in soil solution.
5. In standing crops, apply $FeSO_4$ or Fe chelates (0.5% w/v solution) as foliar sprays. Foliar sprays must be repeated every 10–15 days.

Further reading

Kim, S.A. and Guerinot, M.L. (2007) Mining iron: iron uptake and transport in plants. *FEBS Letters* 581, 2273–2280.

Kruger, C., Berkowitz, O., Stephan, U.W. and Hell, R. (2002) A metal-binding member of the late embryogenesis abundant protein family transports iron in the phloem of *Ricinus communis* L. *Journal of Biological Chemistry* 277, 25062–25069.

Rojas, A.I. and Neptune, A.M.L. (1971) Effect of macronutrients and iron upon the growth and mineral composition of castor bean (*Ricinus communis* L.) cultivated in nutrient solution. *Anais da Escola Superior de Agricultura 'Luiz de Queiroz'* 28, 31–67.

Plate 435. (Photo by Dr Prakash Kumar.).

Plate 436. Pale yellow chlorotic young leaves, veins and their adjacent tissues are dark green. (Photo by Dr Prakash Kumar.)

Plate 437. Stunted plant with small leaves having pale yellow chlorosis. (Photo by Dr Prakash Kumar.)

Plate 438. Severely deficient plant showing chlorotic leaves with dark green broad veins. (Photo by Dr Prakash Kumar.)

CASTOR (*Ricinus communis* Linn.) ZINC (Zn) DEFICIENCY

Symptoms

1. The deficiency of zinc leads to disruption of chloroplasts, disintegration of thylakoids and absence of amyloplasts.
2. Castor appears to be suitable for phytoremediation of highly zinc-contaminated soil.
3. Plant growth is severely restricted. The internodes become short and the size of young leaves is drastically reduced.
4. Zinc deficiency reduces photosynthesis and chlorophyll concentrations in plants.
5. The number and size of capsules are reduced, contributing to poor yields.
6. The deficiency symptoms primarily appear on younger leaves while older leaves remain normally green (Plate 435).
7. Interveinal chlorosis develops on younger leaves.
8. Initially a pale yellow chlorosis becomes evident, but as the severity increases it turns into white chlorotic interveinal areas.

Developmental stages

Stage I: In mild deficiency conditions, plants appear stunted and leaves become smaller.
Stage II: As the symptoms advance, yellow chlorosis develops on younger leaves while veins with adjacent tissues remain dark green (Plates 436 and 437).
Stage III: In prolonged deficient conditions, the yellow chlorosis proceeds rapidly inwards leaving veins broad and dark green (Plate 438).
Stage IV: In severely deficient conditions, yellow chlorotic tissues turn white and then become necrotic.

Likely to occur in

1. Leached, light sandy soils where zinc content is low.
2. Alkaline and calcareous soils, where zinc availability is depressed.
3. Recently levelled soils where subsoil is exposed to cultivation. Plant-available zinc in surface soil is often double that of the subsoil.
4. Soil in which high rates of phosphatic fertilizers are applied, which can hamper zinc uptake by crops.
5. Acid soils having pH below 5.0.
6. Alkaline soils having pH above 7.5.

Integrated nutrient management

1. Analyse the soil before sowing to estimate the amount of plant-available zinc in the soil.
2. Reclaim problematic alkaline soils.
3. Incorporate any organic manure well before sowing.
4. Apply $ZnSO_4$ (commonly at 25–30 kg/ha) or Zn chelates (10 kg/ha) once every 2 years in zinc-deficient soils. Do not mix zinc fertilizers with phosphate fertilizers.
5. If deficiency symptoms appear in standing crops, spray 0.5% w/v $ZnSO_4$ to correct the deficiency.

Further reading

Anderson, W.B. (1972) Zinc in soils and plant nutrition. *Advances in Agronomy* 24, 147–186.

Lavres, J. Jr, Cabral, C.P., Rossi, M.L., Nogueira, T.A.R., Nogueira, N.L. and Malavolta, E. (2012) Deficiency symptoms and uptake of micronutrients by castor bean grown in nutrient solution. *Revista Brasileira de Ciência do Solo* 36, 233–242.

Melton, J.R., Mehtab, S.K. and Swoboda, A.R. (1973) Diffusion of zinc in soils as a function of applied zinc, phosphorus, and soil pH. *Soil Science Society of America Proceedings* 37, 370–380.

Xiaoyi, L. and Chiquan, H. (2005) Environmental behavior of zinc in *Ricinus communis* L. *Environmental Pollution and Control* 27(6), 1–2.

Plate 439. Manganese deficient castor plant. (Photo Dr Prakash Kumar.)

CASTOR (*Ricinus communis* Linn.) MANGANESE (Mn) DEFICIENCY

Plate 440. Yellow chlorotic manganese-deficient leaf. (Photo by Dr Prakash Kumar.)

Plate 441. Yellow mottled leaf showing a fine network of veins. (Photo by Dr Prakash Kumar.)

Plate 442. Dark yellow interveinal tissues and light green veins in a severely deficient leaf. (Photo by Dr Prakash Kumar.)

Symptoms

1. The deficiency of manganese leads to disruption of chloroplasts, disintegration of thylakoids and absence of amyloplasts.
2. Castor has high demand for manganese for fruit yield. The fruit yield is drastically reduced by manganese deficiency.
3. Deficient plants become stunted. The number of branches is reduced. Fewer capsules are produced.
4. Manganese is not readily transferred from older to younger tissues under poor supply conditions, because manganese is an immobile nutrient within plants. Thus, the deficiency symptoms tend to occur first in younger leaves while older leaves remain normally green.
5. Pale green to pale yellow interveinal mottling develops in recently matured young leaves.
6. The dark green main veins and minor veins surrounding the chlorotic interveinal tissues produce a fine netting pattern (Plate 439).
7. With the advancement of symptoms, the chlorosis (mottling) intensifies in interveinal tissues and the veins also become faded green (Plate 440).

Developmental stages

Stage I: In mildly deficient conditions, the veins appear dark green while interveinal tissues become light green (Plate 439).
Stage II: In prolonged deficiency conditions, the interveinal tissues turn pale yellow in small pockets and are surrounded by broad minor veins, showing a fine netting pattern.
Stage III: As symptoms advance, the interveinal areas enlarge whereas the minor veins turn light green and appear fine (Plate 441).
Stage IV: In severe deficiency, the interveinal tissues turn yellow and the main veins also become fine (Plate 442).

Likely to occur in

1. Light-textured sandy soils that have been extensively leached by heavy rainfall or excessive irrigation.
2. Calcareous and alkaline soils where solubility of manganese is very low.
3. Waterlogged peaty soils where organic matter ties up manganese and reduces its availability in solution.
4. Soils derived from parent material that is inherently low in manganese.
5. Acid soils having pH below 5.0.
6. Alkaline soils having pH above 7.5.
7. Soils containing high concentration of iron.

Integrated nutrient management

1. Get the soil analysed before sowing to measure the amount of 'available' manganese in the soil.
2. Problematic alkaline soils should be reclaimed.
3. Add organic manures well before sowing.
4. Apply soluble salts of manganese such as manganese sulphate as basal.
5. If deficiency appears in the standing crop, apply manganese sulphate (0.2 to 0.3% w/v solution) as a foliar spray. Repeated sprays may be required if symptoms reappear.

Further reading

Clarkson, D.T. (1988) The uptake and translocation of manganese by plant roots. In: Graham, R.D., Hannam, R.J. and Uren, N.C. (eds) *Manganese in Soils and Plants*. Kluwer Academic Publishers, Dordrecht, the Netherlands, pp. 101–111.

Lavres, J. Jr, Cabral, C.P., Rossi, M.L., Nogueira, T.A.R., Nogueira, N.L. and Malavolta, E. (2012) Deficiency symptoms and uptake of micronutrients by castor bean grown in nutrient solution. *Revista Brasileira de Ciência do Solo* 36, 233–242.

Reuter, D.J., Alston, A.M. and McFarlane, J.D. (1988) Occurrence and correction of manganese deficiency in plants. In: Graham, R.D., Hannam, R.J. and Uren, N.C. (eds) *Manganese in Soils and Plants*. Kluwer Academic Publishers, Dordrecht, the Netherlands, pp. 205–224.

Schmidke, I. and Stephan, U.W. (1995) Transport of metal micronutrients in the phloem of castor bean (*Ricinus communis*) seedlings. *Plant Physiology* 95, 147–153.

Plate 443. Stunted plant showing bottom leaves dark yellow, middle leaves pale yellow and top leaves light green. (Photo by Dr Prakash Kumar.)

Plate 444. Entire plant appearing chlorotic with more pronounced effect on lower leaves. (Photo by Dr Prakash Kumar.)

Plate 445. Uniformly dark yellow older leaves, pale yellow upper leaf and yellowing appearing on the stem. (Photo by Dr Prakash Kumar.)

Plate 446. Severely nitrogen-deficient whitish yellow old leaf. (Photo by Dr Prakash Kumar.)

SESAME (*Sesamum indicum* Linn.) NITROGEN (N) DEFICIENCY

Symptoms

1. Increased nitrogen nutrition results in increased protein level and decreased oil content.
2. Improved nitrogen nutrition increases non-amino acid forms of nitrogen.
3. The nitrogen-deficient plant shows poor growth. The stem becomes short and thin.
4. The plant has poor branching. The number and size of capsules are drastically reduced and fewer seeds are produced per capsule. Crop yield declines sharply.
5. Paling of the entire plant occurs due to lack of chlorophyll content in the leaves.
6. Nitrogen is fairly mobile within plants, thus it is quickly retranslocated from older to younger tissues when the supply to the plant is restricted.
7. The deficiency symptoms become evident first and more severely on the older leaves, then gradually progress to the upper leaves (Plate 444).
8. The severely deficient bottom leaves die and shed prematurely.

Developmental stages

Stage I: In mild deficiency, the plant lacks green colour and appears pale green.
Stage II: If deficiency continues, the older leaves turn yellow and the upper leaves can be pale green (Plate 445).
Stage III: In severely deficient plants, the old leaves turn dark yellow, the middle leaves become pale yellow and the top leaves appear light green (Plates 443 and 446).
Stage IV: In extremely deficient conditions, the bottom leaves dry and fall off early.

Likely to occur in

1. Soils having low organic matter.
2. Light-textured sandy soils that have been leached by heavy rainfall or excessive irrigation.
3. Soils exhausted by intensive cropping.
4. Waterlogged conditions.
5. Acid soils having pH below 6.0.
6. Alkaline soils having pH above 8.0.

Integrated nutrient management

1. Get the soil analysed before sowing to measure the amount of 'available' nitrogen in the soil.
2. Apply analysis-based recommended quantity of nitrogen as basal by using:
  a. organic manures;
  b. bio-fertilizers;
  c. nitrogenous fertilizers.
3. Top-dress soluble nitrogenous fertilizers such as urea in split doses.
4. For quick recovery, apply urea (2% w/v solution) as a foliar spray in the standing crop. Foliar sprays must be repeated every 10–15 days.

Further reading

Mitchell, G.A., Bingham, F.T. and Yermanos, D.M. (1974) Growth, mineral composition and seed characteristics of sesame as affected by nitrogen, phosphorus, and potassium nutrition. *Soil Science Society of America Journal* 38, 925–931.

Mitchell, G.A., Bingham, F.T., Labanauskas, C.K. and Yermanos, D.M. (1976) Protein and free amino acid composition of sesame meal as affected by nitrogen, phosphorus, and potassium nutrition. *Soil Science Society of America Journal* 40, 64–68.

Sayyad-Amin, P. and Ehsanzadeh, P. (2008) The effect of nitrogen on seed and oil yield of seven sesame (*Sesamum indicum* L.) genotypes in Isfahan. In: *Proceedings of the International Meeting on Soil Fertility Land Management and Agroclimatology, Kusadasi, Turkey, 29 October–1 November 2008*. Adnan Menderes University, Faculty of Agriculture, Aydin, Turkey, pp. 581–586.

Tiwari, R.K., Namdeo, K.N. and Girish Jha (2000) Effect of nitrogen and sulphur on growth, yield and quality of sesame (*Sesamum indicum*) varieties. *Research on Crops* 1, 163–167.

Plate 447. Phosphorus-deficient plant displaying purple discoloration on older leaves. (Photo by Dr Prakash Kumar.)

Plate 448. Purpling of tissues in an older leaf. (Photo by Dr Prakash Kumar.)

Plate 449. Purpling extending over the entire leaf blade. (Photo by Dr Prakash Kumar.)

Plate 450. Severely deficient leaf displaying brown necrosis. (Photo by Dr Prakash Kumar.)

SESAME (*Sesamum indicum* Linn.) PHOSPHORUS (P) DEFICIENCY

Symptoms

1. Restricted supply of phosphorus causes severe reduction in dry matter yield.
2. Deficient plants exhibit stunted growth and produce a lower number of capsules.
3. Phosphorus deficiency adversely affects seed formation but has no noticeable effect on oil content.
4. Phosphorus deficiency increases acid phosphatase activity of sesame.
5. Under short supply conditions, phosphorus is rapidly transferred from lower to upper tissues of the plant (since phosphorus is fairly mobile within plants). Thus the older leaves display deficiency symptoms first.
6. The deficient leaves primarily appear dark green.
7. Distinct purple pigmentation develops on the leaves with a more pronounced effect on the bottom leaves, then works up the plant to the upper leaves (Plate 447).

Developmental stages

Stage I: In mild deficiency, plants appear stunted and leaves turn dark green.
Stage II: As the symptoms advance, the purple pigmentation appears in the interveinal tissues of older leaves (Plate 448).
Stage III: In severe deficiency, the purpling spreads on the entire leaf blade, followed by yellowing of older leaves (Plate 449).
Stage IV: Finally, the entire leaf turns brown necrotic and dies (Plate 450).

Likely to occur in

1. Soils having low organic matter.
2. Alkaline and calcareous soils.
3. Soils exhausted by intensive cropping.
4. Acid soils and highly weathered soils.
5. Soils where topsoil has been removed by erosion.
6. Acid soils having pH below 6.0.
7. Alkaline soils having pH between 7.5 and 8.5.
8. Cool and wet conditions where uptake of phosphorus is reduced.

Integrated nutrient management

1. Get the soil analysed to measure the amount of 'available' phosphate in the soil well before sowing.
2. Apply analysis-based recommended quantity of phosphorus as basal by using:
- **a.** organic manures;
- **b.** phosphate-solubilizing microbial cultures;
- **c.** phosphatic fertilizers.

3. If standing crops show the deficiency, apply soluble phosphatic fertilizers such as ammonium phosphate with irrigation water.

Further reading

Chaplot, P.C. (1996) Effect of sulphur and phosphorus on the growth and yield attributes of sesame (*Sesamum indicum* L.). *International Journal of Tropical Agriculture* 14, 255–258.

Muthusamy, A., Patrick, G.M. and Jayabalan, N. (2001) Effect of phosphorus deficiency on acid phosphatase activity in sesame. *Indian Journal of Plant Physiology* 6, 202–204.

Sen, P.K. and Lahiri, A. (1960) Studies on the nutrition of oilseed crops. IV. Effects of phosphorus and sulphur on the uptake of nitrogen and growth, yield and oil content of sesame (*Sesamum indicum* Linn.). *Indian Agriculturist* 4, 23–26.

Shehu, H.E., Kwari, J.D. and Sandabe, M.K. (2010) Effects of N, P and K fertilizers on yield, content and uptake of N, P and K by sesame (*Sesamum indicum*). *International Journal of Agriculture and Biology* 12, 845–850.

Plate 451. Deficient plant showing yellow marginal chlorosis on lower leaves and small and dark green young leaves. (Photo by Dr Prakash Kumar.)

SESAME (*Sesamum indicum* Linn.) POTASSIUM (K) DEFICIENCY

Plate 452. Marginal and interveinal yellow chlorosis on older leaves. (Photo by Dr Prakash Kumar.)

Plate 453. Marginal yellowing of leaf as very early symptom. (Photo by Dr Prakash Kumar.)

Plate 454. Yellow chlorosis developed along the leaf margin and progressing rapidly into interveinal tissues. (Photo by Dr Prakash Kumar.)

Symptoms

1. Potassium deficiency results in an accumulation of free amino acids and a decrease in protein amino acids because potassium is required for the utilization of amino acids in protein synthesis. Therefore the rate of protein synthesis is reduced in potassium deficiency.
2. The increasing levels of potassium result in increased protein content and decreased oil content in sesame seeds.
3. Potassium-deficient plants can appear as if wilted in hot and dry conditions.
4. Plant growth is retarded. Deficient plants appear stunted and younger leaves become smaller in size.
5. Under short supply conditions potassium is readily mobilized from older to younger leaves because it is mobile within plants. Therefore, the deficiency symptoms occur first in older leaves whereas upper leaves remain green.
6. Symptoms develop initially as marginal chlorosis on lower leaves (Plate 453).
7. The marginal chlorosis then rapidly proceeds inwards (Plates 451 and 454).
8. The chlorotic tissues progressively become necrotic. The leaves eventually die and fall off prematurely.

Developmental stages

Stage I: In mild deficiency, plants show retarded growth and tendency of wilting in water stress and hot conditions.
Stage II: If deficiency persists, visible symptoms appear as marginal and interveinal yellow chlorosis in older leaves (Plate 452).
Stage III: In severe deficiency, yellow chlorosis turns into necrosis beginning from the margins and spreads inwards rapidly.

Likely to occur in

1. Soils formed from low potassium-bearing parent material.
2. Light-textured soils where potassium has been leached by heavy rainfall or excessive irrigation.
3. Soils having low organic matter.
4. Soils with wide Na:K, Mg:K or Ca:K ratio.
5. Acid soils having pH below 6.0.

Integrated nutrient management

1. Analyse the soil before planting to measure the amount of plant-available potassium.
2. Problematic acid/alkaline/saline soils should be reclaimed.
3. Add organic manures well before planting.
4. Apply potassic fertilizers to the soil at or before planting as per soil testing recommendations.
5. In standing crops, apply soluble potassium salts with irrigation water.

Further reading

Crocomo, O.J. and Basso, L.C. (1974) Accumulation of putrescine and related amino acids in potassium deficient *Sesamum*. *Phytochemistry* 13, 2659–2665.

Mitchell, G.A., Bingham, F.T. and Yermanos, D.M. (1974) Growth, mineral composition and seed characteristics of sesame as affected by nitrogen, phosphorus, and potassium nutrition. *Soil Science Society of America Journal* 38, 925–931.

Mitchell, G.A., Bingham, F.T., Labanauskas, C.K. and Yermanos, D.M. (1976) Protein and free amino acid composition of sesame meal as affected by nitrogen, phosphorus, and potassium nutrition. *Soil Science Society of America Journal* 40, 64–68.

Perrenoud, S. (1990) *Potassium and Plant Health. IPI Research Topic No. 3*, 2nd edn. International Potash Institute, Horgen, Switzerland.

Plate 455. Pale yellow chlorotic top leaves and light green lower leaves. (Photo by Dr Prakash Kumar.)

SESAME (*Sesamum indicum* Linn.) SULPHUR (S) DEFICIENCY

Plate 456. Sulphur-sufficient plant showing dark green leaves. (Photo by Dr Prakash Kumar.)

Plate 457. Paling of top leaves in moderate sulphur deficiency. (Photo by Dr Prakash Kumar.)

Plate 458. Severely deficient leaf showing uniform yellowing including veins. (Photo by Dr Prakash Kumar.)

Symptoms

1. Sulphur is regarded as the fourth key nutrient element after nitrogen, phosphorus and potassium. It plays an important role in oil-producing crops.
2. Sufficient sulphur supply can alleviate the harmful effect of water stress on growth, yield components and nutrient content in sesame seeds.
3. Sulphur nutrition enhances both protein and oil content in seeds.
4. Deficient plants show restricted growth and the leaves become smaller.
5. The number and size of flowers are reduced. Flowers shed prematurely, consequently producing fewer capsules.
6. Because sulphur is fairly immobile within the plant, under restricted supply conditions it is not readily retranslocated from older to younger plant tissues. Therefore, the younger leaves display deficiency symptoms first.
7. The younger leaves become pale green to pale yellow, while the lower leaves appear comparatively darker (Plates 455 and 457).
8. The young leaves become evenly chlorotic, including the veins.
9. In acute shortage of sulphur, symptoms progress to lower leaves and the whole plant appears chlorotic.

Developmental stages

Stage I: In early deficiency, paling develops in the top leaves whereas the old leaves stay normally green (Plate 457).
Stage II: If the deficiency continues, younger leaves become pale yellow while lower leaves appear light green (Plate 455).
Stage III: In acute deficiency, the entire plant turns yellow chlorotic (Plate 458).

Likely to occur in

1. Soils having low organic matter.
2. Light-textured sandy soils that have been leached by heavy rainfall or excessive irrigation.
3. Soils exhausted by intensive cropping.
4. Soils derived from low sulphur-bearing parent material.
5. Acid soils having pH below 6.0.

Integrated nutrient management

1. Get the soil analysed before sowing to measure the amount of 'available' sulphate in the soil.
2. Apply analysis-based recommended quantity of sulphur by mixing into the surface soil, well before sowing, either:
   a. elemental sulphur; or
   b. gypsum.
3. In deficient standing crops apply ammonium sulphate, magnesium sulphate or potassium sulphate with irrigation water.

Further reading

Cram, W.J. (1990) Uptake and transport of sulfate. In: Rennenberg, H., Brunold, C., De Kok, L.J. and Stulen, I. (eds) *Sulfur Nutrition and Sulfur Assimilation in Higher Plants: Fundamental, Environmental and Agricultural Aspects*. SPB Academic Publishing, The Hague, the Netherlands, pp 3–11.

Heidari, M., Galavi, M. and Hassani, M. (2011) Effect of sulphur and iron fertilizers on yield, yield components and nutrient uptake in sesame (*Sesamum indicum* L.) under water stress. *African Journal of Biotechnology* 10, 8695–8702.

Lewandowska, M. and Sirko, A. (2008) Recent advances in understanding plant responses to sulphur deficiency stress. *Acta Biochimia Polonica* 55, 457–471.

Tiwari, R.K., Namdeo, K.N. and Girish Jha (2000) Effect of nitrogen and sulphur on growth, yield and quality of sesame (*Sesamum indicum*) varieties. *Research on Crops* 1, 163–167.

Plate 459. Iron-deficient sesame plant (Photo by Dr Prakash Kumar and Dr Manoj Kumar Sharma.)

SESAME (*Sesamum indicum* Linn.) IRON (Fe) DEFICIENCY

Plate 460. Interveinal yellowing in younger leaves. (Photo by Dr Prakash Kumar.)

Plate 461. Top leaves most affected and a gradually decreasing effect on lower leaves. (Photo by Dr Prakash Kumar and Dr Manoj Kumar Sharma.)

Plate 462. Acutely deficient sesame leaf. (Photo by Dr Prakash Kumar.)

Symptoms

1. Foliar application of iron helps in alleviating the harmful effect of water stress on growth, yield components and nutrient content in sesame seeds.
2. Iron-deficient plants appear stunted and the stem becomes elongated. Number and size of capsules are decreased and fewer seeds per capsule are formed. Maturity also gets delayed.
3. Under reduced supply of iron to the plant, it is not easily mobilized from lower to younger leaves because iron is considered fairly immobile within plants.
4. The deficiency symptoms are observed first in younger leaves.
5. The younger leaves become pale green or yellow with prominent green veins. As the symptoms advance, the veins also become faded (Plates 459 and 461).
6. In prolonged deficiency, yellow chlorotic leaves turn white and symptoms progress to the lower leaves of the plant.
7. In acute short supply conditions, the chlorotic interveinal tissues turn into necrotic spots.

Developmental stages

Stage I: In the early stage, pale green interveinal chlorosis develops in younger leaves.
Stage II: If deficiency continues, the pale green chlorosis turns into yellow chlorosis with prominent green veins (Plate 460).
Stage III: In severe deficiency, the entire leaf blade turns white although the veins and adjacent tissues stay green (Plate 462).
Stage IV: In extremely deficient conditions, necrotic spots develop in chlorotic tissues.

Likely to occur in

1. Sandy soils having low total iron content.
2. Alkaline and calcareous soils where solubility of iron is very low.
3. Peat and muck soils where organic matter ties up iron and reduces its availability in soil solution.
4. Acid soils with excessively high levels of soluble zinc, manganese, copper or nickel that can hinder the uptake of iron despite high iron availability.
5. Alkaline soils having pH above 7.5.

Integrated nutrient management

1. Analyse the soil before sowing to measure the 'available' iron in the soil.
2. Reclaim problematic alkaline soils.
3. Add organic manure well before sowing.
4. Apply a basal dose of soluble iron fertilizers such as $FeSO_4$ (commonly at 25 kg/ha) or Fe chelates (10 kg/ha). Use of organic chelates proves to be more promising as they maintain iron in soil solution.
5. In standing crops, apply $FeSO_4$ or Fe chelates (0.5% w/v solution) as foliar sprays. Foliar sprays must be repeated every 10–15 days.

Further reading

Heidari, M., Galavi, M. and Hassani, M. (2011) Effect of sulphur and iron fertilizers on yield, yield components and nutrient uptake in sesame (*Sesamum indicum* L.) under water stress. *African Journal of Biotechnology* 10, 8695–8702.

Kannan, S. (1983) Fe-deficiency stress tolerance and pH reduction: some contrasts in the cultivars of sesame and lentil. *Journal of Plant Nutrition* 6, 1025–1031.

Kurdistani, R., Tohidinejad, E., Mohammadi-Nejad, G. and Zareie, S. (2011) Yield potential evaluation and path analysis of different sesame genotypes under various levels of iron. *African Journal of Plant Sciences* 5, 862–866.

Suresh, K., Rao, J.S.P. and Jagannatham, A. (1994) Effect of iron deficiency on photosynthetic characters, phytomass production and nutrient composition of sesame (*Sesamum indicum* Linn.). *Journal of Agricultural Sciences* 64, 244–246.

Plate 463. Nitrogen-deficient safflower plant. (Photo by Dr Prakash Kumar.)

Plate 464. Deficient leaf uniformly pale yellow. (Photo by Dr Prakash Kumar.)

Plate 465. Dark yellow to orange–yellow leaf. (Photo by Dr Prakash Kumar.)

Plate 466. Leaves showing stages of nitrogen deficiency. (Photo by Dr Prakash Kumar.)

SAFFLOWER (*Carthamus tinctorius* Linn.) NITROGEN (N) DEFICIENCY

Symptoms

1. Safflower is very sensitive to nitrogen deficiency. Nitrogen-deficiency symptoms appear even in mild deficiency conditions.
2. Nitrogen-deficient plants show poor growth. Plants are stunted with spindly stems and pale green to yellow leaves. Deficient plants produce fewer branches and flowers, resulting in a drastic reduction of crop yields.
3. Nitrogen is mobile in plants and under short supply conditions it is easily mobilized from older to younger leaves. The deficiency symptoms appear first and more severely on old leaves (Plate 463).
4. In mild deficiencies or when deficiency occurs in the young plant stage, the entire plant appears uniformly light green to pale yellow in colour. Older leaves are paler than younger.
5. If deficiency persists and becomes more severe, the older leaves show uniform pale green to pale yellow chlorosis (Plate 464).
6. As symptoms advance affected leaves turn to dark yellow to orange–yellow. Yellowing starts from the leaf tip and soon covers the entire leaf (Plate 465).
7. The yellowing is followed by brown necrosis. Eventually, the entire leaf burns and dies (Plate 466).

Developmental stages

Stage I: In mild deficiencies or when deficiency occurs in the young plant stage, the entire plant appears uniformly light green to pale yellow in colour.
Stage II: If the deficiency persists, the older leaves show uniform pale green to pale yellow chlorosis (Plate 464).
Stage III: As the symptoms advance, affected older leaves become dark yellow to orange–yellow. Yellowing starts from leaf tips (Plates 463 and 465).
Stage IV: In the most advanced stage, brown necrosis develops on affected older leaves. Eventually, the entire leaf burns and dies (Plate 466).

Likely to occur in

1. Soils having low organic matter.
2. Light-textured sandy soils that have been leached by heavy rainfall or excessive irrigation.
3. Soils exhausted by intensive cropping.
4. Waterlogged conditions.
5. Acid soils having pH below 6.0.
6. Alkaline soils having pH above 8.0.

Integrated nutrient management

1. Get the soil analysed before sowing to measure the amount of 'available' nitrogen in the soil.
2. Apply analysis-based recommended quantity of nitrogen as basal by using:
  - **a.** organic manures;
  - **b.** bio-fertilizers;
  - **c.** nitrogenous fertilizers.

3. Top-dress soluble nitrogenous fertilizers such as urea in two split doses.
4. For quick recovery, apply urea with irrigation water or as a foliar spray in the standing crop. Foliar sprays must be repeated every 10–15 days.
5. Incorporate legume crops in rotation.

Further reading

Abbadi, J. and Gerendas, J. (2009) Nitrogen use efficiency of safflower as compared to sunflower. *Journal of Plant Nutrition* 32, 929–945.

Abbadi, J., Gerendas, J. and Sattelmacher, B. (2008) Effects of nitrogen supply on growth, yield and yield components of safflower and sunflower. *Plant and Soil* 306, 167–180.

Dordas, C.A. and Sioulas, C. (2008) Safflower yield, chlorophyll content, photosynthesis, and water use efficiency response to nitrogen fertilization under rain fed conditions. *Industrial Crops and Products* 27, 75–85.

Haghighati, A. (2010) Study on the effects of nitrogen and phosphorus fertilizers on the yield and oil content of safflower lines in drylands. *Research Journal of Agronomy* 4(3), 57–62.

Plate 467. Potassium-deficient safflower plant. (Photo by Dr Prakash Kumar.)

Plate 468. Marginal chlorosis starts from the tip of lower leaves. (Photo by Dr Prakash Kumar.)

Plate 469. Yellow chlorosis covering margins. (Photo by Dr Prakash Kumar.)

Plate 470. Chlorosis followed by necrosis around the margin. (Photo by Dr Prakash Kumar.)

SAFFLOWER (*Carthamus tinctorius* Linn.) POTASSIUM (K) DEFICIENCY

Symptoms

1. Safflower has strong hidden hunger to potassium deficiency. Deficiency symptoms appear only in severe deficiency conditions.
2. Safflower has an extraordinary capability to utilize absorbed potassium in low potassium supply conditions. The total potassium requirement of safflower is also very low in comparison with other oilseed crops like sunflower. Being a deep-rooted, drought-resistant, low-input crop, safflower outperforms sunflower in soils having low available potassium.
3. Potassium moves readily from old to young leaves; therefore, deficiency symptoms appear first and more severely on older leaves (Plate 467).
4. The symptoms begin as marginal yellowing of older leaves. Yellowing starts from the leaf tip and advances around the leaf margin (Plates 468 and 469).
5. If deficiency persists, the leaf tip becomes scorched. The scorching spreads around the leaf margin (Plate 470).
6. A yellow band is found between the scorched area and healthy green tissues.
7. In acute deficiency conditions, the symptoms move towards upper leaves.

Developmental stages

Stage I: A pale yellow chlorosis appears on older leaves starting from the tip of the leaf (Plate 468).
Stage II: As the symptoms advance, pale yellow chlorosis covers the margins of the lamina (Plate 469).
Stage III: The leaf tip becomes scorched as the symptoms become severe. Yellow chlorosis is followed by necrosis along the margins of the leaf (Plate 470).
Stage IV: In acute deficiency conditions, the symptoms move towards upper leaves.

Likely to occur in

1. Soils formed from parent material low in potassium.
2. Light-textured soils where potassium has been leached by heavy rainfall or excessive irrigation.
3. Soils low in organic matter.
4. Soils with wide Na:K, Mg:K or Ca:K ratio.
5. Large bicarbonate concentration in irrigation water.
6. Highly acid soils having pH below 6.0.

Integrated nutrient management

1. Get the soil analysed before sowing to measure the amount of 'available' potassium in the soil.
2. Problematic acid soils should be reclaimed.
3. Add organic manures well before sowing, if organic matter is low.
4. Apply potassium nitrate, potassium sulphate or potassium chloride to the soil at or before sowing as per soil testing recommendations.
5. If potassium deficiency appears on the standing crop, apply soluble potassium salts such as potassium nitrate, potassium sulphate or potassium chloride with irrigation water. Foliar application of these salts is usually not recommended because a number of sprays are needed to fulfil crop requirements.

Further reading

Abbadi, J., Gerendas, J. and Sattlemacher, B. (2008) Effect of potassium supply on growth and yield of safflower as compared to sunflower. *Journal of Plant Nutrition and Soil Science* 171, 272–280.

Aslam, M. (1975) Potassium and sodium interrelations in growth and alkali cation content of safflower. *Agronomy Journal* 67, 262–264.

Bailey, L.D. and Soper, R.J. (1985) Potassium nutrition of rape, flax, sunflower and safflower. In: Munson, R.D. (ed.) *Potassium in Agriculture*. American Society of Agronomy, Crop Science Society of America and Soil Science Society of America, Madison, Wisconsin, pp. 765–798.

Gerendas, J., Abbadi, J. and Sattlemacher, B. (2008) Potassium efficiency of safflower (*Carthamus tinctorius* L.) and sunflower (*Helianthus annuus* L.). *Journal of Plant Nutrition and Soil Science* 171, 431–439.

Plate 471. Sulphur-deficient safflower plant. (Photo by Dr Prakash Kumar.)

SAFFLOWER (*Carthamus tinctorius* Linn.) SULPHUR (S) DEFICIENCY

Plate 472. Plant showing pale yellow younger and green older leaves. (Photo by Dr Prakash Kumar.)

Plate 473. Close-up of uniform pale yellow leaves. (Photo by Dr Prakash Kumar and Dr Manoj Kumar Sharma.)

Plate 474. Close-up of an affected leaf. (Photo by Dr Prakash Kumar.)

Symptoms

1. Safflower is a sulphur-responsive oilseed crop. The sulphur-deficient safflower crop lacks vigour and yields poorly. Affected plants are stunted with thin stems and pale green to pale yellow foliage.

2. Sulphur deficiency, even if severe, does not produce any type of necrosis or burning of leaves. The change in leaf colour is the only indicative symptom of sulphur deficiency.

3. In safflower, at the initial stage of growth, sulphur-deficiency symptoms are often confused with those caused by nitrogen deficiency and iron deficiency.

4. Sulphur is immobile in plants and does not mobilize from older to younger leaves in short supply conditions. Thus, just the reverse to nitrogen deficiency, sulphur-deficiency symptoms appear first and more severely on younger leaves while older leaves remain green and healthy.

5. Deficiency symptoms appear as an even and uniform pale green to pale yellow chlorosis across the lamina of young leaves. The colour of the midrib and other veins becomes very similar to that of interveinal areas of the leaf (in the case of iron deficiency, veins remain green and prominent).

6. If deficiency persists and becomes more severe, symptoms eventually move downwards covering more leaves.

Developmental stages

Stage I: Symptoms begin with uniform fading of young leaves from green to a pale green colour. Older leaves remain green and healthy (Plates 471 and 472).

Stage II: If the deficiency persists and becomes more severe, an even and uniform pale green to pale yellow chlorosis develops across the lamina of young leaves (Plates 473 and 474).

Stage III: In acute conditions, the deficiency symptoms move downwards covering more leaves.

Likely to occur in

1. Soils low in organic matter.

2. Light-textured sandy soils that have been leached by heavy rainfall or excessive irrigation.

3. Soils exhausted by intensive cropping.

4. Soils derived from parent material that is inherently low in sulphur (for example, soils formed from volcanic rocks).

5. Acid soils having pH below 6.0.

Integrated nutrient management

1. Get the soil analysed before sowing to measure the amount of 'available' sulphate in the soil and to predict the amount of fertilizer required.

2. Apply analysis-based recommended quantity of sulphur by mixing into the soil, well before sowing, either:

  a. elemental sulphur; or

  b. gypsum (calcium sulphate).

3. In deficient standing crops apply ammonium sulphate, magnesium sulphate or potassium sulphate with irrigation water.

4. Problematic acid soils should be reclaimed.

Further reading

Abbas, M., Tomar, S.S. and Nigan, K.B. (1995) Effect of phosphorus and sulphur fertilization in safflower (*Carthamus tinctorius* L.). *Indian Journal of Agronomy* 40, 243–248.

Babhulkar, P.S., Dinesh, K., Badole, W.P. and Balpande, S.S. (2000) Effect of sulphur and zinc on yield, quality and nutrient uptake by safflower in vertisol. *Journal of the Indian Society of Soil Science* 43, 541–543.

Hegde, D.M. (2008) Sulphur fertilization for safflower in different soil types of India. In: *Proceedings of the 7th International Safflower Confererence, Wagga Wagga, New South Wales, Australia, 3–6 November 2008*. Australian Oilseeds Federation, Royal Exchange, Australia, pp. 1–6.

Ravi, S., Channal, T., Hebsur, N.S. and Dharamtti, P.R. (2008) Effect of sulphur, zinc and iron nutrition on growth, yield, nutrient uptake and quality of safflower (*Carthamus tinctorius* L.). *Karnataka Journal of Agricultural Sciences* 21, 382–385.

Plate 475. Iron-deficient safflower plant. (Photo by Dr Prakash Kumar.)

SAFFLOWER (*Carthamus tinctorius* Linn.) IRON (Fe) DEFICIENCY

Plate 476. Close-up of a leaf showing interveinal chlorosis. (Photo by Dr Prakash Kumar and Dr Manoj Kumar Sharma.)

Plate 477. Advanced stage showing bright yellow top leaves. (Photo by Dr Prakash Kumar.)

Plate 478. Necrosis of affected leaves. (Photo by Dr Prakash Kumar.)

Symptoms

1. Iron is immobile in plants. Deficiency symptoms appear first and more severely on younger leaves. The older leaves remain normal and apparently healthy.
2. Chlorosis develops in interveinal tissues, leaving the veins green and prominent. Interveinal chlorosis of top leaves is the specific symptom of iron deficiency (Plate 475).
3. The advanced stage of iron deficiency in safflower looks very similar to sulphur deficiency. At the initial stage of iron deficiency veins remain green while in the case of sulphur deficiency the entire lamina becomes uniformly pale yellow; but in the advanced stage of iron deficiency the entire lamina may become uniformly bright yellow and may be confused with sulphur deficiency.
4. In the most advanced stage, necrosis develops on affected leaves.

Developmental stages

Stage I: The topmost leaves develop fading of interveinal tissues to a pale yellow colour with prominent green veins (Plates 475 and 476).
Stage II: The affected leaves turn bright yellow (Plate 477).
Stage III: Necrosis develops on affected leaves (Plate 478).

Likely to occur in

1. Sandy soils having low total iron.
2. Calcareous soils where solubility of iron is very low.
3. Peat and muck soils where organic matter ties up iron and reduces its availability in soil solution.
4. Acid soils where, despite high availability of iron in soil solution, excessively high levels of soluble zinc, manganese, copper or nickel hamper iron uptake by the plant.
5. Waterlogged soils.
6. Alkaline soils having pH above 7.5.

Integrated nutrient management

1. Get the soil analysed before sowing to measure the amount of 'available' iron in the soil.
2. As inorganic soluble salts of iron such as iron sulphates or chlorides quickly become insoluble in soil, soil dressing of organic forms of iron such as iron chelates (10 kg/ha) is more effective than inorganic forms of iron. These organic iron forms should be used as per soil pH. If the soil is neutral (around pH 7.0), the most suitable chelates are FeHEDTA and FeDTPA. For acid soils FeEDTA is the most effective chelate, while on alkaline calcareous soils FeEDDHA is recommended.
5. If deficiency appears on standing crops apply organic iron chelates or inorganic soluble salts such as iron sulphates or chlorides (0.5 to 1.0% w/vsolution) as a foliar spray. As iron is immobile in plants, foliar sprays must be repeated every 10–15 days. Two or three sprays are required to correct deficiency.

Further reading

Ravi, S., Channal, T., Hebsur, N.S. and Dharamtti, P.R. (2008) Effect of sulphur, zinc and iron nutrition on growth, yield, nutrient uptake and quality of safflower (*Carthamus tinctorius* L.). *Karnataka Journal of Agricultural Sciences* 21, 382–385.

Sangale, P.B., Patil, G.D. and Daftardar, S.Y. (1981) Effect of foliar application of zinc, iron and boron on yield of safflower. *Journal of Maharashtra Agricultural Universities* 6, 65–66.

Zareie, S., Golkar, P. and Mohammadi-Nejad, G. (2011) Effect of nitrogen and iron fertilizers on seed yield and yield components of safflower genotypes. *African Journal of Agricultural Research* 6, 3924–3929.

Plate 479. Nitrogen-deficient sunflower plant with lower leaves pale green. (Photo by Dr Prakash Kumar.)

SUNFLOWER (*Helianthus annuus* Linn.) NITROGEN (N) DEFICIENCY

Plate 480. Entire plant appearing light green. (Photo by Dr Prakash Kumar.)

Plate 481. Pale yellow chlorosis on an older leaf. (Photo by Dr Prakash Kumar.)

Plate 482. Close-up of a leaf turning from green to pale yellow. (Photo by Dr Prakash Kumar.)

Symptoms

1. Nitrogen-deficient sunflower shows very poor plant growth. Leaves are smaller in size with thin and spindly stems. Nitrogen deficiency decreases root hydraulic conductivity by about half. As a result, nitrogen-deficient leaves are unable to maintain adequate turgor for growth especially during the daytime, when transpiration further affects leaf turgor. Overall inhibition of leaf growth may be up to 75%.

2. Deficient plants mature more slowly than healthy plants. Deficient plants develop small heads with very poor seed setting, resulting in a drastic reduction in crop yields.

3. Deficiency symptoms develop first and more severely on the old leaves because nitrogen is readily transferred from older to younger leaves under deficient conditions. The younger leaves usually remain green and apparently healthy (Plate 479).

4. The older leaves turn pale green and later pale yellow (Plates 480 and 481).

5. A pale yellow chlorosis starts from marginal tissues of older leaves, soon covering the entire leaf. Affected leaves eventually die and hang down.

Developmental stages

Stage I: In mild deficiencies or when deficiency occurs in the young plant stage (six- to eight-leaf stage), the entire plant appears uniformly light green in colour (Plate 480).

Stage II: If the deficiency becomes severe, older leaves turn uniformly pale green (Plate 479).

Stage III: In acute deficiency conditions, a pale yellow chlorosis develops on older leaves (Plates 481 and 482).

Stage IV: In the most advanced stage the affected leaves die, turn brown and hang around the stem.

Likely to occur in

1. Soils having low organic matter.

2. Light-textured sandy soils that have been leached by heavy rainfall or excessive irrigation.

3. Soils exhausted by intensive cropping.

4. Waterlogged conditions.

5. Acid soils having pH below 6.0.

6. Alkaline soils having pH above 8.0.

Integrated nutrient management

1. Get the soil analysed before sowing to measure the amount of 'available' nitrogen in the soil.

2. Apply analysis-based recommended quantity of nitrogen as basal by using:

- **a.** organic manures;
- **b.** bio-fertilizers;
- **c.** nitrogenous fertilizers.

3. Top-dress soluble nitrogenous fertilizers such as urea in split doses.

4. For quick recovery, apply urea (2% w/v solution) as a foliar spray in the standing crop. Foliar sprays must be repeated every 10–15 days.

5. Incorporate legume crops in rotation.

Further reading

Ciompi, S., Gentili, E., Guidi, L. and Soldatini, G.F. (1996) The effect of nitrogen deficiency on leaf gas exchange and chlorophyll fluorescence parameters in sunflower. *Plant Science* 118, 177–184.

Grundon, N.J. (1987) *Hungry Crops: A Guide to Nutrient Deficiencies in Field Crops*. Queensland Department of Primary Industries, Brisbane, Australia.

Lehman, R.H and Rice, E.L. (1972) Effects of deficiencies of nitrogen, potassium and sulphur on chlorogenic acids and scopolin in sunflower. *American Midland Naturalist* 87, 71–80.

Radin, J.W. and Boyer, J.S. (1982) Control of leaf expansion by nitrogen nutrition in sunflower plants. *Plant Physiology* 69, 771–775.

Plate 483. Potassium-deficient sunflower plant. (Photo by Dr Prakash Kumar.)

SUNFLOWER (*Helianthus annuus* Linn.) POTASSIUM (K) DEFICIENCY

Plate 484. Marginal chlorosis starts from the tip of the leaf. (Photo by Dr Prakash Kumar.)

Plate 485. Pale yellow chlorosis followed by brown necrosis. (Photo by Dr Prakash Kumar.)

Plate 486. Leaf showing marginal and interveinal chlorosis. (Photo by Dr Prakash Kumar.)

Symptoms

1. Sunflower is very sensitive to potassium deficiency. Prominent symptoms appear even in mild deficiency conditions. Potassium moves readily from old to young leaves and therefore deficiency symptoms appear first on older leaves (Plate 483).
2. The symptoms begin as curvature of the lamina of older leaves. Affected leaves curve downwards until the leaf tip points towards the ground. This is the typical symptom of potassium deficiency in sunflower and usually occurs before any other symptom.
3. If deficiency persists, a pale yellow chlorosis appears on older leaves, covering the margins of the lamina (Plate 484).
4. Marginal yellow chlorosis and interveinal chlorosis develop on the tip and margins of the leaf which advance towards the base, ultimately covering the entire leaf (Plate 486).
5. The chlorosis is followed by brown necrosis both on marginal and interveinal tissues (Plates 485 and 486).

Developmental stages

Stage I: Affected old leaves curve downwards.
Stage II: A pale yellow chlorosis appears on older leaves, covering the margins of the lamina (Plate 485).
Stage III: Pale yellow chlorosis on margins and interveinal chlorosis between veins develop from the tip and side of the leaf, and proceed in a broad front towards the leaf base. Chlorosis is followed by brown necrosis (Plates 484, 485 and 486).

Likely to occur in

1. Soils formed from parent material low in potassium.
2. Light-textured soils where potassium has been leached by heavy rainfall or excessive irrigation.
3. Soils low in organic matter.
4. Soils with wide Na:K, Mg:K or Ca:K ratio.
5. Large bicarbonate concentration in irrigation water.
6. Highly acid soils having pH below 6.0.

Integrated nutrient management

1. Get the soil analysed before sowing to measure the amount of 'available' potassium in the soil.
2. Problematic acid soils should be reclaimed.
3. Add organic manures well before sowing, if organic matter is low.
4. Apply potassium nitrate, potassium sulphate or potassium chloride to the soil at or before sowing as per soil testing recommendations.
5. If potassium deficiency appears on the standing crop, apply soluble potassium salts such as potassium nitrate, potassium sulphate or potassium chloride with irrigation water. Foliar application of these salts is usually not recommended because a number of sprays are needed to fulfil crop requirements.

Further reading

Eaton, S.V. (1952) Effects of potassium deficiency on growth and metabolism of sunflower plants. *Botanical Gazette* 114, 165–180.

Grundon, N.J. (1987) *Hungry Crops: A Guide to Nutrient Deficiencies in Field Crops*. Queensland Department of Primary Industries, Brisbane, Australia.

Lawal, B.A., Obigbesan, G.O. and Akinrinde, E.A. (2011) Determination of the most limiting mineral nutrient element for growth and development of sunflower (*Helianthus annuus*) in southwestern Nigeria. *Advances in Agriculture and Botanics* 3, 93–103.

Plate 487. Calcium-deficient sunflower plant. (Photo by Dr Prakash Kumar.)

Plate 488. Malformed upper leaves and normal lower leaves. (Photo by Dr Prakash Kumar.)

Plate 489. Torn margins and holes in leaves. (Photo by Dr Prakash Kumar.)

Plate 490. Newly emerging very small leaves. (Photo by Dr Prakash Kumar.)

SUNFLOWER (*Helianthus annuus* Linn.) CALCIUM (Ca) DEFICIENCY

Symptoms

1. Calcium is immobile in the plant body. The deficiency symptoms appear first and more severely on younger leaves (Plate 487).
2. In mild calcium-deficiency conditions, the growing young leaves become malformed and brittle. The lamina becomes torn from the margins and may develop holes within. The veins are less affected and continue to grow, resulting in a claw-shaped deformity of the leaf.
3. As the deficiency becomes more severe, the new emerging leaves become very small (Plate 490). In acute deficiency conditions, black and brown necrotic lesions develop on the top of the stem and extend up to the petioles of upper leaves. Later on, the shoot tops and flower buds turn black and die.

Developmental stages

Stage I: In mild deficiencies, the growing young leaves become malformed, torn and brittle (Plates 488 and 489).
Stage II: If the deficiency becomes severe, new emerging leaves become very small (Plate 490).
Stage III: In acute deficiency conditions, black and brown necrotic lesions develop on the top of the stem (upper internodes) and on the petioles of upper leaves.
Stage IV: In the most advanced stage, the shoot tops and flower buds turn brown or black and die.

Likely to occur in

1. Strongly acid peat and muck soils that are low in total calcium.
2. Acid sandy soils that have been leached by heavy rainfall.
3. Alkaline or sodic soils where exchangeable sodium is very high, thus inhibiting uptake of calcium by the plant.
4. Soils having high soluble aluminium and low exchangeable calcium.
5. Heavy and excess application of nitrogenous fertilizers like urea.
6. Acid soils having pH below 6.5.
7. Alkaline soils having pH above 9.0.

Integrated nutrient management

1. Get the soil analysed before sowing to measure the lime or calcium requirement of the soil.
2. Apply analysis-based recommended quantity of calcium-containing fertilizers well before sowing, using soluble salts such as gypsum (calcium sulphate), calcium nitrate or calcium chloride.
3. Avoid heavy and excess application of nitrogenous fertilizers.
4. In acid soils lime (calcium carbonate) is more suitable, while in alkaline soils gypsum (calcium sulphate) is more suitable as a calcium supplement.
5. In low-pH soils, lime (calcium carbonate) should be applied to correct the soil pH.

Further reading

Chrominski, A., Abia, J.A. and Smith, B.N. (1987) Calcium deficiency and gibberellic acid enhance susceptibility of pumpkin and sunflower seedlings to *Sclerotinia sclerotiorum* infection. *Journal of Plant Nutrition* 10, 2181–2193.

Grundon, N.J. (1987) *Hungry Crops: A Guide to Nutrient Deficiencies in Field Crops*. Queensland Department of Primary Industries, Brisbane, Australia.

Tang, P.M. and Fuente, R.K.D. (1986) The transport of indole-3-acetic acid in boron- and calcium-deficient sunflower hypocotyls segments. *Plant Physiology* 81, 646–650.

Plate 491. Magnesium-deficient sunflower plant. (Photo by Dr Prakash Kumar and Dr Manoj Kumar Sharma.)

Plate 492. Fading of interveinal tissues. (Photo by Dr Prakash Kumar and Dr Manoj Kumar Sharma.)

Plate 493. Magnesium-deficient lower leaf with interveinal chlorosis. (Photo by Dr Prakash Kumar and Dr Manoj Kumar Sharma.)

Plate 494. Close-up of a leaf showing the chequer pattern. (Photo by Dr Prakash Kumar and Dr Manoj Kumar Sharma.)

SUNFLOWER (*Helianthus annuus* Linn.) MAGNESIUM (Mg) DEFICIENCY

Symptoms

1. Deficient plants appear stunted with short and thin stems.
2. Magnesium is mobile in plants; under short supply conditions it is transferred from older to younger leaves.
3. The deficiency symptoms appear first and more severely on the older leaves. The youngest leaves remain green and apparently healthy (Plate 491).
4. A pale yellow to almost white interveinal chlorosis develops on older leaves.
5. All veins (primary and secondary) of the leaf remain green, providing the leaf a green and whitish yellow chequer pattern. This is the specific symptom of magnesium deficiency in sunflower.
6. If the deficiency persists and becomes more severe, these symptoms move up the plant towards upper leaves.
7. In acute deficiency conditions, small brown necrotic lesions develop on the margins of affected leaves.

Developmental stages

Stage I: In mild deficiency the lower leaves develop temporary fading of interveinal tissues with prominent green veins (Plate 492).
Stage II: A pale yellow to almost white interveinal chlorosis develops on older leaves, giving the leaf a green and yellow chequer look (Plates 493 and 494).
Stage III: If the deficiency persists and becomes more severe, these symptoms move up the plant towards upper leaves.
Stage IV: In acute deficiency conditions, small brown necrotic lesions develop on margins of affected leaves.

Likely to occur in

1. Acid sandy soils from which magnesium has been leached by heavy rainfall.
2. Peat and muck soils that are low in total magnesium.
3. Soils derived from parent material that is inherently low in magnesium.
4. Soils with heavy and excess application of potassium fertilizers.
5. Soils with heavy and excess application of lime (calcium carbonate) or other calcium fertilizers.
6. Acid soils having pH below 6.5.
7. Alkaline soils having pH above 8.5.

Integrated nutrient management

1. Get the soil analysed before sowing to estimate the amount of soluble and exchangeable magnesium in the soil.
2. Apply analysis-based recommended quantity of magnesium before sowing using soluble salts such as magnesium sulphate or magnesium chloride.
3. In deficient standing crops, apply soluble magnesium salts such as the sulphate, chloride or nitrate with irrigation water. Foliar sprays of these salts are usually not advised as many sprays at frequent intervals are required to fulfil the crop need.
4. Magnesium deficiency in acid soils may be corrected by applying dolomite (a mixture of calcium carbonate and magnesium carbonate, $CaCO_3{\cdot}MgCO_3$) through broadcasting and mixing in the soil a few months before sowing.
5. Reclamation of problematic acid soils or alkaline soils should be done to regulate a proper supply of magnesium.

Further reading

Grundon, N.J. (1987) *Hungry Crops: A Guide to Nutrient Deficiencies in Field Crops*. Queensland Department of Primary Industries, Brisbane, Australia.

Lasa, B., Frechilla, S., Aleu, M., Moro, B.G., Lamsfus, C. and Tejo, P.M.A. (2000) Effects of low and high levels of magnesium on the response of sunflower plants grown with ammonium and nitrate. *Plant and Soil* 225, 167–174.

Masoni, A., Ercoli, L. and Mariotti, M. (1996) Spectral properties of leaves deficient in iron, sulphur, magnesium and manganese. *Agronomy Journal* 88, 937–943.

Plate 495. Sulphur-deficient sunflower plant. (Photo by Dr Prakash Kumar.)

Plate 496. Pale green sulphur-deficient leaf. (Photo by Dr Prakash Kumar.)

Plate 497. Upper leaves turning pale green to pale yellow. (Photo by Dr Prakash Kumar.)

Plate 498. Close-up of a sulphur-deficient yellow leaf. (Photo by Dr Prakash Kumar.)

SUNFLOWER (*Helianthus annuus* Linn.) SULPHUR (S) DEFICIENCY

Symptoms

1. Sulphur deficiency, even if severe, does not produce any type of necrosis or burning of leaves. The change in leaf colour is the only indicative symptom of sulphur deficiency.
2. At the initial stage sulphur-deficiency symptoms are often confused with those caused by nitrogen deficiency or iron deficiency.
3. In mild deficiency conditions or when the plant is young, the whole plant becomes light green to pale green. At this stage it is difficult to differentiate between sulphur deficiency and nitrogen deficiency. Close observation is required to see whether the older leaves are more dark green and the younger ones are more pale (case of sulphur deficiency), or whether the younger leaves are more dark green and the older ones are more pale (case of nitrogen deficiency).
4. Sulphur is immobile in plants and does not mobilize from older to younger leaves in short supply conditions. Thus, the deficiency symptoms appear first on younger leaves while older leaves remain green and healthy (Plate 495).
5. A uniform pale green to pale yellow chlorosis develops on younger leaves (Plate 497). The veins also become very similar in colour to the interveinal tissues (in the case of iron deficiency, these veins remain green and prominent).
6. As deficiency persists and become mores severe, leaf colour turns to yellow (Plate 498).
7. Plant vigour and size of flower heads reduce drastically, resulting in poor crop yields.

Developmental stages

Stage I: In mild deficiencies or when the plant is young, the whole plant becomes light green to pale green (Plate 495).
Stage II: If the deficiency persists, a uniform pale green to pale yellow chlorosis develops on younger leaves (Plate 496).
Stage III: In severe deficiency conditions affected leaves turn to yellow (Plate 498).
Stage IV: In acute conditions the deficiency symptoms move downwards, covering more leaves.

Likely to occur in

1. Soils low in organic matter.
2. Light-textured sandy soils that have been leached by heavy rainfall or excessive irrigation.
3. Soils exhausted by intensive cropping.
4. Soils derived from parent material that is inherently low in sulphur (for example, soils formed from volcanic rocks).
5. Acid soils having pH below 6.0.

Integrated nutrient management

1. Get the soil analysed before sowing to measure the amount of 'available' sulphate in the soil.
2. Apply analysis-based recommended quantity of sulphur by mixing into the soil, well before sowing, either:
  a. elemental sulphur; or
  b. gypsum (calcium sulphate).
3. In deficient standing crops apply ammonium sulphate, magnesium sulphate or potassium sulphate with irrigation water.

Further reading

Eaton, S.V. (1941) Influence of sulphur deficiency on metabolism of the sunflower. *Botanical Gazette* 102, 536–556.

Grundon, N.J. (1987) *Hungry Crops: A Guide to Nutrient Deficiencies in Field Crops*. Queensland Department of Primary Industries, Brisbane, Australia.

Lehman, R.H and Rice, E.L. (1972) Effects of deficiencies of nitrogen, potassium and sulphur on chlorogenic acids and scopolin in sunflower. *American Midland Naturalist* 87, 71–80.

Masoni, A., Ercoli, L. and Mariotti, M. (1996) Spectral properties of leaves deficient in iron, sulphur, magnesium and manganese. *Agronomy Journal* 88, 937–943.

Plate 499. Iron-deficient sunflower plant. (Photo by Dr Prakash Kumar and Dr Manoj Kumar Sharma.)

Plate 500. Plant showing interveinal chlorosis on top leaves. (Photo by Dr Prakash Kumar and Dr Manoj Kumar Sharma.)

Plate 501. Top leaves bleach to bright papery white. (Photo by Dr Prakash Kumar and Dr Manoj Kumar Sharma.)

Plate 502. Close-up of a leaf with brown necrosis. (Photo by Dr Prakash Kumar and Dr Manoj Kumar Sharma.)

SUNFLOWER (*Helianthus annuus* Linn.) IRON (Fe) DEFICIENCY

Symptoms

1. Sunflower is very susceptible to iron deficiency. Deficiency symptoms appear even in mild deficiency conditions.
2. Iron is immobile in plants. Deficiency symptoms appear first and more severely on the younger leaves. The older leaves remain normal and apparently healthy (Plate 499).
3. A pale yellow chlorosis develops in interveinal tissues leaving the veins green and prominent (Plate 500).
4. If the deficiency persists and becomes more severe, the youngest affected leaves bleach to papery white and show puckered growth (Plate 501).
5. As the symptoms advance, brown necrotic lesions develop in the white interveinal areas (Plate 502). The entire affected leaves burn and die, resulting in complete collapse of apical plant growth.

Developmental stages

Stage I: The topmost leaves develop temporary fading of interveinal tissues to a pale yellow colour with prominent green veins (Plate 500).
Stage II: Interveinal tissues of the affected leaves turn bright yellow with prominent green veins.
Stage III: The affected top leaves bleach to bright papery white (Plate 501).
Stage IV: Brown necrotic lesions develop on affected white leaves. The necrosis spreads and eventually kills the entire leaf (Plate 502).

Likely to occur in

1. Sandy soils having low total iron.
2. Calcareous soils where solubility of iron is very low.
3. Peat and muck soils where organic matter ties up iron and reduces its availability in soil solution.
4. Acid soils where, despite high availability of iron in soil solution, excessively high levels of soluble zinc, manganese, copper or nickel hamper iron uptake by the plant.
5. Waterlogged soils.
6. Alkaline soils having pH above 7.5.

Integrated nutrient management

1. Get the soil analysed before sowing to measure the amount of 'available' iron in the soil.
2. As inorganic soluble salts of iron such as iron sulphates or chlorides quickly become insoluble in soil, soil dressing of organic forms of iron such as iron chelates (10 kg/ha) is more effective than inorganic forms of iron. These organic iron forms should be used as per soil pH. If the soil is neutral (around pH 7.0), the most suitable chelates are FeHEDTA and FeDTPA. For acid soils FeEDTA is the most effective chelate, while on alkaline calcareous soils FeEDDHA is recommended.
3. If deficiency appears on standing crops apply organic iron chelates or inorganic soluble salts such as iron sulphates or chlorides (0.5 to 1.0% w/v solution) as a foliar spray. Foliar sprays must be repeated every 10–15 days. Two or three sprays are required to correct deficiency.

Further reading

Grundon, N.J. (1987) *Hungry Crops: A Guide to Nutrient Deficiencies in Field Crops*. Queensland Department of Primary Industries, Brisbane, Australia.

Masoni, A., Ercoli, L. and Mariotti, M. (1996) Spectral properties of leaves deficient in iron, sulphur, magnesium and manganese. *Agronomy Journal* 88, 937–943.

Romheld, V. and Marschner, H. (1981) Iron deficiency stress induced morphological and physiological changes in root tips of sunflower. *Physiologia Plantarum* 53, 354–360.

Plate 503. Boron-deficient sunflower plant. (Photo by Dr Prakash Kumar.)

SUNFLOWER (*Helianthus annuus* Linn.) BORON (B) DEFICIENCY

Plate 504. Bud leaves develop a bronze colour. (Photo by Dr Prakash Kumar.)

Plate 505. Crowded leaves at the top of a plant. (Photo by Dr Prakash Kumar.)

Plate 506. Leaf margins cup downwards. (Photo by Dr Prakash Kumar.)

Symptoms

1. Sunflower is highly sensitive to boron deficiency. Boron-deficiency symptoms on sunflower are so clear that sunflower is considered as one of the best indicator plants for boron deficiency.
2. Affected plants cease to grow and yield is very low.
3. Boron is immobile in plants and under short supply conditions it is not mobilized from older to younger leaves. Thus, deficiency symptoms first become evident on the younger leaves while the older leaves remain normal and apparently healthy.
4. In mild deficiency conditions, the growing upper internodes fail to elongate so the leaves at the top of the shoot crowd together, providing the plant a rosette appearance (Plate 503).
5. The upper leaves cease expanding and the margins cup downwards giving the leaflets a puckered and malformed appearance (Plate 506).
6. If deficiency persists and become mores severe, the bud leaves develop bronze colour followed by grey necrosis. The grey necrosis starts from basal portions of the leaves, leaving the tips green. The young leaves become brittle and are easily torn.
7. In acute deficiency conditions, the top of the growing shoot dies with brown necrotic lesions on the dead stem.

Developmental stages

Stage I: In mild deficiencies the upper leaves at the top crowd together, providing a rosette appearance to the plant (Plates 503 and 505).
Stage II: The upper leaves cease expanding and the margins cup downwards (Plate 506).
Stage III: If deficiency persists and becomes more severe, the bud leaves develop a bronze colour (Plate 504) followed by a grey necrosis.
Stage IV: In the most advanced stage the top of the growing shoot dies.

Likely to occur in

1. Soils formed from parent material low in boron, such as freshwater sediments or acid igneous rocks.
2. Soils low in organic matter.
3. Light-textured sandy soils that have been leached by heavy rainfall or excessive irrigation.
4. Peat and muck soils with low pH.
5. Lime-containing alkaline soils.
6. Acid soils having pH below 5.0.
7. Alkaline soils having pH between 8.0 and 9.0.

Integrated nutrient management

1. Get the soil analysed before sowing to measure the amount of 'available' boron in the soil.
2. Apply analysis-based recommended quantity of boron as a soil dressing, well before sowing, through the use of boron fertilizers such as borax, boric acid or chelated boron.
3. Boric acid and chelated boron may be applied as a foliar spray at 5–6 weeks after sowing or as soon as deficiency appears in the standing crop.

Further reading

Asad, A., Blamey, F.P.C. and Edwards, D.G. (2003) Effects of boron foliar applications on vegetative and reproductive growth of sunflower. *Annals of Botany* 92, 565–570.

Bergman, W. (1986) *Colour Atlas of Nutritional Disorders in Cultivated Plants*. VEB Gustav Fischer Verlag, Jena, Germany.

Grundon, N.J. (1987) *Hungry Crops: A Guide to Nutrient Deficiencies in Field Crops*. Queensland Department of Primary Industries, Brisbane, Australia.

Schuster, C.E. and Stephenson, R.E. (1940) Sunflower as an indicator plant of boron deficiency in soils. *Journal of the American Society of Agronomy* 32, 607–621.

Plate 507. Nitrogen-deficient plant showing dark yellow old leaves. (Photo by Dr Prakash Kumar.)

Plate 508. Entire plant pale green and stunted. (Photo by Dr Prakash Kumar.)

Plate 509. Uniform yellow old leaves and pale green young leaves. (Photo by Dr Manoj Kumar Sharma.)

Plate 510. Severely deficient uniformly dark yellow old leaf. (Photo by Dr Prakash Kumar.)

GROUNDNUT (*Arachis hypogaea* Linn.) NITROGEN (N) DEFICIENCY

Symptoms

1. Nitrogen deficiency is expressed as stunted growth of the plants.
2. The stem becomes thin and elongated.
3. The entire plant may become light green in appearance (Plate 508).
4. Since nitrogen is a mobile nutrient within plants, it is rapidly translocated from older to younger leaves (if the plant is not supplied with sufficient nitrogen).
5. The deficiency symptoms appear primarily on older leaves. In prolonged deficiency, the symptoms also become more severe on lower leaves.
6. Older leaves become uniformly yellow while young leaves may remain light green (Plate 509).
7. The stem may turn reddish in appearance.

Developmental stages

Stage I: In the early stage or in mild deficiency, the entire plant appears uniformly pale green (Plate 508).
Stage II: In prolonged deficiency, the older leaves turn uniformly yellow and the upper leaves appear pale green (Plate 509).
Stage III: In severe deficiency, the entire plant turns yellow (Plate 507).
Stage IV: In acute deficiency, the old leaves turn dark yellow and fall off prematurely (Plate 510).

Likely to occur in

1. Soils having low organic matter.
2. Light-textured sandy soils that have been leached by heavy rainfall or excessive irrigation.
3. Soils exhausted by intensive cropping.
4. Waterlogged conditions.
5. Acid soils having pH below 6.0.
6. Alkaline soils having pH above 8.0.

Integrated nutrient management

1. Get the soil analysed before sowing to measure the amount of 'available' nitrogen in the soil.
2. Apply analysis-based recommended quantity of nitrogen as basal by using:
  a. organic manures;
  b. bio-fertilizers;
  c. nitrogenous fertilizers.
3. Top-dress soluble nitrogenous fertilizers such as urea.
4. For quick recovery, apply urea (2% w/v solution) as a foliar spray in the standing crop. Foliar sprays must be repeated every 10–15 days.

Further reading

Burkhart, L. and Collins, E.R. (1941) Mineral nutrients in peanut plant growth. *Soil Science Society of America Proceedings* 6, 272–280.

Cox, F.R. and Perry, A. (1989) Groundnut (peanut). In: Plucknett, D.L. and Sprague, H.B. (eds) *Detecting Mineral Nutrient Deficiencies in Tropical and Temperate Crops*. Westview, Boulder, Colorado, pp. 137–144.

Smith, D.H., Wells, M.A., Porter, D.M. and Cox, F.R. (1993) Peanuts. In: Bennett, W.F. (ed.) *Nutrient Deficiencies and Toxicities in Crop Plants*. APS Press, St Paul, Minnesota, pp. 105–110.

Thornton, G.D. and Broadbent, F.E. (1948) Preliminary greenhouse studies of the influence of nitrogen fertilization of peanut on nodulation, yield, and gynophores absorption of this element. *Journal of the American Society of Agronomy* 40, 64–69.

Plate 511. Potassium-deficient plant showing various stages of leaf symptoms. (Photo by Dr Prakash Kumar and Dr Manoj Kumar Sharma.)

Plate 512. Yellowing of leaf margins and tips with some necrosis at tips of older leaves. (Photo by Dr Prakash Kumar and Dr Manoj Kumar Sharma.)

Plate 513. Yellow chlorosis progressing towards the base along the border and some chlorosis into interveinal tissues. (Photo by Dr Prakash Kumar and Dr Manoj Kumar Sharma.)

Plate 514. Brown necrosis and scorching of leaf tip and margins. (Photo by Dr Prakash Kumar and Dr Manoj Kumar Sharma.)

GROUNDNUT (*Arachis hypogaea* Linn.) POTASSIUM (K) DEFICIENCY

Symptoms

1. Potassium deficiency results in stunted plants and poor growth.
2. The branches become small and weak.
3. Potassium is a mobile nutrient within plants and it is readily moved from older leaves to younger parts of the plant. So the symptoms typically appear first on older leaves and progress up the plant with time. The young leaves usually remain dark green.
4. Symptoms begin as chlorosis at the tip and along the leaf margins of old leaves; some chlorosis may also occur in interveinal areas (Plate 511).
5. Old leaves turn brown and become scorched from the tips and along the margins (Plate 514).
6. Eventually, the old leaves dry and fall off early.

Developmental stages

Stage I: In mild deficiency, the plant exhibits poor growth and small branches.
Stage II: When deficiency becomes severe, yellowing develops along the margins and at the tips of older leaves (Plates 512 and 513).
Stage III: When deficiency persists, the tips and margins of old leaves turn brown necrotic or scorched (Plate 514).
Stage IV: In the later stage, old leaves dry or scorch and shed prematurely.

Likely to occur in

1. Soils formed from parent material low in potassium.
2. Light-textured soils where potassium has been leached by heavy rainfall or excessive irrigation.
3. Soils low in organic matter.
4. Soils with wide Na:K, Mg:K or Ca:K ratio.
5. Acid soils having pH below 6.0.

Integrated nutrient management

1. Analyse the soil before planting to measure the amount of plant-available potassium.
2. Problematic acid/alkaline soils should be reclaimed.
3. Add organic manures well before sowing.
4. Apply potassium chloride, potassium sulphate or potassium nitrate to the soil at or before planting as per soil testing recommendations.
5. In standing crops, apply soluble potassium salts with irrigation water.

Further reading

Burkhart, L. and Collins, E.R. (1941) Mineral nutrients in peanut plant growth. *Soil Science Society of America Proceedings* 6, 272–280.

Cox, F.R., Adams, F. and Tucker, B.B. (1982) Liming, fertilization and mineral nutrition. In: Patte, H.E. and Young, C.T. (eds) *Peanut Science and Technology*. American Peanut Research and Education Society, Yoakum, Texas, pp. 139–163.

Harris, H.C. (1949) The effect on the growth of peanuts of nutrient deficiencies in the roots and pegging zone. *Plant Physiology* 24, 150–161.

Smith, D.H., Wells, M.A., Porter, D.M. and Cox, F.R. (1993) Peanuts. In: Bennett, W.F. (ed.) *Nutrient Deficiencies and Toxicities in Crop Plants*. APS Press, St Paul, Minnesota, pp. 105–110.

Plate 515. Plant showing interveinal chlorosis on older leaves and dark green young leaves. (Photo by Dr Prakash Kumar.)

GROUNDNUT (*Arachis hypogaea* Linn.) MAGNESIUM (Mg) DEFICIENCY

Plate 516. Old chlorotic leaves on a magnesium-deficient plant. (Photo by Dr Prakash Kumar.)

Plate 517. Yellow chlorosis advancing towards the midrib of an older leaflet. (Photo by Dr Prakash Kumar.)

Plate 518. Necrotic areas developed on older leaflets. (Photo by Dr Prakash Kumar.)

Symptoms

1. Magnesium is considered a mobile nutrient within plants, so under poor supply conditions it is rapidly moved from older tissues to younger leaves.
2. The visual deficiency symptoms are first observed on older leaves and severity increases with time, if deficiency persists.
3. Symptoms are expressed on the plant as interveinal chlorosis in older leaves showing prominent dark green veins (Plate 516).
4. Initially, the chlorosis begins at the leaf margins and then proceeds towards the mid-vein (Plate 515).
5. In severely deficient conditions, the chlorotic older leaves may develop necrotic lesions in the interveinal tissues.
6. Eventually, the deficient leaves shed prematurely.

Developmental stages

Stage I: In early stage of deficiency, the branches become weak and prone to disease attack.
Stage II: In prolonged deficiency conditions, chlorosis develops between the veins of old leaves leaving the veins dark green (Plate 517).
Stage III: As the symptoms advance, necrotic lesions develop in the interveinal tissues of chlorotic older leaves (Plate 518).
Stage IV: In the later stage, the affected leaves shed prematurely.

Likely to occur in

1. Acid sandy soils that have been leached by heavy rainfall.
2. Peat and muck soils that are low in total magnesium.
3. Soils having higher quantity of calcium or potassium.
4. Soils derived from parent material that is inherently low in magnesium.
5. Acid soils having pH below 6.5.
6. Alkaline soils having pH above 8.5.

Integrated nutrient management

1. Get the soil analysed before sowing to estimate the amount of 'available' magnesium in the soil.
2. Apply analysis-based recommended quantity of magnesium before sowing using soluble salts such as magnesium sulphate or magnesium chloride.
3. In deficient standing crops, apply soluble magnesium salts such as magnesium sulphate, chloride or nitrate with irrigation water.

Further reading

Bledsoe, R.W., Harris, H.C. and Tisdale, W.B. (1946) Leafspot of peanut associated with magnesium deficiency. *Plant Physiology* 21, 237–240.

Cox, F.R. and Perry, A. (1989) Groundnut (peanut). In: Plucknett, D.L. and Sprague, H.B. (eds) *Detecting Mineral Nutrient Deficiencies in Tropical and Temperate Crops*. Westview, Boulder, Colorado, pp. 137–144.

Cox, F.R., Adams, F. and Tucker, B.B. (1982) Liming, fertilization and mineral nutrition. In: Patte, H.E. and Young, C.T. (eds) *Peanut Science and Technology*. American Peanut Research and Education Society, Yoakum, Texas, pp. 139–163.

Smith, D.H., Wells, M.A., Porter, D.M. and Cox, F.R. (1993) Peanuts. In: Bennett, W.F. (ed.) *Nutrient Deficiencies and Toxicities in Crop Plants*. APS Press, St Paul, Minnesota, pp. 105–110.

Plate 519. Sulphur-deficient plant stunted and with pale yellow young leaves. (Photo by Dr Prakash Kumar.)

Plate 520. Uniformly chlorotic young leaves. (Photo by Dr Prakash Kumar.)

Plate 521. Pale green sulphur-deficient young leaves (left) compared with dark green normal leaves (right). (Photo by Dr Prakash Kumar.)

Plate 522. Severely deficient yellow young leaves. (Photo by Dr Prakash Kumar.)

GROUNDNUT (*Arachis hypogaea* Linn.) SULPHUR (S) DEFICIENCY

Symptoms

1. Sulphur deficiency results in stunted plants and reduced branching.
2. The deficient crop gets delayed in maturity.
3. In restricted supply conditions, sulphur is not easily transferred from lower to upper leaves because sulphur is immobile in the plant body. Consequently the deficiency symptoms are observed first on the younger, upper leaves.
4. Initially, the whole plant appears light green.
5. The younger leaves become pale green or pale yellow (Plates 519 and 521).
6. The uniform chlorosis develops over the entire leaf, covering both the veins and the interveinal tissues (Plate 520).
7. In acute deficiency conditions, the entire plant turns yellow.

Developmental stages

Stage I: In mild deficiency, general yellowing develops on the entire plant; however, old leaves look darker (Plate 520).
Stage II: If the deficiency becomes severe, the younger leaves turn yellow (Plate 522).
Stage III: In acutely deficient conditions, the whole plant turns yellow.

Likely to occur in

1. Soils having low organic matter.
2. Light-textured sandy soils that have been leached by heavy rainfall or excessive irrigation.
3. Soils exhausted by intensive cropping.
4. Soils derived from parent material that is inherently low in sulphur.
5. Acid soils having pH below 6.0.

Integrated nutrient management

1. Get the soil analysed before sowing to measure the amount of 'available' sulphate in the soil.
2. Apply analysis-based recommended quantity of sulphur by mixing into the surface soil, well before sowing, either:
  a. elemental sulphur; or
  b. gypsum.
3. In deficient standing crops apply ammonium sulphate, magnesium sulphate or potassium sulphate with irrigation water.

Further reading

Cox, F.R. and Perry, A. (1989) Groundnut (peanut). In: Plucknett, D.L. and Sprague, H.B. (eds) *Detecting Mineral Nutrient Deficiencies in Tropical and Temperate Crops*. Westview, Boulder, Colorado, pp. 137–144.

Cox, F.R., Adams, F. and Tucker, B.B. (1982) Liming, fertilization and mineral nutrition. In: Patte, H.E. and Young, C.T. (eds) *Peanut Science and Technology*. American Peanut Research and Education Society, Yoakum, Texas, pp. 139–163.

Paricha, N.S. and Aulakh, M.S. (1986) Role of sulphur in the nutrition of groundnut. *Fertilizer News* September issue, 17–21.

Smith, D.H., Wells, M.A., Porter, D.M. and Cox, F.R. (1993) Peanuts. In: Bennett, W.F. (ed.) *Nutrient Deficiencies and Toxicities in Crop Plants*. APS Press, St Paul, Minnesota, pp. 105–110.

Plate 523. Iron-deficient plant showing chlorotic younger leaves and green old leaves. (Photo by Dr Prakash Kumar.)

Plate 524. Interveinal chlorosis on younger leaves. (Photo by Dr Prakash Kumar.)

Plate 525. Leaves completely devoid of chlorophyll. (Photo by Dr Prakash Kumar.)

Plate 526. Papery white leaves. (Photo by Dr Prakash Kumar.)

GROUNDNUT (*Arachis hypogaea* Linn.) IRON (Fe) DEFICIENCY

Symptoms

1. Groundnut is sensitive to iron deficiency.
2. Deficient plants show poor growth and the leaves become smaller in size.
3. Iron is not rapidly translocated from older tissues to younger parts of the plant. Therefore, the visible deficiency symptoms are observed primarily on younger leaves (Plate 523).
4. A pale yellow chlorosis develops in interveinal tissues while the veins remain green and prominent; this chlorosis extends the full length of the leaves (Plate 524).
5. As the chlorosis advances, veins also become chlorotic and the entire leaf may appear pale yellow (Plate 524).
6. In the later stage, leaves turn almost white and may become necrotic.

Developmental stages

Stage I: In mild deficiencies, pale yellow interveinal chlorosis develops in interveinal areas leaving dark green and prominent veins (Plate 524).
Stage II: When deficiency persists for long, pale green young leaves turn yellow with faded veins.
Stage III: In acutely deficient conditions, the entire leaf blade becomes bleached and papery white (Plates 525 and 526).
Stage IV: In the later stage, leaves may develop brown or necrotic lesions.

Likely to occur in

1. Sandy soils having low total iron.
2. Alkaline and calcareous soils where solubility of iron is very low.
3. Peat and muck soils where organic matter ties up iron and reduces its availability in soil solution.
4. Acid soils with excessively high levels of soluble zinc, manganese, copper or nickel that can hinder the uptake of iron despite high iron availability.
5. Alkaline soils having pH above 7.5.

Integrated nutrient management

1. Analyse the soil before sowing to measure the amount of 'available' iron in the soil.
2. Reclaim problematic alkaline soils.
3. Add organic manure well before sowing.
4. Apply a basal dose of soluble iron fertilizers such as $FeSO_4$ (commonly at 25 kg/ha) or Fe chelates (10 kg/ha). Use of organic chelates proves to be more promising as they maintain iron in soil solution.
5. In standing crops, apply 0.5% w/v $FeSO_4$ + 0.1% w/v citric acid solution, or 2% w/v ferric citrate solution, as a foliar spray. Foliar sprays must be repeated every 10–15 days.

Further reading

Cox, F.R. and Perry, A. (1989) Groundnut (peanut). In: Plucknett, D.L. and Sprague, H.B. (eds) *Detecting Mineral Nutrient Deficiencies in Tropical and Temperate Crops*. Westview, Boulder, Colorado, pp. 137–144.

Cox, F.R., Adams, F. and Tucker, B.B. (1982) Liming, fertilization and mineral nutrition. In: Patte, H.E. and Young, C.T. (eds) *Peanut Science and Technology*. American Peanut Research and Education Society, Yoakum, Texas, pp. 139–163.

Lachover, D., Fichman, M. and Hartzook, A. (1970) The use of iron chelate to correct chlorosis in peanuts under field conditions. *Oleagineux* 25, 85–88.

Zaharieva, T., Kasabov, D. and Romheld, V. (1988) Responses of peanuts to iron–manganese interaction in calcareous soil. *Journal of Plant Nutrition* 11, 1015–1024.

Plate 527. Zinc-deficient stunted plant showing white chlorotic leaves. (Photo by Dr Prakash Kumar.)

GROUNDNUT (*Arachis hypogaea* Linn.) ZINC (Zn) DEFICIENCY

Plate 528. Leaves showing white chlorotic areas. (Photo by Dr Prakash Kumar and Dr Manoj Kumar Sharma.)

Plate 529. White chlorotic areas proceeding towards the mid-vein. (Photo by Dr Prakash Kumar and Dr Manoj Kumar Sharma.)

Plate 530. White chlorotic leaflets almost devoid of chlorophyll. (Photo by Dr Prakash Kumar.)

Symptoms

1. Zinc-deficient plants become stunted.
2. The size of internodes is reduced and the young leaves become small.
3. Since zinc is only partly mobile within plants, it is not rapidly transferred from older to younger parts of the plant.
4. The deficiency symptoms first become evident on recently matured leaves, while older leaves remain usually green and healthy.
5. The chlorosis develops in interveinal areas of recently matured young leaves (Plate 527).
6. In prolonged deficient conditions, leaves may become reddish brown and fall off.
7. The deficient plants can develop reddish pigmentation on stems, petioles and leaf veins.

Developmental stages

Stage I: The early deficiency symptom in plants is expressed as stunted growth and small leaves.
Stage II: The recently matured leaves develop interveinal chlorosis (Plates 527, 528, 529 and 530).
Stage III: As the deficiency becomes severe, chlorotic leaves turn reddish brown and fall off.

Likely to occur in

1. Leached, light sandy soils where zinc content is low.
2. Alkaline and calcareous soils, where zinc availability is depressed.
3. Recently levelled soils where subsoil is exposed to cultivation. Plant-available zinc in surface soil is often double that of the subsoil.
4. Soils in which high rates of phosphatic fertilizers are applied, which can hamper zinc uptake by crops.
5. Acid soils having pH below 5.0.
6. Alkaline soils having pH above 7.5.

Integrated nutrient management

1. Analyse the soil before sowing to estimate the amount of plant-available zinc in the soil.
2. Reclaim problematic alkaline soils.
3. Incorporate any organic manure well before sowing.
4. Apply $ZnSO_4$ (commonly at 25–30 kg/ha) or Zn chelates (10 kg/ha) once every 2 years in zinc-deficient soils.
5. In standing crops, spray 5 kg of zinc sulphate plus 2.5 kg of unslaked lime in 500 l of water.

Further reading

Bell, R.H., Kirk, G., Plaskett, D. and Loneragan, J.F. (1990) Diagnosis of zinc deficiency in peanut (*Arachis hypogaea* L.) by plant analysis. *Communications in Soil Science and Plant Analysis* 21, 273–285.

Cox, F.R. and Perry, A. (1989) Groundnut (peanut). In: Plucknett, D.L. and Sprague, H.B. (eds) *Detecting Mineral Nutrient Deficiencies in Tropical and Temperate Crops*. Westview, Boulder, Colorado, pp. 137–144.

Cox, F.R., Adams, F. and Tucker, B.B. (1982) Liming, fertilization and mineral nutrition. In: Patte, H.E. and Young, C.T. (eds) *Peanut Science and Technology*. American Peanut Research and Education Society, Yoakum, Texas, pp. 139–163.

Smith, D.H., Wells, M.A., Porter, D.M. and Cox, F.R. (1993) Peanuts. In: Bennett, W.F. (ed.) *Nutrient Deficiencies and Toxicities in Crop Plants*. APS Press, St Paul, Minnesota, pp. 105–110.

Plate 531. Manganese-deficient plant showing interveinal chlorosis on new leaves. (Photo by Dr Prakash Kumar.)

Plate 532. Chlorotic interveinal tissues with dark green broad veins on young leaves. (Photo by Dr Manoj Kumar Sharma.)

Plate 533. In early deficiency, leaflets are light green with green veins. (Photo by Dr Manoj Kumar Sharma.)

Plate 534. Severely deficient leaflets with yellow chlorotic interveinal areas and green veins. (Photo by Dr Manoj Kumar Sharma.)

GROUNDNUT (*Arachis hypogaea* Linn.) MANGANESE (Mn) DEFICIENCY

Symptoms

1. Groundnut is considered relatively less susceptible to manganese deficiency.
2. Deficient plants show poor growth and stems appear slender.
3. The number of pods is reduced markedly and there are fewer seeds (kernels) per pod.
4. Manganese is fairly immobile within plants and it is not easily moved from older to younger leaves.
5. The deficiency symptoms become evident first and more pronounced on younger leaves.
6. The symptoms begin as interveinal chlorosis on young leaves, the tissues between the veins become chlorotic while the veins and adjacent tissues remain dark green (Plate 533).
7. With the advancement of symptoms, the chlorosis intensifies in interveinal tissues leaving the veins light green (Plates 531 and 534).

Developmental stages

Stage I: In mild deficiencies, the leaves become chlorotic with broad green veins (Plate 532).
Stage II: In prolonged deficiency conditions, the interveinal tissues become yellow chlorotic with light green narrow veins (Plate 534).
Stage III: If deficiency becomes severe, bronzing may develop on leaf margins.
Stage IV: In acute deficiency, the older leaves become brown necrotic and fall off prematurely.

Likely to occur in

1. Light-textured sandy soils that have been leached by heavy rainfall or excessive irrigation.
2. Calcareous and alkaline soils where solubility of manganese is very low.
3. Waterlogged peaty soils where organic matter ties up manganese and reduces its availability in solution.
4. Soils derived from parent material that is inherently low in manganese.
5. Acid soils having pH below 5.0.
6. Alkaline soils having pH above 7.5.

Integrated nutrient management

1. Get the soil analysed before sowing to measure the amount of 'available' manganese in the soil.
2. Problematic alkaline soils should be reclaimed.
3. Add organic manures well before sowing.
4. Apply soluble salts of manganese such as manganese sulphate as basal.
5. In deficient standing crops, apply manganese sulphate (0.2 to 0.3% w/v solution) as a foliar spray. Repeated sprays may be required if symptoms reappear.

Further reading

Lombin, G.L. and Bates, T.E. (1982) Comparative responses of peanuts, alfalfa and soybeans to varying rates of boron and manganese on two calcareous Ontario soils. *Canadian Journal of Soil Science* 62, 1–9.

Cox, F.R. and Perry, A. (1989) Groundnut (peanut). In: Plucknett, D.L. and Sprague, H.B. (eds) *Detecting Mineral Nutrient Deficiencies in Tropical and Temperate Crops*. Westview, Boulder, Colorado, pp. 137–144.

Parker, M.B. and Walker, M.E. (1986) Soil pH and manganese effects on manganese nutrition of peanut. *Agronomy Journal* 78, 614–620.

Smith, D.H., Wells, M.A., Porter, D.M. and Cox, F.R. (1993) Peanuts. In: Bennett, W.F. (ed.) *Nutrient Deficiencies and Toxicities in Crop Plants*. APS Press, St Paul, Minnesota, pp. 105–110.

Plate 535. Entire plant pale green with pale yellow old leaves. (Photo by Dr Manoj Kumar Sharma.)

Plate 536. Pale green nitrogen-deficient leaf (left) compared with a dark green normal leaf (right). (Photo by Dr Manoj Kumar Sharma.)

Plate 537. Pale yellow severely nitrogen-deficient leaves. (Photo by Dr Manoj Kumar Sharma.)

Plate 538. Dark yellow and necrotic severely nitrogen-deficient leaf. (Photo by Dr Prakash Kumar.)

SOYBEAN (*Glycine max* Linn.) NITROGEN (N) DEFICIENCY

Symptoms

1. Nitrogen-deficient plants become short in height with slender and spindly stems.
2. Fewer pods are produced.
3. The number and size of seeds per pod are reduced, producing poor grain yields.
4. The entire plant becomes light green in colour.
5. Leaves become smaller in size and the number of branches gets reduced.
6. Nitrogen is a fairly mobile nutrient in plants and it is readily translocated from older to younger leaves as its supply becomes restricted.
7. The deficiency symptoms appear primarily on older leaves and progress rapidly to upper leaves (Plate 535).
8. Initially, the old leaves become pale green to pale yellow (Plate 536).
9. In the later stage, leaves turn dark yellow or almost white (Plate 538).
10. Eventually leaves turn brown, die and drop early.

Developmental stages

Stage I: In the early stage of deficiency, the entire plant can become uniformly pale green.
Stage II: If the deficiency becomes severe, the lower leaves become uniformly pale yellow (Plate 537).
Stage III: As the symptoms advance, the entire leaf becomes dark yellow and then almost white (Plate 538).
Stage IV: In acutely deficient conditions, old leaves turn brown and die.

Likely to occur in

1. Soils having low organic matter.
2. Light-textured sandy soils that have been leached by heavy rainfall or excessive irrigation.
3. Soils exhausted by intensive cropping.
4. Waterlogged conditions.
5. Acid soils having pH below 6.0.
6. Alkaline soils having pH above 8.0.
7. Soils low in nitrogen and limiting moisture conditions.

Integrated nutrient management

1. Get the soil analysed before sowing to measure the amount of 'available' nitrogen in the soil.
2. Apply analysis-based recommended quantity of nitrogen as basal by using:
- **a.** organic manures;
- **b.** bio-fertilizers;
- **c.** nitrogenous fertilizers.

3. Top-dress soluble nitrogenous fertilizers such as urea.
4. For quick recovery, apply urea (2% w/v solution) as a foliar spray in the standing crop. Foliar sprays must be repeated every 10–15 days.

Further reading

Grundon, N.J. (1987) *Hungry Crops: A Guide to Nutrient Deficiencies in Field Crops*. Queensland Department of Primary Industries, Brisbane, Australia, pp. 161–183.

Rapper, C.C. Jr and Kramer, P.J. (1987) Stress physiology. In: Wilcox, J.R. (ed.) *Soybean: Improvement, Production and Uses*, 2nd edn. American Society of Agronomy, Madison, Wisconsin, pp. 590–641.

Sinclair, J.B. (1993) Soybeans. In: Bennett, W.F. (ed.) *Nutrient Deficiencies and Toxicities in Crop Plants*. APS Press, St Paul, Minnesota, pp. 99–103.

Sprague, H.B. (ed.) (1964) *Hunger Signs in Crops – A Symposium*, 3rd edn. David McKay, New York.

Plate 539. Purple pigmentation appearing on older trifoliate of a phosphorus-deficient plant. (Photo by Dr Manoj Kumar Sharma.)

SOYBEAN (*Glycine max* Linn.) PHOSPHORUS (P) DEFICIENCY

Plate 540. Deficient old leaves are small and dark green. (Photo by Dr Manoj Kumar Sharma.)

Plate 541. In continued deficiency, leaves become dull green. (Photo by Dr Manoj Kumar Sharma.)

Plate 542. Older leaflet displaying dark brown necrotic spots in interveinal areas. (Photo by Dr Prakash Kumar.)

Symptoms

1. Deficient plants appear unhealthy and stunted.
2. The stem becomes thin and spindly and has reduced branching.
3. Both flowering and maturity are delayed in phosphorus-deficient plants.
4. A smaller number of pods and fewer seeds are formed, both contributing to poor yields.
5. Leaves become smaller in size and appear dark green to bluish green in colour (Plate 540).
6. Phosphorus is fairly mobile in plants and in poor supply conditions it is easily translocated from older to younger tissues.
7. The deficiency symptoms first become evident on the lower leaves and then progress to the upper leaves.
8. Older leaves exhibit dark brown necrotic spots in interveinal tissues.
9. Purple pigmentation often develops on the leaflets of lower leaves, working up the plant to the upper leaves (Plate 539).

Developmental stages

Stage I: In the early stage of deficiency, the plant becomes stunted and its leaves are small and dark green (Plates 540 and 541).
Stage II: Purple pigmentation often develops on the lower leaves, beginning from the margins and rapidly proceeding inwards (Plate 539).
Stage III: If the deficiency is severe, the old leaves develop dark brown necrotic lesions in interveinal tissues (Plate 542).
Stage IV: In the later stage, leaves appear dark yellow with dark brown necrotic lesions in interveinal tissues; eventually the leaves turn dark brown and fall off.

Likely to occur in

1. Soils having low organic matter.
2. Alkaline and calcareous soils.
3. Soils exhausted by intensive cropping.
4. Acid soils and highly weathered soils.
5. Soils where topsoil has been removed by erosion.
6. Acid soils having pH below 6.0.
7. Alkaline soils having pH between 7.5 and 8.5.
8. Cool and wet conditions where uptake of phosphorus is reduced.

Integrated nutrient management

1. Get the soil analysed to measure the amount of 'available' phosphate in the soil well before sowing.
2. Apply analysis-based recommended quantity of phosphorus as basal by using:
- **a.** organic manures;
- **b.** phosphate-solubilizing microbial cultures;
- **c.** phosphatic fertilizers.

3. If standing crops show the deficiency, apply soluble phosphatic fertilizers such as ammonium phosphate with irrigation water.

Further reading

Grundon, N.J. (1987) *Hungry Crops: A Guide to Nutrient Deficiencies in Field Crops*. Queensland Department of Primary Industries, Brisbane, Australia.

Hambidge, G. (1941) *Hunger Signs in Crops*. American Society of Agronomy and National Fertilizer Council, Washington, DC.

Sinclair, J.B. (1993) Soybeans. In: Bennett, W.F. (ed.) *Nutrient Deficiencies and Toxicities in Crop Plants*. APS Press, St Paul, Minnesota, pp. 99–103.

Sprague, H.B. (ed.) (1964) *Hunger Signs in Crops – A Symposium*, 3rd edn. David McKay, New York.

Plate 543. Plant showing marginal yellow chlorosis on old leaves. (Photo by Dr Prakash Kumar.)

Plate 544. Early symptom appearing as pale yellow marginal chlorosis (Photo by Dr Manoj Kumar Sharma.)

Plate 545. Severely deficient leaves showing yellow chlorosis followed by necrosis. (Photo by Dr Manoj Kumar Sharma.)

Plate 546. Dark yellow chlorotic stripes along the margins advancing into interveinal tissues. (Photo by Dr Prakash Kumar.)

SOYBEAN (*Glycine max* Linn.) POTASSIUM (K) DEFICIENCY

Symptoms

1. Potassium deficiency results in poor plant growth and delayed maturity.
2. Deficient plants have poor branching with the development of fewer pods.
3. The number and size of seeds per pod are reduced, resulting in low production.
4. The potassium requirement of the soybean crop is high and it is required most during rapid vegetative growth.
5. Deficient plants have weak stems and are susceptible to diseases.
6. Potassium is highly mobile within plants and it is quickly translocated from older to younger tissues. As a consequence, the older leaves show deficiency symptoms first while the upper leaves remain dark green.
7. Symptoms develop as a pale yellow marginal chlorosis beginning at the tip and proceeding towards the base of old leaves (Plate 544).
8. In a continuous short supply condition, the pale yellow chlorosis becomes more pronounced along the margins and progresses interveinally towards the midrib (Plate 545).
9. As the deficiency becomes severe, the marginal chlorosis is followed by necrosis and both spread inwards. The necrotic tissues then drop away, which results in a ragged appearance of the leaves.

Developmental stages

Stage I: In mild deficiency, plants show stunted growth, thin stems and small green foliage.
Stage II: When deficiency becomes more severe, pale yellow marginal chlorosis develops on older leaves (Plate 544).
Stage III: Marginal chlorosis advances into interveinal areas (Plates 543, 545 and 546).
Stage IV: In acute deficiency, the yellow chlorosis turns into necrosis.

Likely to occur in

1. Soils formed from parent material low in potassium.
2. Light-textured soils where potassium has been leached by heavy rainfall or excessive irrigation.
3. Soils low in organic matter.
4. Soils with wide Na:K, Mg:K or Ca:K ratio.
5. Acid soils having pH below 6.0.

Integrated nutrient management

1. Analyse the soil before planting to measure the amount of plant-available potassium.
2. Problematic acid/alkaline soils should be reclaimed.
3. Add organic manures well before sowing.
4. Apply potassium chloride, potassium sulphate or potassium nitrate to the soil at or before planting as per soil testing recommendations.
5. In standing crops, apply soluble potassium salts with irrigation water.

Further reading

Grundon, N.J. (1987) *Hungry Crops: A Guide to Nutrient Deficiencies in Field Crops*. Queensland Department of Primary Industries, Brisbane, Australia, pp. 161–183.

Hambidge, G. (1941) *Hunger Signs in Crops*. American Society of Agronomy and National Fertilizer Council, Washington, DC.

Rapper, C.C. Jr and Kramer, P.J. (1987) Stress physiology. In: Wilcox, J.R. (ed.) *Soybean: Improvement, Production and Uses*, 2nd edn. American Society of Agronomy, Madison, Wisconsin, pp. 590–641.

Sinclair, J.B. (1993) Soybeans. In: Bennett, W.F. (ed.) *Nutrient Deficiencies and Toxicities in Crop Plants*. APS Press, St Paul, Minnesota, pp. 99–103.

Plate 547. Magnesium-deficient plant, light green in colour, with interveinal chlorosis on lower leaves. (Photo by Dr Manoj Kumar Sharma.)

Plate 548. Pale yellow interveinal chlorosis on old leaves. (Photo by Dr Manoj Kumar Sharma.)

Plate 549. Chlorotic leaf showing prominent green veins. (Photo by Dr Manoj Kumar Sharma.)

Plate 550. Yellow mottling in interveinal tissues under prolonged deficiency. (Photo by Dr Manoj Kumar Sharma.)

SOYBEAN (*Glycine max* Linn.) MAGNESIUM (Mg) DEFICIENCY

Symptoms

1. Magnesium is essential for chlorophyll synthesis and also plays an important role in nitrogen fixation.
2. Deficient plants become small and have a pale green appearance.
3. The stem becomes thin and spindly.
4. Deficient plants produce a lower number of pods containing fewer small seeds, resulting in poor grain yields.
5. Magnesium is mobile in plants, so under poor supply conditions it is rapidly translocated from older to younger leaves.
6. The visual deficiency symptoms appear first and become more severe on older leaves. If deficiency persists, symptoms spread rapidly to younger leaves.
7. Pale yellow interveinal mottling develops on older leaves. Leaves then turn deep yellow (Plate 550).
8. In severely deficient conditions, pale brown necrotic lesions develop in the interveinal tissues of old leaves.

Developmental stages

Stage I: In the early stage of deficiency, the plants are short and pale green in appearance.
Stage II: If the deficiency is prolonged, pale yellow interveinal chlorosis occurs on old leaves (Plates 547, 548 and 549).
Stage III: As the symptoms advance, pale yellow chlorosis turns into pale brown necrosis in interveinal regions.

Likely to occur in

1. Acid sandy soils that have been leached by heavy rainfall.
2. Peat and muck soils that are low in total magnesium.
3. Soils having higher quantity of ammonium, calcium or potassium.
4. Soils derived from parent material that is inherently low in magnesium.
5. Acid soils having pH below 6.5.
6. Alkaline soils having pH above 8.5.
7. Soils highly fertilized with potassium and manganese that hinder root uptake of magnesium.

Integrated nutrient management

1. Get the soil analysed before sowing to estimate the amount of 'available' magnesium in the soil.
2. Apply analysis-based recommended quantity of magnesium before sowing using soluble salts such as magnesium sulphate or magnesium chloride.
3. In deficient standing crops, apply soluble magnesium salts such as magnesium sulphate, chloride or nitrate with irrigation water.

Further reading

Grundon, N.J. (1987) *Hungry Crops: A Guide to Nutrient Deficiencies in Field Crops*. Queensland Department of Primary Industries, Brisbane, Australia, pp. 161–183.

Heenan, D.P. and Campbell, L.C. (1981) Influence of potassium and manganese on growth and uptake of magnesium by soybeans (*Glycine max* (L.) Merr. cv. Bragg). *Plant and Soil* 61, 447–456.

Rapper, C.C. Jr and Kramer, P.J. (1987) Stress physiology. In: Wilcox, J.R. (ed.) *Soybean: Improvement, Production and Uses*, 2nd edn. American Society of Agronomy, Madison, Wisconsin, pp. 590–641.

Sprague, H.B. (ed.) (1964) *Hunger Signs in Crops – A Symposium*, 3rd edn. David McKay, New York.

Plate 551. Sulphur-deficient plant showing yellow young leaves and green old leaves. (Photo by Dr Manoj Kumar Sharma.)

SOYBEAN (*Glycine max* Linn.)
SULPHUR (S) DEFICIENCY

Plate 552. Light green whole plant with pale yellow upper leaves. (Photo by Dr Prakash Kumar.)

Plate 553. Pale yellow chlorotic young leaves. (Photo by Dr Prakash Kumar.)

Plate 554. Severely deficient yellow chlorotic young leaf. (Photo by Dr Manoj Kumar Sharma.)

Symptoms

1. Plant height and number of branches are drastically reduced.
2. The stem becomes thin and elongated; stem elongation is a typical symptom of sulphur deficiency.
3. The appearance of small and yellowish green youngest leaves is the most striking symptom of sulphur deficiency.
4. Sulphur deficiency results in a decreased number of pods with poor filling and thereby causes a significant reduction in yields.
5. Sulphur is an immobile nutrient in plants and under restricted supply conditions it is not easily transferred from older to younger leaves. Therefore, youngest leaves show the deficiency symptoms first.
6. The whole plant appears light green while the youngest leaves become yellow.
7. The chlorosis develops evenly over the entire leaf, covering both the veins and the interveinal tissues uniformly (Plates 553 and 554).
8. In mature crops, the young leaves become pale yellow to yellow while the old leaves remain green.

Developmental stages

Stage I: In mild deficiency conditions, the whole plant becomes pale green although the old leaves appear darker (Plate 552).
Stage II: If the deficiency becomes severe, the younger leaves turn pale yellow to yellow while the old leaves remain green (Plate 551).

Likely to occur in

1. Soils having low organic matter.
2. Light-textured sandy soils that have been leached by heavy rainfall or excessive irrigation.
3. Soils exhausted by intensive cropping.
4. Soils derived from parent material that is inherently low in sulphur.
5. Acid soils having pH below 6.0.

Integrated nutrient management

1. Get the soil analysed before sowing to measure the amount of 'available' sulphate in the soil.
2. Apply analysis-based recommended quantity of sulphur by mixing into the surface soil, well before sowing, either:
- **a.** elemental sulphur; or
- **b.** gypsum.

3. In deficient standing crops apply ammonium sulphate, magnesium sulphate or potassium sulphate with irrigation water.

Further reading

Grundon, N.J. (1987) *Hungry Crops: A Guide to Nutrient Deficiencies in Field Crops*. Queensland Department of Primary Industries, Brisbane, Australia, pp. 161–183.

Sinclair, J.B. (1993) Soybeans. In: Bennett, W.F. (ed.) *Nutrient Deficiencies and Toxicities in Crop Plants*. APS Press, St Paul, Minnesota, pp. 99–103.

Sprague, H.B. (ed.) (1964) *Hunger Signs in Crops – A Symposium*, 3rd edn. David McKay, New York.

Tandon, H.L.S. (1991) *Sulphur Research and Agricultural Production in India*, 3rd edn. The Sulphur Institute, Washington, DC.

Plate 555. Interveinal chlorosis on young leaves while old leaves are dark green. (Photo by Dr Prakash Kumar and Dr Manoj Kumar Sharma.)

Plate 556. Pale yellow upper leaves with prominent green veins. (Photo by Dr Prakash Kumar and Dr Manoj Kumar Sharma.)

Plate 557. Papery white bleached leaves with green mid-vein only. (Photo by Dr Prakash Kumar and Dr Manoj Kumar Sharma.)

Plate 558. Entire leaflet yellow to white with almost disappeared veins. (Photo by Dr Prakash Kumar and Dr Manoj Kumar Sharma.)

SOYBEAN (*Glycine max* Linn.) IRON (Fe) DEFICIENCY

Symptoms

1. Deficient plants show poor growth and have thin and spindly stems.
2. Poor pod formation and poor pod filling result in significant yield reductions.
3. Iron is immobile within plants and under restricted supply conditions it is not easily mobilized from older to younger parts of the plant. Thus, the deficiency symptoms appear first and become more severe on younger leaves while older leaves remain green (Plate 555).
4. The symptoms begin as a pale yellow interveinal chlorosis with sharp green veins on younger leaves (Plate 556).
5. Under prolonged deficient conditions, the pale yellow chlorotic leaves turn to dark yellow and the veins also become faded.
6. Eventually, the entire leaf turns almost white (Plate 558) and brown necrotic lesions may develop near the margins.
7. The symptoms usually become more severe under low or high temperatures and under high intensity of sunlight.

Developmental stages

Stage I: In mild deficiency, the uppermost leaves develop pale green interveinal chlorosis with dark green veins. If the iron supply is restored the plant can recover its normal appearance.
Stage II: When deficiency persists, the pale green young leaves turn to pale yellow with prominent green veins (Plate 556).
Stage III: As the deficiency advances, the prominent green veins also fade and become light green (Plate 557).
Stage IV: In the later stage, pale brown necrotic lesions develop on the margins of young leaves.

Likely to occur in

1. Sandy soils having low total iron content.
2. Alkaline and calcareous soils where solubility of iron is very low.
3. Peat and muck soils where organic matter ties up iron and reduces its availability in soil solution.
4. Acid soils with excessively high levels of soluble zinc, manganese, copper or nickel that can hinder the uptake of iron despite high iron availability.
5. Alkaline soils having pH above 7.5.

Integrated nutrient management

1. Analyse the soil before sowing to measure the amount of 'available' iron in the soil.
2. Reclaim problematic alkaline soils.
3. Add organic manure well before sowing.
4. Apply a basal dose of soluble iron fertilizers such as $FeSO_4$ (commonly at 25 kg/ha) or Fe chelates (10 kg/ha). Use of organic chelates proves to be more promising as they maintain iron in soil solution.
5. In standing crops, apply $FeSO_4$ or Fe chelates (0.5% w/v solution) as a foliar spray. Foliar sprays must be repeated every 10–15 days.

Further reading

Grundon, N.J. (1987) *Hungry Crops: A Guide to Nutrient Deficiencies in Field Crops*. Queensland Department of Primary Industries, Brisbane, Australia, pp. 161–183.

Rapper, C.C. Jr and Kramer, P.J. (1987) Stress physiology. In: Wilcox, J.R. (ed.) *Soybean: Improvement, Production and Uses*, 2nd edn. American Society of Agronomy, Madison, Wisconsin, pp. 590–641.

Sinclair, J.B. (1993) Soybeans. In: Bennett, W.F. (ed.) *Nutrient Deficiencies and Toxicities in Crop Plants*. APS Press, St Paul, Minnesota, pp. 99–103.

Sprague, H.B. (ed.) (1964) *Hunger Signs in Crops – A Symposium*, 3rd edn. David McKay, New York.

Plate 559. Deficient plant showing pale green young leaves with dark green old leaves. (Photo by Dr Manoj Kumar Sharma.)

SOYBEAN (*Glycine max* Linn.) MANGANESE (Mn) DEFICIENCY

Plate 560. Pale green leaflets showing interveinal mottling. (Photo by Dr Prakash Kumar and Dr Manoj Kumar Sharma.)

Plate 561. Small leaflets with chequered interveinal mottling. (Photo by Dr Prakash Kumar and Dr Manoj Kumar Sharma.)

Plate 562. Mottled leaflet showing broad green veins and sub-veins. (Photo by Dr Prakash Kumar and Dr Manoj Kumar Sharma.)

Symptoms

1. Symptoms easily appear on the deficient crop because soybean is sensitive to manganese deficiency.
2. Deficient plants appear small and the stem becomes short and thin.
3. Manganese deficiency badly affects the branching and number of pods.
4. The pods fill with fewer seeds, resulting in poor grain yields.
5. Manganese is an immobile nutrient within plants and it is not readily moved from older to younger tissues under reduced supply conditions.
6. The deficiency symptoms become first evident on younger leaves while older leaves remain normally green.
7. Pale green to white interveinal mottling develops in younger leaves while veins remain prominently green (Plates 561 and 562).
8. In severe deficiency, brown necrotic spots develop in the interveinal areas.

Developmental stages

Stage I: In mild deficiencies, the leaves become light green with pale yellow interveinal chlorosis (Plates 559 and 560).
Stage II: In severely deficient conditions, brown necrotic lesions develop in interveinal tissues.
Stage III: If deficiency prolongs, the affected young leaves readily fall off.

Likely to occur in

1. Light-textured sandy soils that have been leached by heavy rainfall or excessive irrigation.
2. Calcareous and alkaline soils where solubility of manganese is very low.
3. Waterlogged peaty soils where organic matter ties up manganese and reduces its availability in solution.
4. Soils derived from parent material that is inherently low in manganese.
5. Acid soils having pH below 5.0.
6. Alkaline soils having pH above 7.5.
7. Soils containing higher concentration of iron.

Integrated nutrient management

1. Get the soil analysed before sowing to measure the amount of 'available' manganese in the soil.
2. Problematic alkaline soils should be reclaimed.
3. Add organic manures well before sowing.
4. Apply soluble salts of manganese such as manganese sulphate as basal.
5. In deficient standing crops, apply manganese sulphate (0.2 to 0.3% w/v solution) as a foliar spray. Repeated sprays may be required if symptoms reappear.

Further reading

Davis, V. and Fernandez, F. (2009) Yellow soybeans in high pH soils: signs of Mn deficiency. *The Bulletin* issue 15(2). University of Illinois Extension, Urbana, Illinois.

Grundon, N.J. (1987) *Hungry Crops: A Guide to Nutrient Deficiencies in Field Crops*. Queensland Department of Primary Industries, Brisbane, Australia, pp. 161–183.

Sinclair, J.B. (1993) Soybeans. In: Bennett, W.F. (ed.) *Nutrient Deficiencies and Toxicities in Crop Plants*. APS Press, St Paul, Minnesota, pp. 99–103.

Sprague, H.B. (ed.) (1964) *Hunger Signs in Crops – A Symposium*, 3rd edn. David McKay, New York.

Plate 563. Stunted plant showing dark yellow old leaves and pale green new leaves. (Photo by Dr Prakash Kumar.)

MUSTARD (*Brassica campestris* Linn.) NITROGEN (N) DEFICIENCY

Plate 564. Chlorotic bottom leaf turned yellow in persistent deficiency. (Photo by Dr Prakash Kumar.)

Symptoms

1. Nitrogen enhances yield by influencing the growth parameters.
2. Nitrogen deficiency results in retarded plant growth.
3. Deficient plants have short and thin stems.
4. Only a few branches are produced and a smaller number of pods are formed.
5. Since nitrogen is mobile within plants, it is rapidly transferred (as the deficiency occurs) from older to younger tissues.
6. The deficiency symptoms are first observed and become more severe on older leaves (Plate 566), then work up the plant to younger leaves.
7. The older leaves become yellow chlorotic because the plant is unable to form sufficient chlorophyll (Plates 564 and 565).
8. The yellow chlorotic leaves then turn brown and become necrotic.

Plate 565. Pale green leaf becoming yellow from the tip and proceeding to the base, along with reddening starting at the margins. (Photo by Dr Prakash Kumar.)

Developmental stages

Stage I: If deficiency is mild or early, symptoms expressed as pale green leaves are more pronounced in lower leaves.
Stage II: In prolonged deficiency, the older leaves turn evenly yellow and the upper leaves appear pale green (Plate 563).
Stage III: In severe deficiency, the old leaves turn dark yellow to brown and easily detach from the plant.

Likely to occur in

1. Soils having low organic matter.
2. Light-textured sandy soils that have been leached by heavy rainfall or excessive irrigation.
3. Soils exhausted by intensive cropping.
4. Waterlogged conditions.
5. Acid soils having pH below 6.0.
6. Alkaline soils having pH above 8.0.

Integrated nutrient management

1. Get the soil analysed before sowing to measure the amount of 'available' nitrogen in the soil.
2. Apply analysis-based recommended quantity of nitrogen as basal by using:
- **a.** organic manures;
- **b.** bio-fertilizers;
- **c.** nitrogenous fertilizers.

3. Top-dress soluble nitrogenous fertilizers such as urea in split doses.
4. For quick recovery, apply urea (2% w/v solution) as a foliar spray in the standing crop. Foliar sprays must be repeated every 10–15 days.

Plate 566. Severely nitrogen-deficient evenly dark yellow old leaf. (Photo by Dr Prakash Kumar.)

Further reading

Allen, E.J. and Morgan, D.J. (1972) A quantitative analysis of the effects of nitrogen on the growth, development and yield of oilseed rape. *Journal of Agricultural Science* 78, 315–324.

Bhatty, R.S. (1964) Influence of nitrogen fertilization on the yield, protein and oil content of two varieties of rape. *Canadian Journal of Plant Science* 44, 215–217.

Grant, C.A. and Bailey, L.D. (1993) Fertility management in canola production. *Canadian Journal of Plant Science* 73, 651–670.

Nuttall, W.F., Ukrainetz, H., Stewart, J.W.B. and Spurr, D.T. (1987) The effect of nitrogen, sulphur and boron on yield and quality of rapeseed (*Brassica napus* L. and *B. campestris* L.). *Canadian Journal of Soil Science* 67, 545–559.

Plate 567. Stunted plant, lower leaves and stem showing purple and reddish violet discoloration. (Photo by Dr Prakash Kumar.)

Plate 568. Purple discoloration on bottom leaves and progressing to upper leaves. (Photo by Dr Prakash Kumar.)

Plate 569. Purpling started at leaf tip and along the margins, then spreading to the entire lamina. (Photo by Dr Prakash Kumar.)

Plate 570. Deep purple discoloration occurring on the entire leaf. (Photo by Dr Prakash Kumar.)

MUSTARD (*Brassica campestris* Linn.) PHOSPHORUS (P) DEFICIENCY

Symptoms

1. Rapid uptake of phosphorus occurs in the early stages of plant growth.
2. In deficient conditions, plant growth is restricted. The stem becomes thin and short. Root development is adversely affected.
3. Plants appear stunted having few branches and small leaves.
4. When phosphorus supply is restricted to the plant, it is rapidly translocated from older to younger leaves (since phosphorus is mobile within the plant); therefore deficiency symptoms typically appear in older leaves first.
5. Old leaves become dark bluish green with purple or red tints (Plate 567).
6. Purple or red discoloration occurs on the tips and margins of old leaves and then advances into interveinal tissues (Plate 569).
7. The stem can also develop purple or red discoloration.

Developmental stages

Stage I: Plants show mild symptoms as poorer growth with short stem and restricted root development.
Stage II: In prolonged deficiency conditions, the older leaves show purple tints on the tips and margins and into interveinal tissues (Plates 568 and 569).
Stage III: In acutely deficient conditions, the entire leaf turns purple or red (Plate 570).

Likely to occur in

1. Soils having low organic matter.
2. Alkaline and calcareous soils.
3. Soils exhausted by intensive cropping.
4. Acid soils and highly weathered soils.
5. Soils where topsoil has been removed by erosion.
6. Acid soils having pH below 6.0.
7. Alkaline soils having pH between 7.5 and 8.5.

Integrated nutrient management

1. Get the soil analysed to measure the amount of 'available' phosphate in the soil well before sowing.
2. Apply analysis-based recommended quantity of phosphorus as basal by using:
- **a.** organic manures;
- **b.** phosphate-solubilizing microbial cultures;
- **c.** phosphatic fertilizers.

3. If standing crops show the deficiency, apply soluble phosphatic fertilizers such as ammonium phosphate with irrigation water.

Further reading

Bidwell, R.G.S. (1979) *Plant Physiology*, 2nd edn. Macmillan Publishing Co., Inc., New York.
Grant, C.A. and Bailey, L.D. (1993) Fertility management in canola production. *Canadian Journal of Plant Science* 73, 651–670.
Racz, G.J., Webber, M.D., Soper, R.J. and Hedlin, R.A. (1965) Phosphorus and nitrogen utilization by rape, flax, and wheat. *Agronomy Journal* 57, 335–337.
Ukrainetz, H., Soper, R.J. and Nyborg, M. (1975) Plant nutrient requirements of oilseed and pulse crops. In: Harapiak, J.T. (ed.) *Oilseed and Pulse Crops in Western Canada – A Symposium*. Western Cooperative Fertilizers Ltd, Calgary, Canada, pp. 314–325.

Plate 571. Severely deficient bottom leaf showing scorching on the leaf borders and yellow–brown interveinal chlorosis. (Photo by Dr Manoj Kumar Sharma.)

MUSTARD (*Brassica campestris* Linn.) POTASSIUM (K) DEFICIENCY

Plate 572. Beginning of yellowing along the margins. (Photo by Dr Prakash Kumar and Dr Manoj Kumar Sharma.)

Plate 573. Yellow–brown marginal chlorosis spreading inwards. (Photo by Dr Manoj Kumar Sharma.)

Plate 574. Prominent yellowish brown chlorosis developed along the leaf border and spreading into interveinal tissues. (Photo by Dr Manoj Kumar Sharma.)

Symptoms

1. The deficient crop appears as if wilted in dry and hot weather conditions.
2. The overall growth of deficient plants is reduced. Internodes become shorter. The younger leaves appear smaller.
3. Potassium is mobile within plants, therefore it is readily redistributed from older to younger tissues. The deficiency symptoms are observed first in older leaves.
4. Marginal scorching of older leaves is a typical symptom of potassium deficiency (Plate 571).
5. Symptoms develop as a yellow–brown marginal and interveinal chlorosis.
6. As the symptoms advance, necrosis starts from the leaf margins.
7. Eventually leaves become necrotic and die.

Developmental stages

Stage I: In mild deficiency, growth is poorer and plants show wilting in dry and hot season.
Stage II: If deficiency persists, yellow marginal chlorosis develops in older leaves (Plates 572 and 573).
Stage III: In prolonged deficient conditions, marginal chlorosis rapidly spreads inwards (Plate 574).
Stage IV: In very severe conditions, the leaf margins become scorched (Plate 571), then leaves may die and fall off.

Likely to occur in

1. Soils formed from parent material low in potassium.
2. Light-textured soils where potassium has been leached by heavy rainfall or excessive irrigation.
3. Soils low in organic matter.
4. Soils with wide Na:K, Mg:K or Ca:K ratio.
5. Acid soils having pH below 6.0.

Integrated nutrient management

1. Analyse the soil before planting to measure the amount of plant-available potassium.
2. Problematic acid/alkaline/saline soils should be reclaimed.
3. Add organic manures well before planting.
4. Apply potassium chloride, potassium sulphate or potassium nitrate to the soil at or before planting as per soil testing recommendations.
5. In standing crops, apply soluble potassium salts with irrigation water.

Further reading

Bailey, L.D. and Soper, R.J. (1985) Potassium nutrition of rape, flax, sunflower, and safflower. In: Munson, R.D. (ed.) *Potassium in Agriculture*. American Society of Agronomy, Madison, Wisconsin, pp. 765–798.

Grant, C.A. and Bailey, L.D. (1993) Fertility management in canola production. *Canadian Journal of Plant Science* 73, 651–670.

Perrenoud, S. (1990) *Potassium and Plant Health. IPI Research Topic No. 3*, 2nd edn. International Potash Institute, Horgen, Switzerland.

Pissarek, H.P. (1973) The development of potassium deficiency symptoms in summer rape. *Zeitschrift für Pflanzenernährung und Bodenkunde* 136, 1–19.

Plate 575. Chlorotic new leaves and normal green old leaves. (Photo by Dr Prakash Kumar.)

MUSTARD (*Brassica campestris* Linn.) SULPHUR (S) DEFICIENCY

Plate 576. Uniform yellowing occurring in the top leaves. (Photo by Dr Prakash Kumar.)

Plate 577. Yellowing developed on an entire lamina with marginal red or purple pigmentation. (Photo by Dr Prakash Kumar.)

Plate 578. Severely deficient, spoon-shaped, dark yellow young leaves having purple pigmentation on the margins. (Photo by Dr Prakash Kumar.)

Symptoms

1. Sulphur-deficient plants appear chlorotic as sulphur is involved in chlorophyll synthesis.
2. Plant growth is reduced. Leaves become small. The plant appears rigid and erect.
3. Flowering is delayed. The number and size of pods are decreased. Fewer seeds are formed per pod.
4. Sulphur is fairly immobile within plants, so in restricted supply conditions it is not easily mobilized from older to younger leaves. Thus, the deficiency symptoms tend to appear first on younger leaves (Plate 576).
5. Initially, the younger leaves become pale green to pale yellow.
6. The chlorosis also proceeds from upper to lower leaves, with more pronounced chlorosis on the younger leaves.
7. The chlorosis appears uniform on the entire leaf, including the veins (Plate 575).
8. In severe deficiency of sulphur, some varieties clearly show pink, red or purple pigmentation on the leaf edges which spreads into chlorotic areas (Plate 577).

Developmental stages

Stage I: If deficiency is mild, yellowing develops in the youngest leaves whereas the older leaves remain normally green (Plate 575).
Stage II: If deficiency persists, the youngest leaves turn more chlorotic and the symptoms advance to the lower leaves.
Stage III: Severely deficient younger leaves become spoon-shaped and turn yellow with red or purple pigmentation on the margins (Plate 578).
Stage IV: Later, leaves become dark yellow and develop red or purple pigmentation into chlorotic areas.

Likely to occur in

1. Soils having low organic matter.
2. Light-textured sandy soils that have been leached by heavy rainfall or excessive irrigation.
3. Soils exhausted by intensive cropping.
4. Soils derived from parent material that is inherently low in sulphur.
5. Acid soils having pH below 6.0.

Integrated nutrient management

1. Get the soil analysed before sowing to measure the amount of 'available' sulphate in the soil.
2. Apply analysis-based recommended quantity of sulphur by mixing into the surface soil, well before sowing, either:
  a. elemental sulphur; or
  b. gypsum.
3. In deficient standing crops apply ammonium sulphate, magnesium sulphate or potassium sulphate with irrigation water.

Further reading

Cram, W.J. (1990) Uptake and transport of sulfate. In: Rennenberg, H., Brunold C., De Kok L.J. and Stulen, I. (eds) *Sulfur Nutrition and Sulfur Assimilation in Higher Plants: Fundamental, Environmental and Agricultural Aspects*. SPB Academic Publishing, The Hague, the Netherlands, pp. 3–11.

Grant, C.A. and Bailey, L.D. (1993) Fertility management in canola production. *Canadian Journal of Plant Science* 73, 651–670.

Schnug, E. and Haneklaus, S. (1994) Sulphur deficiency in *Brassica napus*. *Landbauforschung Völkenrode/Sonderheft* 44, 1–31.

Stuiver, C.E.E., De Kok, L.J. and Westermann, S. (1997) Sulfur deficiency in *Brassica oleracea* L.: development, biochemical characterization, and sulfur/nitrogen interactions. *Russian Journal of Plant Physiology* 44, 505–513.

Plate 579. Iron deficient mustard plant showing interveinal chlorosis on younger leaves. (Photo by Dr Prakash Kumar and Dr Manoj Kumar Sharma.)

MUSTARD (*Brassica campestris* Linn.) IRON (Fe) DEFICIENCY

Plate 580. Interveinal chlorosis on younger leaves. (Photo by Dr Prakash Kumar and Dr Manoj Kumar Sharma.)

Plate 581. Bleached leaf lamina. (Photo by Dr Prakash Kumar and Dr Manoj Kumar Sharma.)

Plate 582. Necrotic interveinal lesions. (Photo by Dr Prakash Kumar and Dr Manoj Kumar Sharma.)

Symptoms

1. Iron deficiency reduces the aconitase activity in mustard leaves.
2. Interveinal chlorosis of young leaves is a typical symptom of iron deficiency. Retarded plant growth is a more sensitive indicator of iron deficiency in the mustard crop.
3. Soil microbial activity is essential to supply required iron from soil to the crop.
4. Leaves become small. The number and size of pods are reduced and pods also appear chlorotic.
5. Iron is considered an immobile nutrient within plants, therefore deficiency symptoms are first observed and become more severe on younger leaves.
6. The younger leaves become chlorotic. The chlorosis extends the full length of the leaf, starting from the basal part of the leaf and proceeding towards the apical part.
7. In severe deficiency, the leaves become almost white or bleached. The newly emerging leaves can also be completely bleached.

Developmental stages

Stage I: In mild deficiencies, pale yellow interveinal chlorosis develops with prominent dark green veins (Plates 579 and 580).
Stage II: If deficiency persists, pale yellow young leaves turn to white–yellow with faded veins.
Stage III: In acutely deficient conditions, the entire leaf blace becomes bleached or papery white (Plate 581).
Stage IV: In the later stage, the papery white leaves become necrotic (Plate 582).

Likely to occur in

1. Sandy soils having low total iron content.
2. Alkaline and calcareous soils where solubility of iron is very low.
3. Peat and muck soils where organic matter ties up iron and reduces its availability in soil solution.
4. Acid soils with excessively high levels of soluble zinc, manganese, copper or nickel that can hinder the uptake of iron despite high iron availability.
5. Alkaline soils having pH above 7.5.

Integrated nutrient management

1. Analyse the soil before sowing to measure the amount of 'available' iron in the soil.
2. Reclaim problematic alkaline soils.
3. Add organic manure well before sowing.
4. Apply a basal dose of soluble iron fertilizers such as $FeSO_4$ (commonly at 25 kg/ha) or Fe chelates (10 kg/ha). Use of organic chelates proves to be more promising as they maintain iron in soil solution.
5. In standing crops, apply $FeSO_4$ or Fe chelates (0.5% w/v solution) as a foliar spray. Foliar sprays must be repeated every 10–15 days.

Further reading

Bacon, J.S.D., Dekock, P.C. and Palmer, M.J. (1961) Aconitase levels in the leaves of iron-deficient mustard plants (*Sinapis alba*). *Biochemistry Journal* 80, 64–70.

Holmes, M.R.J. (1980) *Nutrition of the Oil Seed Rape Crop*. Applied Science Publishers, London.

Miller, G.W., Pushnik, J.C. and Welkie, G.W. (1984) Iron chlorosis, a world wide problem, the relation of chlorophyll biosynthesis to iron. *Journal of Plant Nutrition* 7, 1–22.

Rroco, E., Kosegarten, H., Harizaj, F., Imani, J. and Mengel, K. (2003) The importance of soil microbial activity for the supply of iron to sorghum and rape. *European Journal of Agronomy* 19, 487–493.

PART IV

Nutrient Deficiencies in Cash Crops

Plate 583. Yellowing of lower leaves. (Photo by Dr Prakash Kumar.)

COTTON (*Gossypium hirsutum* Linn.) NITROGEN (N) DEFICIENCY

Plate 584. Entire plant appearing light green and stunted. (Photo by Dr Prakash Kumar and Dr Manoj Kumar Sharma.)

Plate 585. Field view of a nitrogen-deficient cotton crop. (Photo by Dr Prakash Kumar.)

Plate 586. Entire leaf uniformly pale yellow. (Photo by Dr Prakash Kumar and Dr Manoj Kumar Sharma.)

Symptoms

1. Deficient plants appear stunted with short, thin stems.
2. A bright red pigmentation often develops on the lower parts of the stem.
3. The entire plant exhibits a light green appearance (Plate 584).
4. Younger leaves become smaller in size and the number of branches is reduced.
5. Nitrogen is a mobile nutrient in plants and it is rapidly moved from older to younger parts of the plants when its supply is reduced.
6. The deficiency symptoms appear primarily on older leaves and become more severe with time (Plates 583 and 585).
7. Initially, old leaves become pale green, then yellow and finally develop brown necrosis, usually in interveinal regions.
8. Eventually, the affected leaves die and fall off early.

Developmental stages

Stage I: In mild deficiencies, the entire plant appears uniformly light green in colour (Plate 584).
Stage II: If the deficiency becomes severe, the whole leaf becomes uniformly pale yellow (Plate 586).
Stage III: As the symptoms advance, the entire leaf becomes pale brown and dies.

Likely to occur in

1. Soils having low organic matter.
2. Light-textured sandy soils that have been leached by heavy rainfall or excessive irrigation.
3. Soils exhausted by intensive cropping.
4. Waterlogged conditions.
5. Acid soils having pH below 6.0.
6. Alkaline soils having pH above 8.0.

Integrated nutrient management

1. Get the soil analysed before sowing to measure the amount of 'available' nitrogen in the soil.
2. Apply analysis-based recommended quantity of nitrogen as basal by using:
  a. organic manures;
  b. bio-fertilizers;
  c. nitrogenous fertilizers.
3. Top-dress soluble nitrogenous fertilizers such as urea in split doses.
4. For quick recovery, apply urea (3% w/v solution) as a foliar spray in the standing crop. Foliar sprays must be repeated every 10–15 days.
5. Incorporate legume crops in rotation.

Further reading

Cassman, K.G. (1993) Cotton. In: Bennett, W.F. (ed.) *Nutrient Deficiencies and Toxicities in Crop Plants*. APS Press, St Paul, Minnesota, pp. 111–119.

Donald, L. (1964) Nutrient deficiencies in cotton. In: Sprague, H.B. (ed.) *Hunger Signs in Crops – A Symposium*, 3rd edn. David McKay, New York, pp. 59–98.

Grundon, N.J. (1987) *Hungry Crops: A Guide to Nutrient Deficiencies in Field Crops*. Queensland Department of Primary Industries, Brisbane, Australia, pp. 215–239.

Tucker, T.C. (1984) Diagnosis of nitrogen deficiency in plants. In: Hauck, R.D. (ed.) *Nitrogen in Crop Production*. American Society of Agronomy, Madison, Wisconsin, pp. 249–262.

Plate 587. Phosphorus-deficient cotton plant showing stunted growth and purple pigmentation on old leaves. (Photo by Dr Prakash Kumar and Dr Manoj Kumar Sharma.)

Plate 588. Plant showing small and dark green leaves with purple discoloration on lower leaves. (Photo by Dr Prakash Kumar and Dr Manoj Kumar Sharma.)

Plate 589. Purpling of leaf along the margins and into interveinal areas. (Photo by Dr Prakash Kumar and Dr Manoj Kumar Sharma.)

Plate 590. Acutely deficient leaf. (Photo by Dr Prakash Kumar and Dr Manoj Kumar Sharma.)

COTTON (*Gossypium hirsutum* Linn.) PHOSPHORUS (P) DEFICIENCY

Symptoms

1. The distinct visible symptoms of phosphorus do not appear easily. However, the deficient plants become stunted and have short, thin stems.
2. Leaves appear small in size with dark green colour.
3. The number of branches is reduced severely.
4. The flowering gets delayed and fewer flowers are produced.
5. Boll formation is retarded, which results in poor yields.
6. Phosphorus is mobile within plants and it is readily mobilized from older to younger leaves under short supply conditions.
7. The deficiency symptoms become evident first and more severe on older leaves.
8. A purple pigmentation often develops on the lower leaves, working up the plant to the upper leaves (Plates 587 and 588).
9. The phosphorus-deficient crop gets delayed in maturity.

Developmental stages

Stage I: In the early stage of deficiency, the plant becomes stunted having small and dark green leaves (Plate 588).
Stage II: Purple pigmentation often develops on the leaf margins and rapidly proceeds into interveinal tissues (Plate 589).
Stage III: If the deficiency becomes severe, brown necrosis develops on the leaf margins and rapidly spreads into interveinal tissues.
Stage IV: Eventually, leaves die and fall off prematurely (Plate 590).

Likely to occur in

1. Soils having low organic matter.
2. Alkaline and calcareous soils.
3. Soils exhausted by intensive cropping.
4. Acid soils and highly weathered soils.
5. Soils where topsoil has been removed by erosion.
6. Acid soils having pH below 6.0.
7. Alkaline soils having pH between 7.5 and 8.5.

Integrated nutrient management

1. Get the soil analysed to measure the amount of 'available' phosphate in the soil well before sowing.
2. Apply analysis-based recommended quantity of phosphorus as basal by using:
  a. organic manures;
  b. phosphate-solubilizing microbial cultures;
  c. phosphatic fertilizers.
3. If standing crops show the deficiency, apply soluble phosphatic fertilizers such as ammonium phosphate with irrigation water, or a foliar application of diammonium phosphate at 12 kg/ha per spray or granular ammonium polyphosphate at 9 kg/ha per spray.

Further reading

Donald, L. (1964) Nutrient deficiencies in cotton. In: Sprague, H.B. (ed.) *Hunger Signs in Crops – A Symposium*, 3rd edn. David McKay, New York, pp. 59–98.
Grundon, N.J. (1987) *Hungry Crops: A Guide to Nutrient Deficiencies in Field Crops*. Queensland Department of Primary Industries, Brisbane, Australia, pp. 215–239.
Hambidge, G. (1941) *Hunger Signs in Crops*. American Society of Agronomy and National Fertilizer Council, Washington, DC.
Jones, U.S. and Bardsley, C.E. (1968) Phosphorus nutrition. In: Elliot, F.C., Hoower, M. and Porter, W.K. Jr (eds) *Advances in Production Utilization of Quality Cotton: Principles and Practices*. Iowa State University Press, Ames, Iowa, pp. 212–253.
Venugopalan, M.V., Tarhalkar, P.P. and Ragvir Singh (1995) Efficacy of phosphate carriers as foliar fertilizer on rainfed upland cotton (*Gossypium hirsutum*). *Indian Journal of Agricultural Science* 65, pp. 328–333.

Plate 591. Potassium deficient cotton plant. (Photo by Dr Prakash Kumar and Dr Manoj Kumar Sharma.)

Plate 592. Beginning of yellowing along the margins. (Photo by Dr Prakash Kumar.)

Plate 593. Marginal chlorosis spreading inwards. (Photo by Dr Prakash Kumar.)

Plate 594. Advanced stage of marginal chlorosis. (Photo by Dr Prakash Kumar and Dr Manoj Kumar Sharma.)

COTTON (*Gossypium hirsutum* Linn.) POTASSIUM (K) DEFICIENCY

Symptoms

1. Potassium deficiency delays crop maturity.
2. Plant growth is retarded and the number of flowers is reduced remarkably.
3. The number and size of bolls are decreased severely, which results in low yields and poor quality produce.
4. Low potassium content in the plants causes premature crop senescence.
5. Under deficient conditions, potassium is rapidly translocated from older to younger leaves. Therefore, the deficiency symptoms appear first on lower leaves.
6. Symptoms start as a pale yellow interveinal chlorosis on older leaves (Plate 591).
7. In continuous deficiency, the pale yellow chlorosis turns into a pale brown necrosis that spreads from the leaf edges (Plate 592).
8. If deficiency becomes severe, the necrosis advances rapidly into interveinal areas.
9. If the crop suffers from potassium deficiency in the maturing stage, the deficiency symptoms first become evident on younger mature leaves as interveinal chlorosis, rapidly turning bronze and necrotic, spreading from the margins.

Developmental stages

Stage I: Mild deficiency causes stunted growth, thin stems and small green foliage.
Stage II: When deficiency becomes more severe, marginal chlorosis develops on older leaves (Plates 593 and 594).
Stage III: Marginal chlorosis is followed by pale brown necrosis and both advance into interveinal areas (Plate 592).
Stage IV: In the later stage, affected leaves become bronze and drop off prematurely.

Likely to occur in

1. Soils formed from parent material low in potassium.
2. Light-textured soils where potassium has been leached by heavy rainfall or excessive irrigation.
3. Soils low in organic matter.
4. Soils with wide Na:K, Mg:K or Ca:K ratio.
5. Acid soils having pH below 6.0.

Integrated nutrient management

1. Analyse the soil before planting to measure the amount of plant-available potassium.
2. Problematic acid/alkaline soils should be reclaimed.
3. Add organic manures well before planting.
4. Apply potassium chloride, potassium sulphate or potassium nitrate to the soil at or before planting as per soil testing recommendations.
5. In standing crops, apply soluble potassium salts with irrigation water.

Further reading

Grundon, N.J. (1987) *Hungry Crops: A Guide to Nutrient Deficiencies in Field Crops*. Queensland Department of Primary Industries, Brisbane, Australia, pp. 215–239.

Kerby, T.A. and Adams, F. (1985) Potassium nutrition of cotton. In: Munson, R.D. (ed.) *Potassium in Agriculture*. American Society of Agronomy, Madison, Wisconsin, pp. 843–860.

Maples, R.L., Thompson, W.R. and Varvil, J. (1988) Potassium deficiency in cotton takes on a new look. *Better Crops with Plant Food* 73(1), 6–9. Potash & Phosphate Institute, Atlanta, Georgia.

Weir, B.L., Kerby, T.A., Roberts, B.A., Mikkelsen, D.S. and Garber, R.H. (1986) Potassium deficiency syndrome of cotton. *California Agriculture* 40(5–6), 13–14.

Wright, P.R. (1999) Premature senescence of cotton (*Gossypium hirsutum* L.) – predominantly a potassium disorder caused by an imbalance of source and sink. *Plant and Soil* 211, 231–239.

Plate 595. Magnesium-deficient plant showing reddening of lower leaves. (Photo by Dr Prakash Kumar.)

COTTON (*Gossypium hirsutum* Linn.) MAGNESIUM (Mg) DEFICIENCY

Plate 596. Reddening of leaf margins advancing into interveinal areas while veins are green. (Photo by Dr Prakash Kumar.)

Plate 597. Reddening intensified into interveinal tissues while veins look prominent and green. (Photo by Dr Prakash Kumar.)

Plate 598. Reddening turns into brown necrosis in interveinal areas. (Photo by Dr Prakash Kumar.)

Symptoms

1. Deficient plants appear stunted, showing thin and short stems.
2. The yields of seeds and fibres are reduced severely.
3. Magnesium is fairly mobile within plants; under reduced supply conditions it is easily mobilized from older to younger parts of the plant.
4. The deficiency symptoms are clearly visible first on older leaves.
5. Old leaves become purple to red in colour (reddish purple discoloration of interveinal tissues), leaving veins green and prominent (Plates 595 and 596).
6. In some varieties, pale yellow interveinal chlorosis can deve op on older leaves.
7. The youngest leaves usually remain unaffected and look apparently healthy.
8. Eventually, affected leaves die and can drop prematurely.

Developmental stages

Stage I: In early stage of deficiency, a reddening develops on old leaves leaving veins distinct.
Stage II: If the deficiency becomes severe, the reddening intensifies into interveinal tissues while veins remain green and prominent (Plate 597).
Stage III: As the symptoms advance, reddening turns into brown necrosis in interveinal regions (Plate 598).
Stage IV: In acute conditions, the affected old leaves die and fall off prematurely.

Likely to occur in

1. Acid sandy soils that have been leached by heavy rainfall.
2. Peat and muck soils that are low in total magnesium.
3. Soils having higher quantity of calcium or potassium.
4. Soils derived from parent material that is inherently low in magnesium.
5. Acid soils having pH below 6.5.
6. Alkaline soils having pH above 8.5.
7. Soils fertilized with a high quantity of potassic fertilizers.

Integrated nutrient management

1. Get the soil analysed before sowing to estimate the amount of 'available' magnesium in the soil.
2. Apply analysis-based recommended quantity of magnesium before sowing using soluble salts such as magnesium sulphate or magnesium chloride.
3. In deficient standing crops, apply soluble magnesium salts such as magnesium sulphate, chloride or nitrate with irrigation water. Foliar spray of 3% w/v $MgSO_4$ solution corrects the deficiency of magnesium.

Further reading

Cassman, K.G. (1993) Cotton. In: Bennett, W.F. (ed.) *Nutrient Deficiencies and Toxicities in Crop Plants*. APS Press, St Paul, Minnesota, pp. 111–119.
Grundon, N.J. (1987) *Hungry Crops: A Guide to Nutrient Deficiencies in Field Crops*. Queensland Department of Primary Industries, Brisbane, Australia, pp. 215–239.
Page, A.L. and Bingham, F.T. (1965) Potassium–magnesium interrelationship in cotton. *California Agriculture* 19(11), 6–7.
Ramesh, K., Thirumurugan, V. and Kumar, P.S.S. (1999) Magnesium nutrition of cotton. *Micronutrient News* 13(10), 2–3.

Plate 599. Pale yellow young leaves and darker old leaves. (Photo by Dr Prakash Kumar.)

Plate 600. Whole plant light green with pale yellow upper leaves. (Photo by Dr Prakash Kumar and Dr Manoj Kumar Sharma.)

Plate 601. Pale yellow chlorotic young leaf. (Photo by Dr Prakash Kumar.)

Plate 602. Severely deficient yellowish young leaf. (Photo by Dr Prakash Kumar and Dr Manoj Kumar Sharma.)

COTTON (*Gossypium hirsutum* Linn.) SULPHUR (S) DEFICIENCY

Symptoms

1. Deficient plants appear stunted and the stems become short and thin.
2. The number of branches is reduced and fewer flowers are produced.
3. Fewer bolls are formed and consequently lint and seed yields are decreased.
4. Sulphur is an immobile nutrient within plants; under low availability or poor supply conditions it is not rapidly mobilized from older to younger leaves. Thus, the deficiency symptoms are first evident and more pronounced on younger leaves while older leaves remain green.
5. Initially, the entire plant appears pale green.
6. As the symptoms advance, the pale green youngest leaves then turn pale yellow.
7. Although the whole plant appears light green to pale yellow, the youngest leaves are the palest (Plate 602).
8. The uniform paleness appears on the entire leaf, affecting both the veins and the interveinal tissues uniformly (Plate 601).

Developmental stages

Stage I: In mild deficiencies, all leaves on the plant become pale green although the old leaves remain darker (Plate 600).
Stage II: If the deficiency becomes severe, the younger leaves turn pale yellow while the old leaves remain pale green (Plate 599).
Stage III: In advanced stage of deficiency, the young leaves may develop pale brown necrosis around the margins.

Likely to occur in

1. Soils having low organic matter.
2. Light-textured sandy soils that have been leached by heavy rainfall or excessive irrigation.
3. Soils exhausted by intensive cropping.
4. Soils derived from parent material that is inherently low in sulphur.
5. Acid soils having pH below 6.0.

Integrated nutrient management

1. Get the soil analysed before sowing to measure the amount of 'available' sulphate in the soil.
2. Apply analysis-based recommended quantity of sulphur by mixing into the surface soil, well before sowing, either:
  a. elemental sulphur; or
  b. gypsum.
3. In deficient standing crops apply ammonium sulphate, magnesium sulphate or potassium sulphate with irrigation water.

Further reading

Cassman, K.G. (1993) Cotton. In: Bennett, W.F. (ed.) *Nutrient Deficiencies and Toxicities in Crop Plants*. APS Press, St Paul, Minnesota, pp. 111–119.

Cram, W.J. (1990) Uptake and transport of sulfate. In: Rennenberg, H., Brunold C., De Kok L.J. and Stulen, I. (eds) *Sulfur Nutrition and Sulfur Assimilation in Higher Plants: Fundamental, Environmental and Agricultural Aspects*. SPB Academic Publishing, The Hague, the Netherlands, pp. 3–11.

Ergle, D.R. and Eaton, F.M. (1951) Sulphur nutrition of cotton. *Plant Physiology* 26, 639–654.

Grundon, N.J. (1987) *Hungry Crops: A Guide to Nutrient Deficiencies in Field Crops*. Queensland Department of Primary Industries, Brisbane, Australia, pp. 215–239.

Hearn, A.B. (1981) Cotton nutrition. *Field Crop Abstracts* 34, 13–34.

Plate 603. Interveinal chlorosis on top leaves while older leaves are dark green. (Photo by Dr Prakash Kumar.)

Plate 604. Pale yellow upper leaves and prominent green veins. (Photo by Dr Prakash Kumar.)

Plate 605. Pale yellow leaf showing fading of veins. (Photo by Dr Prakash Kumar.)

Plate 606. Dark yellow top leaf with faded veins. (Photo by Dr Prakash Kumar.)

COTTON (*Gossypium hirsutum* Linn.) IRON (Fe) DEFICIENCY

Symptoms

1. Deficient plants show poor growth and yields are reduced.
2. Iron is an immobile nutrient in plants and it is not rapidly moved from older to younger parts of the plant under reduced supply conditions. Therefore, the visible deficiency symptoms appear primarily on the younger leaves.
3. Initially, the younger leaves develop pale green to pale yellow interveinal chlorosis while the older leaves remain green and normal (Plate 603).
4. A pale yellow chlorosis begins in interveinal tissues, leaving the veins distinctly dark green; while in prolonged short supply conditions, the pale yellow leaves turn dark yellow with visibly faded veins (Plate 606).
5. The chlorosis extends the full length of the leaves.
6. Severely deficient leaves then turn almost white and the veins also disappear.

Developmental stages

Stage I: In mild deficiencies, the uppermost leaves develop pale green interveinal chlorosis with dark green veins (Plate 603). If the iron supply is restored the plant can recover its normal appearance.
Stage II: When deficiency persists, pale green upper leaves turn pale yellow with prominent green veins (Plate 604).
Stage III: As the deficiency advances, the prominent green veins also fade and become light green (Plate 605).
Stage IV: In the later stage, leaves turn dark yellow with faded veins (Plate 606).

Likely to occur in

1. Sandy soils having low total iron content.
2. Alkaline and calcareous soils where solubility of iron is very low.
3. Peat and muck soils where organic matter ties up iron and reduces its availability in soil solution.
4. Acid soils with excessively high levels of soluble zinc, manganese, copper or nickel that can hinder the uptake of iron despite high iron availability.
5. Alkaline soils having pH above 7.5.

Integrated nutrient management

1. Analyse the soil before sowing to measure the amount of 'available' iron in the soil.
2. Reclaim problematic alkaline soils.
3. Add organic manure well before sowing.
4. Apply a basal dose of soluble iron fertilizers such as $FeSO_4$ (commonly at 25 kg/ha) or Fe chelates (10 kg/ha). Use of organic chelates proves to be more promising as they maintain iron in soil solution.
5. In standing crops, apply $FeSO_4$ or Fe chelates (0.5% w/v solution) as a foliar spray. Foliar sprays must be repeated every 10–15 days.

Further reading

Donald, L. (1964) Nutrient deficiencies in cotton. In: Sprague, H.B. (ed.) *Hunger Signs in Crops – A Symposium*, 3rd edn. David McKay, New York, pp. 59–98.

Grundon, N.J. (1987) *Hungry Crops: A Guide to Nutrient Deficiencies in Field Crops*. Queensland Department of Primary Industries, Brisbane, Australia, pp. 215–239.

Hinkle, D.A. and Brown, A.L. (1968) Secondary and micronutrients. In: Elliot, F.C., Hoower, M. and Porter, W.K. Jr (eds) *Advances in Production and Utilization of Quality Cotton: Principles and Practices*. Iowa State University Press, Ames, Iowa, pp. 281–320.

Vretta-Kouskoleka, H. and Kallinis, T.L. (1968) Iron deficiency in cotton in relation to growth and nutrient balance. *Soil Science Society of America Proceedings* 32, 253–257.

Plate 607. Zinc deficient cotton plant. (Photo by Dr Prakash Kumar.)

COTTON (*Gossypium hirsutum* Linn.) ZINC (Zn) DEFICIENCY

Plate 608. Chlorosis occurring between secondary veins. (Photo by Dr Prakash Kumar.)

Plate 609. Interveinal chlorosis becomes more pronounced. (Photo by Dr Prakash Kumar.)

Plate 610. Entire leaf covered with interveinal yellowing. (Photo by Dr Prakash Kumar.)

Symptoms

1. The growth of zinc-deficient plants is adversely affected and plants appear stunted.
2. Zinc deficiency reduces photosynthesis, respiration, carbonic anhydrase activity and chlorophyll concentrations in plants.
3. The deficient plants have short and thin stems and small leaves.
4. Lower numbers of flowers and bolls are produced, leading to poorer yields.
5. Zinc is only partly mobile within plants. Under poor supply conditions, the visible symptoms become evident first on the middle leaves.
6. Middle leaves develop pronounced interveinal chlorosis or mottling between the secondary veins (Plate 609).
7. Zinc deficiency results in reduced size of younger leaves, which often become malformed and produce a rosette appearance at the top of the shoot.

Developmental stages

Stage I: Affected leaves develop interveinal chlorosis (Plate 607).
Stage II: Interveinal chlorosis then spreads into the areas of secondary veins (Plate 608).
Stage III: As the symptoms advance, the interveinal chlorosis turns into yellowing (Plate 610).

Likely to occur in

1. Leached, light sandy soils where zinc content is low.
2. Alkaline and calcareous soils, where zinc availability is depressed.
3. Recently levelled soils where subsoil is exposed to cultivation. Plant-available zinc in surface soil is often double that of the subsoil.
4. Soil in which high rates of phosphatic fertilizers are applied, which can hamper zinc uptake by crops.
5. Acid soils having pH below 5.0.
6. Alkaline soils having pH above 7.5.

Integrated nutrient management

1. Analyse the soil before sowing to estimate the amount of plant-available zinc in the soil.
2. Reclaim problematic alkaline soils.
3. Incorporate any organic manure well before sowing.
4. Apply $ZnSO_4$ (commonly at 25–30 kg/ha) or Zn chelates (10 kg/ha) once every 2 years in zinc-deficient soils. Do not mix zinc fertilizers with phosphate fertilizers.
5. If deficiency symptoms appear in standing crops, spray 0.5% w/v $ZnSO_4$ to correct the deficiency.

Further reading

Brown, J.C. and Jones, W.E. (1977) Fitting plants nutritionally to soils. II. Cotton. *Agronomy Journal* 69, 405–409.
Cassman, K.G. (1993) Cotton. In: Bennett, W.F. (ed.) *Nutrient Deficiencies and Toxicities in Crop Plants*. APS Press, St Paul, Minnesota, pp. 111–119.
Grundon, N.J. (1987) *Hungry Crops: A Guide to Nutrient Deficiencies in Field Crops*. Queensland Department of Primary Industries, Brisbane, Australia, pp. 215–239.
Ohki, K. (1976) Effect of zinc nutrition on photosynthesis and carbonic anhydrase activity in cotton. *Physiologia Plantarum* 36, 300–304.

Plate 611. Nitrogen-deficient plant showing dark yellow old leaves and light green younger leaves. (Photo by Dr Manoj Kumar Sharma.)

SUGARCANE (*Saccharum officinarum* Linn.) NITROGEN (N) DEFICIENCY

Plate 612. Die-back of an old leaf. (Photo by Dr Manoj Kumar Sharma.)

Plate 613. Necrosis occurring on leaf tip and proceeding towards the base. (Photo by Dr Prakash Kumar.)

Plate 614. Nitrogen-deficient plants with short and slender stalks. (Photo by Dr Prakash Kumar.)

Symptoms

1. In the nitrogen-deficient crop, stalks become short and slender (Plate 614).
2. Fewer tillers are produced and overall growth of the plant is reduced.
3. The entire plant may become light green in appearance.
4. Nitrogen is a mobile nutrient in plants and in poor supply conditions it is easily mobilized from older to younger leaves.
5. The deficiency symptoms are primarily observed on older leaves (Plate 611).
6. Later, the entire plant turns chlorotic.
7. Older leaves may become uniformly pale green to yellow (Plate 611).
8. Nitrogen deficiency causes die-back of older leaves (Plates 612 and 613).
9. Necrosis occurs on the tips and margins of recently matured old leaves.

Developmental stages

Stage I: In the early plant stage or mild deficiency, the entire plant appears uniformly light green.
Stage II: In prolonged deficiency, the older leaves turn uniformly yellow to dark yellow (Plate 611).
Stage III: In acute deficiency conditions, the older leaves turn brown and die.

Likely to occur in

1. Soils having low organic matter.
2. Light-textured sandy soils that have been leached by heavy rainfall or excessive irrigation.
3. Soils exhausted by intensive cropping.
4. Waterlogged conditions.
5. Acid soils having pH below 6.0.
6. Alkaline soils having pH above 8.0.

Integrated nutrient management

1. Get the soil analysed before sowing to measure the amount of 'available' nitrogen in the soil.
2. Apply analysis-based recommended quantity of nitrogen as basal by using:
- **a.** organic manures;
- **b.** bio-fertilizers;
- **c.** nitrogenous fertilizers.

3. Top-dress soluble nitrogenous fertilizers such as urea in split doses.
4. For quick recovery, apply urea (2% w/v solution) as a foliar spray in the standing crop. Foliar sprays must be repeated every 10–15 days.
5. Incorporate legume crops in rotation.

Further reading

Anderson, D.L. and Bowen, J.E. (1990) Identification key to the malnutrition of sugarcane. In: *Sugarcane Nutrition*. Foundation for Agronomic Research and Potash and Phosphate Institute, Atlanta, Georgia, pp. 24–25.

Gascho, G.J. and Taha, F.A. (1972) *Nutrient Deficiency Symptoms of Sugarcane. Circular S-221*. University of Florida Agricultural Experiment Station, Gainesville, Florida.

Gascho, G.J., Anderson, D.L. and Bowen, J.E. (1993) Sugarcane. In: Bennett, W.F. (ed.) *Nutrient Deficiencies and Toxicities in Crop Plants*. APS Press, St Paul, Minnesota, pp. 37–42.

Plate 615. Phosphorus-deficient sugarcane plant. (Photo by Dr Prakash Kumar.)

Plate 616. Purpling on old leaf of a phosphorus-deficient sugarcane plant. (Photo by Dr Prakash Kumar.)

Plate 617. Purpling begins on the leaf tip and along the margins. (Photo by Dr Prakash Kumar.)

Plate 618. Purpling proceeding inwards and towards the base. (Photo by Dr Prakash Kumar.)

SUGARCANE (*Saccharum officinarum* Linn.) PHOSPHORUS (P) DEFICIENCY

Symptoms

1. Phosphorus-deficient plants have retarded growth.
2. The stalk becomes short and slender and the length of internodes is reduced.
3. The number of tillers is reduced.
4. Crop maturity is delayed.
5. Phosphorus is mobile in plants, thus it is rapidly transferred from older to younger tissues.
6. The deficiency symptoms are first observed on older leaves and become severe with time.
7. Old leaves become dark green to bluish green (Plate 615).
8. Purple discoloration occurs on the tips and margins of old leaves (Plates 616 and 617).

Developmental stages

Stage I: In the early stage of deficiency or mildly deficient conditions, the overall growth of plant gets reduced.
Stage II: As deficiency becomes severe, the older leaves turn bluish green or purple (Plates 615 and 618).
Stage III: In acutely deficient conditions, the older leaves turn yellow and die back from the tips and margins.

Likely to occur in

1. Soils having low organic matter.
2. Alkaline and calcareous soils.
3. Soils exhausted by intensive cropping.
4. Acid soils and highly weathered soils.
5. Soils where topsoil has been removed by erosion.
6. Acid soils having pH below 6.0.
7. Alkaline soils having pH between 7.5 and 8.5.

Integrated nutrient management

1. Get the soil analysed to measure the amount of 'available' phosphate in the soil well before sowing.
2. Apply analysis-based recommended quantity of phosphorus as basal by using:
- **a.** organic manures;
- **b.** phosphate-solubilizing microbial cultures;
- **c.** phosphatic fertilizers.

3. If standing crops show the deficiency, apply soluble phosphatic fertilizers such as ammonium phosphate with irrigation water.

Further reading

Anderson, D.L. and Bowen, J.E. (1990) Identification key to the malnutrition of sugarcane. In: *Sugarcane Nutrition*. Foundation for Agronomic Research and Potash and Phosphate Institute, Atlanta, Georgia, pp. 26–27.

Gascho, G.J. and Taha, F.A. (1972) *Nutrient Deficiency Symptoms of Sugarcane. Circular S-221*. University of Florida Agricultural Experiment Station, Gainesville, Florida.

Gascho, G.J., Anderson, D.L. and Bowen, J.E. (1993) Sugarcane. In: Bennett, W.F. (ed.) *Nutrient Deficiencies and Toxicities in Crop Plants*. APS Press, St Paul, Minnesota, pp. 37–42.

Plate 619. Potassium-deficient sugarcane plant. (Photo by Dr Prakash Kumar.)

SUGARCANE (*Saccharum officinarum* Linn.) POTASSIUM (K) DEFICIENCY

Plate 620. Yellow–orange chlorosis on tips and leaf margins of older leaves. (Photo by Dr Prakash Kumar.)

Plate 621. Yellow–orange chlorotic leaf borders. (Photo by Dr Prakash Kumar.)

Plate 622. Brown necrosis and scorching of leaf tip and margins. (Photo by Dr Prakash Kumar.)

Symptoms

1. The potassium-deficient plants become highly susceptible to drought and diseases.
2. Potassium deficiency causes reduced growth of the plants and stalks become slender.
3. Potassium deficiency in plant leaves results in decreased translocation of photosynthates.
4. Since potassium is a mobile nutrient within plants, it is readily translocated from older to younger leaves. Therefore, the older leaves show the deficiency symptoms first while the upper leaves appear dark green.
5. Symptoms develop as a yellow–orange chlorosis at the leaf tip and margins that progresses towards the base of old leaves (Plates 619 and 620).
6. Older leaves become brown and scorched from the tips and along the margins (Plate 622).
7. Older leaves may show red discoloration on the upper surface of the leaf midrib.
8. Red discoloration on the upper midrib surface in older leaves is the typical symptom of potassium deficiency.

Developmental stages

Stage I: In mild deficiency, plant growth is poor and stalks are slender.
Stage II: When deficiency becomes severe, yellow–orange marginal chlorosis develops on older leaves (Plate 621).
Stage III: Tips and margins of old leaves turn brown necrotic or scorched (Plate 622).
Stage IV: In the later stage, old leaves turn entirely brown and die.

Likely to occur in

1. Soils formed from parent material low in potassium.
2. Light-textured soils where potassium has been leached by heavy rainfall or excessive irrigation.
3. Soils low in organic matter.
4. Soils with wide Na:K, Mg:K or Ca:K ratio.
5. Acid soils having pH below 6.0.

Integrated nutrient management

1. Analyse the soil before planting to measure the amount of plant-available potassium.
2. Problematic acid/alkaline/saline soils should be reclaimed.
3. Add organic manures well before planting.
4. Apply potassium chloride, potassium sulphate or potassium nitrate to the soil at or before planting as per soil testing recommendations.
5. In standing crops, apply soluble potassium salts with irrigation water.

Further reading

Anderson, D.L. and Bowen, J.E. (1990) Identification key to the malnutrition of sugarcane. In: *Sugarcane Nutrition*. Foundation for Agronomic Research and Potash and Phosphate Institute, Atlanta, Georgia, pp. 28–29.

Gascho, G.J. and Taha, F.A. (1972) *Nutrient Deficiency Symptoms of Sugarcane. Circular S-221*. University of Florida Agricultural Experiment Station, Gainesville, Florida.

Gascho, G.J., Anderson, D.L. and Bowen, J.E. (1993) Sugarcane. In: Bennett, W.F. (ed.) *Nutrient Deficiencies and Toxicities in Crop Plants*. APS Press, St Paul, Minnesota, pp. 37–42.

Hartt, C.E. (1970) Effect of potassium deficiency upon translocation of ^{14}C in detached blades of sugarcane. *Plant Physiology* 45, 183–187.

Plate 623. Young leaves remain rolled and joined together at their tips. (Photo by Dr Manoj Kumar Sharma.)

Plate 624. 'Ladder-like' appearance. (Photo by Dr Manoj Kumar Sharma.)

Plate 625. Hooking of leaves. (Photo by Dr Manoj Kumar Sharma.)

Plate 626. Die-back of a young leaf. (Photo by Dr Manoj Kumar Sharma.)

SUGARCANE (*Saccharum officinarum* Linn.) CALCIUM (Ca) DEFICIENCY

Symptoms

1. In calcium-deficient plants, stalks become weak and slender. The growth of roots and tops of the plants is restricted.
2. Because calcium is immobile within plants, in poor supply conditions it is not easily mobilized from older to younger tissues.
3. When calcium supply becomes restricted in soil, the deficiency symptoms are evident first on the youngest leaves while older leaves remain unaffected.
4. The youngest leaves become distorted and then necrotic.
5. The leaves show die-back of tips and margins.
6. In some cases, the tips of young leaves remain joined together and look like hooks (Plates 624 and 625).
7. Old leaves may show a rusty appearance and die prematurely.

Developmental stages

Stage I: In the early stage of deficiency, the stalk becomes slender and the plant shows poor growth.
Stage II: As the symptoms advance, the young leaves become distorted and remain joined together (Plate 623).
Stage III: In acute deficiency, the newly emerging leaves die back (Plate 626).

Likely to occur in

1. Acid sandy soils that have been leached by heavy rainfall.
2. Sodic soils that are rich in exchangeable sodium.
3. Soils having high soluble aluminium and low exchangeable calcium.
4. Strongly acid peat and muck soils that are low in calcium.
5. Acid soils having pH below 6.5.
6. Alkaline soils having pH above 8.5.

Integrated nutrient management

1. Get the soil analysed before sowing to measure the lime or calcium requirement of the soil.
2. Apply analysis-based recommended quantity of calcium-containing fertilizers well before sowing, using soluble salts such as gypsum (calcium sulphate), calcium nitrate or calcium chloride.
3. In low-pH soils, lime (calcium carbonate) should be applied to correct the soil pH.

Further reading

Anderson, D.L. and Bowen, J.E. (1990) Identification key to the malnutrition of sugarcane. In: *Sugarcane Nutrition*. Foundation for Agronomic Research and Potash and Phosphate Institute, Atlanta, Georgia, pp. 10–11.

Evans, H. (1959) Elements other than nitrogen, potassium and phosphorus in the mineral nutrition of sugar cane. In: Humbert, R. (ed.) *Proceedings of the 10th Congress of the International Society of Sugarcane Technologists, Hawaii, 1959*. Elsevier Publishing Co., Amsterdam, pp. 473–507.

Gascho, G.J. and Taha, F.A. (1972) *Nutrient Deficiency Symptoms of Sugarcane. Circular S-221*. University of Florida Agricultural Experiment Station, Gainesville, Florida.

Gascho, G.J., Anderson, D.L. and Bowen, J.E. (1993) Sugarcane. In: Bennett, W.F. (ed.) *Nutrient Deficiencies and Toxicities in Crop Plants*. APS Press, St Paul, Minnesota, pp. 37–42.

Plate 627. Sulphur-deficient sugarcane plant. (Photo by Dr Manoj Kumar Sharma.)

Plate 628. Entire plant showing pale yellow chlorotic leaves. (Photo by Dr Manoj Kumar Sharma.)

Plate 629. Paleness appearing uniformly on the entire leaf. (Photo by Dr Manoj Kumar Sharma.)

Plate 630. Chlorotic leaves with distinct purpling. (Photo by Dr Prakash Kumar.)

SUGARCANE (*Saccharum officinarum* Linn.) SULPHUR (S) DEFICIENCY

Symptoms

1. Sulphur-deficient plants show stunted growth and stalks become slender.
2. Sulphur is immobile in plants, hence it is not readily translocated from older to younger parts of the plant.
3. In a situation of reduced supply to the plant, the deficiency symptoms first become visible on the younger leaves.
4. The whole plant appears pale green to pale yellow while the youngest leaves become more chlorotic.
5. The paleness appears on the entire leaf, affecting both the veins and the interveinal tissues uniformly (Plates 628 and 629).
6. Leaves become narrow and short and also show a purplish tinge on the margins (Plate 630).

Developmental stages

Stage I: In mild deficiency, the whole plant appears pale green although old leaves remain darker.
Stage II: If the deficiency becomes severe, the younger leaves turn pale yellow while the old leaves remain pale green with a purplish tinge (Plate 630).
Stage III: In acute deficiency, the youngest leaves turn pale yellow to white but without necrosis (Plate 627).

Likely to occur in

1. Soils having low organic matter.
2. Light-textured sandy soils that have been leached by heavy rainfall or excessive irrigation.
3. Soils exhausted by intensive cropping.
4. Soils derived from parent material that is inherently low in sulphur.
5. Acid soils having pH below 6.0.

Integrated nutrient management

1. Get the soil analysed before sowing to measure the amount of 'available' sulphate in the soil.
2. Apply analysis-based recommended quantity of sulphur by mixing into the surface soil, well before sowing:
  a. elemental sulphur; or
  b. gypsum.
3. In deficient standing crops apply ammonium sulphate, magnesium sulphate or potassium sulphate with irrigation water.

Further reading

Dutt, A.K. (1962) Sulphur deficiency in sugarcane. *Empire Journal of Experimental Agriculture* 30(119), 257–262.

Gascho, G.J. and Taha, F.A. (1972) *Nutrient Deficiency Symptoms of Sugarcane. Circular S-221*. University of Florida Agricultural Experiment Station, Gainesville, Florida.

Gosnell, J.M. and Long, A.C. (1969) A sulphur deficiency in sugarcane. *Proceedings of the South African Sugar Technologists' Association. Annual Congress South African Sugar Technologists' Association* No. 43, 26–29.

Sedl, J.M. (1968) The sulphur nutrition of sugarcane. In: *Proceedings of the Queensland Society of Sugarcane Technologists 35th Congress, 1968*. Watson Ferguson, Brisbane, Australia, pp. 131–135.

Plate 631. Iron-deficient plant showing chlorotic younger leaves and green old leaves. (Photo by Dr Prakash Kumar.)

Plate 632. Deficient leaf showing interveinal yellowing and prominent green veins. (Photo by Dr Prakash Kumar.)

Plate 633. In prolonged deficiency the entire plant appears chlorotic. (Photo by Dr Prakash Kumar and Dr Manoj Kumar Sharma.)

Plate 634. Severely deficient papery white sugarcane leaf. (Photo by Dr Prakash Kumar and Dr Manoj Kumar Sharma.)

SUGARCANE (*Saccharum officinarum* Linn.) IRON (Fe) DEFICIENCY

Symptoms

1. Iron is an immobile nutrient within plants, therefore it is not easily transferred from older to younger tissues if its supply to the plant gets reduced.
2. The deficiency symptoms are first observed on younger leaves while older leaves remain green (Plate 631).
3. A pale yellow chlorosis develops in interveinal tissues, while the veins remain green and prominent (Plate 632).
4. The interveinal chlorosis develops from the tips to the base of leaves. Necrosis may occur on younger leaves.
5. In severely deficient conditions, the whole plant may become chlorotic (Plate 633).
6. The newly emerged leaves may appear completely devoid of chlorophyll.
7. Iron inactivation in plant tissues may be responsible for lime-induced iron chlorosis.

Developmental stages

Stage I: In mild deficiency conditions, the young leaves develop pale green interveinal chlorosis with dark green and prominent veins. If the iron supply is restored the plant becomes normal.
Stage II: When deficiency persists, pale green young leaves turn to pale yellow with prominent green veins (Plate 631).
Stage III: As the deficiency becomes severe, the prominent green veins also fade and become light green.
Stage IV: In acute deficiency conditions, the entire leaf blade becomes bleached and papery white (Plate 634).

Likely to occur in

1. Sandy soils having low total iron content.
2. Alkaline and calcareous soils where solubility of iron is very low.
3. Peat and muck soils where organic matter ties up iron and reduces its availability in soil solution.
4. Acid soils with excessively high levels of soluble zinc, manganese, copper or nickel that can hinder the uptake of iron despite high iron availability.
5. Alkaline soils having pH above 7.5.

Integrated nutrient management

1. Analyse the soil before sowing to measure the amount of 'available' iron in the soil.
2. Reclaim problematic alkaline soils.
3. Add organic manure well before sowing.
4. Apply a basal dose of soluble iron fertilizers such as $FeSO_4$ (commonly at 25 kg/ha) or Fe chelates (10 kg/ha). Use of organic chelates proves to be more promising as they maintain iron in soil solution.
5. In standing crops, apply $FeSO_4$ or Fe chelates (0.5% w/v solution) as a foliar spray. Foliar sprays must be repeated every 10–15 days.

Further reading

Anderson, D.L. and Bowen, J.E. (1990) Identification key to the malnutrition of sugarcane. In: *Sugarcane Nutrition*. Foundation for Agronomic Research and Potash and Phosphate Institute, Atlanta, Georgia, pp. 16–17.

Evans, H. (1959) Elements other than nitrogen, potassium and phosphorus in the mineral nutrition of sugar cane. In: Humbert, R. (ed.) *Proceedings of the 10th Congress of the International Society of Sugarcane Technologists, Hawaii, 1959*. Elsevier Publishing Co., Amsterdam, pp. 473–507.

Gascho, G.J. and Taha, F.A. (1972) *Nutrient Deficiency Symptoms of Sugarcane. Circular S-221*. University of Florida Agricultural Experiment Station, Gainesville, Florida.

Yadav, D.V. and Singh, K. (1988) Lime-induced iron chlorosis in sugarcane. *Fertilizer Research* 16, 119–136.

Plate 635. Zinc-deficient plant showing a white chlorotic young leaf. (Photo by Dr Manoj Kumar Sharma.)

Plate 636. White chlorotic regions developed at the basal part of a leaf. (Photo by Dr Manoj Kumar Sharma.)

Plate 637. White chlorotic stripes proceeding from the base towards the tip. (Photo by Dr Manoj Kumar Sharma.)

Plate 638. Well-developed white chlorotic tissues between green midrib and margins of a sugarcane leaf. (Photo by Dr Manoj Kumar Sharma.)

SUGARCANE (*Saccharum officinarum* Linn.) ZINC (Zn) DEFICIENCY

Symptoms

1. Zinc deficiency causes retardation of plant growth.
2. The development of internodes is restricted and they become smaller in size.
3. The number of tillers is drastically reduced.
4. Zinc deficiency reduces the yield of millable canes and the sucrose concentration in cane juice.
5. Zinc-deficient leaves show increased activity of acid phosphatase and peroxidase enzymes.
6. Decreased zinc concentration in leaves results in reduced activity of carbonic anhydrase.
7. Zinc is fairly immobile within plants and is not readily translocated from older to younger parts of the plant (if the plant is poorly supplied in zinc nutrition). Therefore, the deficiency symptoms first become visible on middle or younger leaves while older leaves remain unaffected.
8. Yellow or white chlorosis develops between the green mid-vein and margins at the base of young leaves (Plate 635).

Developmental stages

Stage I: In mild deficiency, the young leaves develop chlorosis beginning at the base and proceeding towards the tip (Plate 636).
Stage II: When deficiency persists, yellow or white stripes develop between the mid-vein and leaf margins (Plates 637 and 638).
Stage III: As the deficiency becomes severe, the yellow or white stripes turn brown and may become necrotic.

Likely to occur in

1. Leached, light sandy soils where zinc content is low.
2. Alkaline and calcareous soils, where zinc availability is depressed.
3. Recently levelled soils where subsoil is exposed to cultivation. Plant-available zinc in surface soil is often double that of the subsoil.
4. Soil in which high rates of phosphatic fertilizers are applied, which can hamper zinc uptake by crops.
5. Acid soils having pH below 5.0.
6. Alkaline soils having pH above 7.5.

Integrated nutrient management

1. Analyse the soil before sowing to estimate the amount of plant-available zinc in the soil.
2. Reclaim problematic alkaline soils.
3. Incorporate any organic manure well before sowing.
4. Apply $ZnSO_4$ (commonly at 25–30 kg/ha) or Zn chelates (10 kg/ha) once every 2 years in zinc-deficient soils.
5. In standing crops, spray 0.5% w/v $ZnSO_4$ solution.

Further reading

Chatterjee, C., Jain, R., Dube, B.K. and Nautiyal, N. (1998) Use of carbonic anhydrase for determining zinc status of sugar cane. *Tropical Agriculture* 75, 480–483.

Evans, H. (1959) Elements other than nitrogen, potassium and phosphorus in the mineral nutrition of sugar cane. In: Humbert, R. (ed.) *Proceedings of the 10th Congress of the International Society of Sugarcane Technologists, Hawaii, 1959*. Elsevier Publishing Co., Amsterdam, pp. 473–507.

Gascho, G.J. and Taha, F.A. (1972) *Nutrient Deficiency Symptoms of Sugarcane. Circular S-221*. University of Florida Agricultural Experiment Station, Gainesville, Florida.

Gascho, G.J., Anderson, D.L. and Bowen, J.E. (1993) Sugarcane. In: Bennett, W.F. (ed.) *Nutrient Deficiencies and Toxicities in Crop Plants*. APS Press, St Paul, Minnesota, pp. 37–42.

PART V

Nutrient Deficiencies in Tuber Crops

Plate 639. Nitrogen-deficient plant showing dark yellow old leaves. (Photo by Dr Prakash Kumar.)

POTATO (*Solanum tuberosum* Linn.) NITROGEN (N) DEFICIENCY

Plate 640. Entire plant pale green and stunted. (Photo by Dr Prakash Kumar.)

Plate 641. Yellow old leaves and small pale green young leaves. (Photo by Dr Prakash Kumar.)

Plate 642. Severely deficient dark yellow old leaf. (Photo by Dr Prakash Kumar.)

Symptoms

1. The entire plant may become light green to pale yellow in appearance (Plate 640).
2. Nitrogen is mobile within plants and it is readily mobilized from older to younger tissues when nitrogen supply to the plant is restricted.
3. The deficiency symptoms appear primarily on older leaves and then move to the younger leaves.
4. Older leaves become uniformly yellow while young leaves may remain light green (Plate 641).
5. In prolonged deficiency, yellow older leaves turn dark yellow then brown.
6. Eventually, the leaves become necrotic then dry and fall off early.

Developmental stages

Stage I: In the early stage or in mild deficiency, the entire plant appears uniformly pale green (Plate 640).
Stage II: When deficiency persists, the older leaves turn uniformly yellow and the upper leaves appear pale green (Plate 641).
Stage III: In severe deficiency, the old leaves turn dark yellow (Plates 639 and 642).
Stage IV: In acute deficiency, the dark yellow old leaves turn brown, then become necrotic and fall off prematurely.

Likely to occur in

1. Soils having low organic matter.
2. Light-textured sandy soils that have been leached by heavy rainfall or excessive irrigation.
3. Soils exhausted by intensive cropping.
4. Waterlogged conditions.
5. Acid soils having pH below 6.0.
6. Alkaline soils having pH above 8.0.

Integrated nutrient management

1. Get the soil analysed before sowing to measure the amount of 'available' nitrogen in the soil.
2. Apply analysis-based recommended quantity of nitrogen as basal by using:
- **a.** organic manures;
- **b.** bio-fertilizers;
- **c.** nitrogenous fertilizers.

3. Top-dress soluble nitrogenous fertilizers such as urea in split doses.
4. For quick recovery, apply urea (2% w/v solution) as a foliar spray in the standing crop. Foliar sprays must be repeated every 10–15 days.

Further reading

Ulrich, A. (1993) Potato. In: Bennett, W.F. (ed.) *Nutrient Deficiencies and Toxicities in Crop Plants*. APS Press, St Paul, Minnesota, pp. 149–156.

Ulrich, A. and Fong, K.H. (1973) Nitrogen nutrition of White Rose potato in relation to vegetative growth and mineral content of leaves and roots. *Communications in Soil Science and Plant Analysis* 4, 413–426.

Wallace, T. (1961) *The Diagnosis of Mineral Deficiencies in Plants by Visual Symptoms, A Colour Atlas and Guide*, 3rd edn. Chemical Publishing Co., Inc., New York.

Weir, R.G. and Cresswell, G.C. (1993) *Plant Nutrient Disorders 3: Vegetable Crops*. Inkata Press, Melbourne, Australia.

Plate 643. Deficient plant showing stunted growth and dark green leaves. (Photo by Dr Prakash Kumar.)

POTATO (*Solanum tuberosum* Linn.) PHOSPHORUS (P) DEFICIENCY

Plate 644. Leaves appearing dull bluish green. (Photo by Dr Prakash Kumar.)

Symptoms

1. Phosphorus deficiency induces a restriction of lateral bud growth and leaf expansion as well as a reduction in leaf dry matter.
2. The root efficiency of phosphorus acquisition is enhanced under phosphorus stress conditions.
3. Leaf content of phosphorus exhibits a high correlation with tuber yield in potato.
4. Deficient plants appear unhealthy and stunted. Leaves become dark green or bluish green.
5. Since phosphorus is considered a mobile nutrient within plants, in poor supply conditions it is readily transferred from older to younger tissues. Thus, the deficiency symptoms are typically observed first on older leaves.
6. Severe deficiency results in rolling and upward curling of leaves.
7. Leaf margins become scorched and leaves fall off prematurely.

Developmental stages

Stage I: In mild deficiency conditions, plants appear stunted and leaves become dark green (Plates 643 and 646).
Stage II: As the symptoms advance, leaves turn dull bluish green (Plates 644 and 645).
Stage III: In severe deficiency conditions, leaves roll and curl upward.
Stage IV: In the later stage, older leaves exhibit marginal scorching and fall off.

Plate 645. Leaves showing dark spots in interveinal tissues. (Photo by Dr Prakash Kumar.)

Likely to occur in

1. Soils having low organic matter.
2. Alkaline and calcareous soils.
3. Soils exhausted by intensive cropping.
4. Acid soils and highly weathered soils.
5. Soils where topsoil has been removed by erosion.
6. Acid soils having pH below 6.0.
7. Alkaline soils having pH between 7.5 and 8.5.
8. Cool and wet conditions where uptake of phosphorus is reduced.

Integrated nutrient management

1. Get the soil analysed to measure the amount of 'available' phosphate in the soil well before sowing.
2. Apply analysis-based recommended quantity of phosphorus as basal by using:
- **a.** organic manures;
- **b.** phosphate-solubilizing microbial cultures;
- **c.** phosphatic fertilizers.

3. If standing crops show the deficiency, apply soluble phosphatic fertilizers such as ammonium phosphate with irrigation water.

Plate 646. Dark coloured phosphorus-deficient leaf (left) compared with a normal leaf (right). (Photo by Dr Prakash Kumar.)

Further reading

McArthur, D. and Knowles, N.R. (1993) Influence of species of vesicular-arbuscular mycorrhizal fungi and phosphorus nutrition on growth, development, and mineral nutrition of potato (*Solanum tuberosum* L.). *Plant Physiology* 102, 771–782.

Singh, J.P. (1987) Role of phosphorus and potassium content of leaf in maximizing potato yield. *Indian Journal of Agricultural Sciences* 57, 565–566.

Ulrich, A. and Fong, K.H. (1970) Phosphorus nutrition of White Rose potato in relation to growth and minerals of leaf and root tissues. *Communications in Soil Science and Plant Analysis* 1, 141–154.

Wallace, T. (1943) *The Diagnosis of Mineral Deficiencies in Plants by Visual Symptoms. A Colour Atlas and Guide.* HM Stationery Office, London.

Plate 647. Deficient plant exhibits yellow marginal and slight interveinal chlorosis in older leaves. (Photo by Dr Prakash Kumar.)

Plate 648. Necrosis appearing on margins and in interveinal tissues. (Photo by Dr Prakash Kumar.)

Plate 649. Early leaf symptom starting as yellow chlorosis along the margins. (Photo by Dr Prakash Kumar.)

Plate 650. Yellow chlorosis appearing along the margins and proceeding inwards. (Photo by Dr Prakash Kumar.)

POTATO (*Solanum tuberosum* Linn.) POTASSIUM (K) DEFICIENCY

Symptoms

1. Potassium-deficient plants appear short and thick.
2. Low potassium status of plant triggers expression of high-affinity K^+ transporters, up-regulates some K^+ channels and activates signalling cascades. The potassium deprivation also triggers developmental responses in roots.
3. The leaf content of potassium exhibits a high correlation with tuber yield in potato.
4. Frost damage in potato is inversely related to the available potassium content of the soil and the potassium concentration in potato leaves.
5. Potassium is considered a highly mobile nutrient within plants, so under restricted supply conditions it is rapidly mobilized from older to younger tissues. Consequently, the deficiency symptoms tend to occur first on older leaves.
6. The older leaves display slight interveinal chlorosis (Plate 647).
7. As the deficiency becomes severe, the marginal scorching that occurs on older leaves rapidly proceeds inwards in the form of necrotic tissues.
8. The bronzing of leaves occurs as brown spots appear in interveinal areas.
9. Eventually, the older leaves die and drop off early.

Developmental stages

Stage I: In the early stage, plants have thick and short stems and a bushy appearance.
Stage II: In prolonged deficient conditions, pale yellow marginal chlorosis and slight interveinal chlorosis develop on older leaves (Plates 647, 649 and 650).
Stage III: Severely deficient plants display marginal scorching and necrosis in interveinal tissues of older leaves (Plate 648).
Stage IV: In the later stage, the leaves die and fall off.

Likely to occur in

1. Soils formed from parent material low in potassium.
2. Light-textured soils where potassium has been leached by heavy rainfall or excessive irrigation.
3. Soils low in organic matter.
4. Soils with wide Na:K, Mg:K or Ca:K ratio.
5. Acid soils having pH below 6.0.

Integrated nutrient management

1. Analyse the soil before planting to measure the amount of plant-available potassium.
2. Problematic acid/alkaline soils should be reclaimed.
3. Add organic manures well before planting.
4. Apply potassium chloride, potassium sulphate or potassium nitrate to the soil at or before planting as per soil testing recommendations.
5. In standing crops, apply soluble potassium salts with irrigation water.

Further reading

Ashley, M.K., Grant, M. and Grabov, A. (2006) Plant responses to potassium deficiencies: a role for potassium transport proteins. *Journal of Experimental Botany* 57, 425–436.

Grewal, J.S. and Singh, S.N. (1980) Effect of potassium nutrition on frost damage and yield of potato plants on alluvial soils of the Punjab (India). *Plant and Soil* 57, 105–110.

Singh, J.P. (1987) Role of phosphorus and potassium content of leaf in maximizing potato yield. *Indian Journal of Agricultural Sciences* 57, 565–566.

Wallace, T. (1943) *The Diagnosis of Mineral Deficiencies in Plants by Visual Symptoms. A Colour Atlas and Guide*. His Majesty's Stationery Office, London.

Plate 651. Light green sulphur-deficient plant. (Photo by Dr Prakash Kumar.)

Plate 652. Upper leaves pale yellow while older leaves remain darker. (Photo by Dr Manoj Kumar Sharma.)

Plate 653. Pale yellow sulphur-deficient plant (left) compared with a dark green normal plant (right). (Photo by Dr Prakash Kumar.)

Plate 654. Severely deficient yellow young leaf (left) compared with a normal leaf (right). (Photo by Dr Manoj Kumar Sharma.)

POTATO (*Solanum tuberosum* Linn.) SULPHUR (S) DEFICIENCY

Symptoms

1. Sulphur nutrition improves tuber yields and quality. Good sulphur status in the plant also provides resistance against *Rhizoctonia solani* infection.

2. In cases where sulphur supply is reduced, it is not easily transferred from older to younger leaves because sulphur is an immobile nutrient in plants.

3. The deficiency symptoms become first evident and more severe on younger leaves.

4. Sulphur deficiency results in general chlorosis on the entire plant in the early stages.

5. As the symptoms advance, the younger leaves become pale green or pale yellow (Plate 652).

6. The chlorosis occurs evenly over the entire leaf blade, including the veins and interveinal tissues (Plate 654).

7. In severe deficiency conditions, the leaves become yellow and cup upward slightly.

Developmental stages

Stage I: In mild deficiency or in the early stage, the entire plant appears light green (Plate 651).

Stage II: If deficiency persists, the entire plant turns pale yellow (Plate 653).

Stage III: If the deficiency is severe, the younger leaves turn yellow (Plate 654).

Stage IV: In continued severe conditions, there may be upward cupping of leaves.

Likely to occur in

1. Soils having low organic matter.

2. Light-textured sandy soils that have been leached by heavy rainfall or excessive irrigation.

3. Soils exhausted by intensive cropping.

4. Soils derived from parent material that is inherently low in sulphur.

5. Acid soils having pH below 6.0.

Integrated nutrient management

1. Get the soil analysed before sowing to measure the amount of 'available' sulphate in the soil.

2. Apply analysis-based recommended quantity of sulphur by mixing into the surface soil, well before sowing, either:

- **a.** elemental sulphur; or
- **b.** gypsum.

3. In deficient standing crops apply ammonium sulphate, magnesium sulphate or potassium sulphate with irrigation water.

Further reading

Klikocka, H., Haneklaus, S., Bloem, E. and Schnug, E. (2005) Influence of sulphur fertilization on infection of potato tubers with *Rhizoctonia solani* and *Streptomyces scabies*. *Journal of Plant Nutrition* 28, 819–833.

McLachlan, K.D. (1978) *An Atlas of Sulphur Deficiency in Commercial Plants*. CSIRO Publishing, Melbourne, Australia.

Wallace, T. (1961) *The Diagnosis of Mineral Deficiencies in Plants by Visual Symptoms, A Colour Atlas and Guide*, 3rd edn. Chemical Publishing Co., Inc., New York.

Weir, R.G. and Cresswell, G.C. (1993) *Plant Nutrient Disorders 3: Vegetable Crops*. Inkata Press, Melbourne, Australia.

Plate 655. White younger leaves at the growing point of the plant. (Photo by Dr Prakash Kumar and Dr Manoj Kumar Sharma.)

POTATO (*Solanum tuberosum* Linn.) IRON (Fe) DEFICIENCY

Plate 656. Pale yellow chlorotic young leaves while old leaves remain normally green. (Photo by Dr Prakash Kumar and Dr Manoj Kumar Sharma.)

Plate 657. Yellow chlorotic young leaves with netted green veins. (Photo by Dr Prakash Kumar and Dr Manoj Kumar Sharma.)

Plate 658. Papery white younger leaves completely devoid of chlorophyll. (Photo by Dr Prakash Kumar and Dr Manoj Kumar Sharma.)

Symptoms

1. Iron-deficient plants have retarded growth and decreased chlorophyll concentration.
2. Iron deficiency causes yield reduction and poor quality of tubers.
3. The tuber quality deteriorates as the concentrations of sugars, starch and protein nitrogen are reduced whereas the accumulation of non-protein nitrogen and phenols is increased.
4. Iron plays a role in potato–*Phytophthora infestans* interaction.
5. Since iron is immobile in plants, it is not readily mobilized from older to younger tissues when its supply becomes restricted to the plant.
6. The visible symptoms typically appear first and become more severe on younger leaves.
7. The interveinal chlorosis becomes distinct in young leaves whereas older leaves stay normally green (Plate 656).
8. A yellowing develops in interveinal tissues while the veins remain prominent green, and may produce a netting look (Plate 657).
9. As the deficiency advances, yellow chlorotic leaves turn white (Plate 655).

Developmental stages

Stage I: In mild deficiency, interveinal tissues become chlorotic with dark green veins.
Stage II: In prolonged deficient conditions, chlorotic young leaves turn yellow with netted green veins (Plate 657).
Stage III: In acute deficiency, the entire leaf blade becomes bleached and papery white although the leaf tips stay green for a long time (Plate 658).
Stage IV: In the later stage, white leaves sometimes develop brown necrosis.

Likely to occur in

1. Sandy soils having low total iron content.
2. Alkaline and calcareous soils where solubility of iron is very low.
3. Peat and muck soils where organic matter ties up iron and reduces its availability in soil solution.
4. Acid soils with excessively high levels of soluble zinc, manganese, copper or nickel that can hinder the uptake of iron despite high iron availability.
5. Alkaline soils having pH above 7.5.

Integrated nutrient management

1. Get the soil analysed before sowing to measure the 'available' iron status of the soil.
2. Reclaim problematic alkaline soils.
3. Add organic manure well before sowing.
4. Apply a basal dose of soluble iron fertilizers such as $FeSO_4$ (commonly at 25 kg/ha) or Fe chelates (10 kg/ha). Use of organic chelates proves to be more promising as they maintain iron in soil solution.
5. In standing crops, apply $FeSO_4$ or Fe chelates (0.5% w/v solution) as a foliar spray. Foliar sprays must be repeated every 10–15 days.

Further reading

Chatterjee, C., Gopal, R. and Dube, B.K. (2006) Impact of iron stress on biomass, yield, metabolism and quality of potato (*Solanum tuberosum* L.). *Sciencia Horticulturae* 108, 1–6.

Mata, C.G., Lamattina, L. and Cassia, R.O. (2001) Involvement of iron and ferritin in the potato–*Phytophthora infestans* interaction. *European Journal of Plant Pathology* 107, 557–562.

Ulrich, A. (1993) Potato. In: Bennett, W.F. (ed.) *Nutrient Deficiencies and Toxicities in Crop Plants*. APS Press, St Paul, Minnesota, pp. 149–156.

Wallace, T. (1943) *The Diagnosis of Mineral Deficiencies in Plants by Visual Symptoms. A Colour Atlas and Guide*. His Majesty's Stationery Office, London.

Plate 659. Zinc-deficient stunted potato crop. (Photo by Dr Prakash Kumar.)

Plate 660. Small-sized young leaves: 'fern leaf'. (Photo by Dr Prakash Kumar.)

Plate 661. Upward rolling of young leaf blades. (Photo by Dr Prakash Kumar.)

Plate 662. Zinc-deficient 'fern leaves' (left) compared with normal leaf blades (right). (Photo by Dr Prakash Kumar.)

POTATO (*Solanum tuberosum* Linn.) ZINC (Zn) DEFICIENCY

Symptoms

1. Zinc-deficient plants appear stunted and exhibit poor growth.
2. The internodes become shorter and the size of leaves is reduced markedly.
3. The leaves cup upwards from the edges and give a typical appearance known as 'fern leaf' of potato (Plate 660).
4. When zinc supply to the plant is restricted, the deficiency symptoms are more pronounced on recently matured leaves. This is because zinc is partly mobile in plants and is not readily moved from lower to younger tissues.
5. The symptoms progress to the lower leaves as the deficiency persists.
6. The chlorosis appears on developing young leaves.
7. In severe deficiency, bronze-coloured necrotic spots develop on young leaves.
8. Eventually, the leaves die.

Developmental stages

Stage I: The early deficiency symptom on plants appears as stunted growth and small leaves (Plate 659).
Stage II: Upward rolling or cupping of smaller young leaves ('fern leaf') occurs (Plates 660, 661 and 662).
Stage III: As the deficiency becomes severe, the leaves become chlorotic and then develop bronze-coloured irregular necrotic spots.
Stage IV: Later, the necrotic leaves die.

Likely to occur in

1. Leached, light sandy soils where zinc content is low.
2. Alkaline and calcareous soils where zinc availability is depressed.
3. Recently levelled soils where subsoil is exposed to cultivation. Plant-available zinc in surface soil is often double that of the subsoil.
4. Soil in which high rates of phosphatic fertilizers are applied, which can hamper zinc uptake by crops.
5. Acid soils having pH below 5.0.
6. Alkaline soils having pH above 7.5.

Integrated nutrient management

1. Analyse the soil before sowing to estimate the amount of plant-available zinc in the soil.
2. Reclaim problematic alkaline soils.
3. Incorporate any organic manure well before sowing.
4. Apply $ZnSO_4$ (commonly at 25–30 kg/ha) or Zn chelates (10 kg/ha) once every 2 years in zinc-deficient soils.
5. In standing crops, spray 5 kg of $ZnSO_4$ plus 2.5 kg of unslaked lime in 500 l of water.

Further reading

Boawn, L.C. and Leggett, G.E. (1963) Zinc deficiency of the Russet Burbank potato. *Soil Science* 95, 137–141.

Hooker, W.J. (ed.) (1981) *Compendium of Potato Diseases*. American Phytopathological Society, St Paul, Minnesota.

Piper, C.S. (1940) The symptoms and diagnosis of minor-element deficiencies in agricultural and horticultural crops. II. Copper. Zinc. Molybdenum. *Empire Journal of Experimental Agriculture* 8, 199–206.

Wallace, T. (1943) *The Diagnosis of Mineral Deficiencies in Plants by Visual Symptoms. A Colour Atlas and Guide*. His Majesty's Stationery Office, London.

Plate 663. Manganese-deficient plant showing chlorosis on young leaves. (Photo by Dr Prakash Kumar and by Dr Manoj Kumar Sharma.)

POTATO (*Solanum tuberosum* Linn.) MANGANESE (Mn) DEFICIENCY

Plate 664. Chlorotic mottling with visible secondary veins. (Photo by Dr Prakash Kumar.)

Symptoms

1. Deficient plants exhibit poor growth.
2. Increased mobility of manganese within the plant is a function of potassium.
3. Manganese is considered an immobile nutrient in plants and has a tendency not to be easily transferred from older to younger tissues if deficiency occurs.
4. The visible symptoms are first observed on younger leaves.
5. The younger leaves become chlorotic.
6. The symptoms develop as interveinal chlorosis on young leaves (Plate 664).
7. The chlorotic leaves then develop brown to black necrotic spots in interveinal tissues.

Developmental stages

Stage I: In mild deficiencies, the younger leaves become chlorotic (Plate 663).
Stage II: In prolonged deficiency conditions, the interveinal tissues become yellow chlorotic with light green veins (Plates 664, 665 and 666).
Stage III: If deficiency becomes severe, brown or black necrotic spots may develop in interveinal areas.
Stage IV: In acute deficiency, the leaves become cupped and the necrotic spots turn darker with increased numbers occurring along the veins.

Plate 665. Leaf showing fading of sub-veins. (Photo by Dr Prakash Kumar.)

Likely to occur in

1. Light-textured sandy soils that have been leached by heavy rainfall or excessive irrigation.
2. Calcareous and alkaline soils where solubility of manganese is very low.
3. Waterlogged peaty soils where organic matter ties up manganese and reduces its availability in soil solution.
4. Soils derived from parent material that is inherently low in manganese.
5. Acid soils having pH below 5.0.
6. Alkaline soils having pH above 7.5.

Integrated nutrient management

1. Get the soil analysed before sowing to measure the amount of 'available' manganese in the soil.
2. Reclaim problematic alkaline soils.
3. Add organic manures well before sowing.
4. Apply soluble salts of manganese such as manganese sulphate as basal.
5. If deficiency occurs in the standing crop, apply manganese sulphate (0.2 to 0.3% w/v solution) as a foliar spray. Repeated sprays may be required if symptoms reappear.

Plate 666. Leaf with yellow chequered growth. (Photo by Dr Prakash Kumar.)

Further reading

Bolle-Jones, E.W. (1955) The effect of varied nutrient levels on the concentration and distribution of manganese within the potato plant. *Plant and Soil* 6, 45–60.

Ulrich, A. (1993) Potato. In: Bennett, W.F. (ed.) *Nutrient Deficiencies and Toxicities in Crop Plants*. APS Press, St Paul, Minnesota, pp. 149–156.

Wallace, T. (1943) *The Diagnosis of Mineral Deficiencies in Plants by Visual Symptoms. A Colour Atlas and Guide*. His Majesty's Stationery Office, London.

Weir, R.G. and Cresswell, G.C. (1993) *Plant Nutrient Disorders 3: Vegetable Crops*. Inkata Press, Melbourne, Australia.

Plate 667. Oldest leaf turns pale yellow while younger leaves appear pale green. (Photo by Dr Manoj Kumar Sharma.)

Plate 668. Bright yellow old leaves. (Photo by Dr Prakash Kumar and Dr Manoj Kumar Sharma.)

Plate 669. Pale yellow chlorotic leaf. (Photo by Dr Prakash Kumar and Dr Manoj Kumar Sharma.)

Plate 670. Yellow chlorotic leaf. (Photo by Dr Prakash Kumar and Dr Manoj Kumar Sharma.)

SWEET POTATO (*Ipomoea batatas* Linn.) NITROGEN (N) DEFICIENCY

Symptoms

1. Insufficient nitrogen supply results in sparse crop cover with slow growth.
2. Nitrogen is mobile within plants; if nitrogen supply is reduced, then it is rapidly redistributed from older to younger leaves. Therefore, the deficiency symptoms are observed primarily in the older leaves.
3. If nitrogen supply is adequate initially and becomes deficient during growth of the crop, the plants usually appear normal except for yellowing and premature shedding of older leaves. The oldest leaves turn uniformly yellow and show symptoms of wilting. Light brown necrosis starts from the leaf tip or margins.
4. If nitrogen supply is low from the beginning, the entire plant turns uniformly pale green. The leaves become small. The number of branches is reduced and plants have a dull appearance.
5. In severe deficiency, purple-coloured flecks or ring spots can occur on the surface of older leaves in some cultivars.

Developmental stages

Stage I: Early deficiency symptoms are expressed as uniform pale green leaves on the entire plant.
Stage II: If deficiency persists, the lower leaves turn pale yellow while the upper leaves appear pale green (Plates 667 and 669).
Stage III: In severe deficiency, the older leaves turn uniformly yellow (Plate 670).
Stage IV: In acute deficiency conditions, yellow old leaves become brown necrotic and the necrosis starts from the tip or margins (Plate 668).

Likely to occur in

1. Soils having low organic matter.
2. Light-textured sandy soils that have been leached by heavy rainfall or excessive irrigation.
3. Soils exhausted by intensive cropping.
4. Waterlogged conditions.
5. Acid soils having pH below 6.0.
6. Alkaline soils having pH above 8.0.

Integrated nutrient management

1. Get the soil analysed before sowing to measure the amount of 'available' nitrogen in the soil.
2. Apply analysis-based recommended quantity of nitrogen as basal by using:
- **a.** organic manures;
- **b.** bio-fertilizers;
- **c.** nitrogenous fertilizers.

3. Top-dress soluble nitrogenous fertilizers such as urea in split doses.
4. For quick recovery, apply urea (2% w/v solution) as a foliar spray in the standing crop. Foliar sprays must be repeated every 10–15 days.
5. Waterlogging adversely affects the growth and yield of the crop. Therefore, good drainage conditions should be maintained regardless of nitrogen supply to the crop.

Further reading

Bourke, R.M. (1985) Influence of nitrogen and potassium fertiliser on growth of sweet potato (*Ipomoea batatas*) in Papua New Guinea. *Field Crops Research* 12, 363–375.

Constantin, R.J., Hernandez, T.P. and Jones, L.G. (1974) Effects of irrigation and nitrogen fertilization on quality of sweetpotatoes. *Journal of the American Society for Horticultural Science* 99, 308–310.

Halavatau, S., Asher, C.J. and Bell, L.C. (1996) Soil fertility and sweet potato research in Tonga – nitrogen and phosphorus. In: Craswell, E.T., Asher, C.J. and O'Sullivan, J.N. (ed) *ACIAR Proceedings No. 65: Mineral Nutrient Disorders of Root Crops in the Pacific*. Australian Centre for International Agricultural Research (ACIAR), Canberra, pp. 58–64.

Plate 671. Phosphorus deficient sweet potato leaf. (Photo by Dr Manoj Kumar Sharma.)

Plate 672. Purpling along the margins. (Photo by Dr Manoj Kumar Sharma.)

Plate 673. Purpling spreading inward. (Photo by Dr Prakash Kumar.)

Plate 674. Purpling of entire leaf. (Photo by Dr Prakash Kumar.)

SWEET POTATO (*Ipomoea batatas* Linn.) PHOSPHORUS (P) DEFICIENCY

Symptoms

1. Plants do not manifest easily recognizable symptoms in mild to moderate deficiency conditions.
2. In mildly deficient conditions, the leaves turn dark bluish green in appearance.
3. Phosphorus is fairly mobile within plants and under deficient conditions it is readily translocated from older to younger tissues. The deficiency symptoms tend to occur first in the older leaves.
4. The symptoms in young crops become evident only in severity. If deficiency is not severe, the deficiency symptoms can develop in the later stage when the crop matures.
5. The primary symptom of phosphorus deficiency is generally premature senescence of the older leaves.
6. In most of the cultivars, purple pigmentation appears first (Plates 672 and 673) followed by yellowing of older leaves. The yellowing may advance from distinct interveinal patches.
7. Necrotic spots develop in the chlorotic areas as irregular patches. Finally, the entire leaf blade turns brown and dies.

Developmental stages

Stage I: In mild deficiency, the plants become stunted and entire leaves turn dark green.
Stage II: As deficiency advances, the dark green leaves turn dull bluish green (Plate 674).
Stage III: If deficiency becomes more severe, the older leaves develop purple pigmentation (Plate 671) followed by yellowing.
Stage IV: In the later stage, the leaves turn brown necrotic and die.

Likely to occur in

1. Soils having low organic matter.
2. Alkaline and calcareous soils.
3. Soils exhausted by intensive cropping.
4. Acid soils and highly weathered soils.
5. Soils where topsoil has been removed by erosion.
6. Acid soils having pH below 6.0.
7. Alkaline soils having pH between 7.5 and 8.5.

Integrated nutrient management

1. Get the soil analysed to measure the amount of 'available' phosphate in the soil well before sowing.
2. Apply analysis-based recommended quantity of phosphorus as basal by using:
  a. organic manures;
  b. phosphate-solubilizing microbial cultures;
  c. phosphatic fertilizers.
3. Rock phosphate is a cheaper alternative source of phosphorus for use in acidic soils only, and should be incorporated well into the soil.
3. If standing crops show the deficiency, apply soluble phosphatic fertilizers such as ammonium phosphate with irrigation water.

Further reading

Halavatau, S., Asher, C.J. and Bell, L.C. (1996) Soil fertility and sweet potato research in Tonga – nitrogen and phosphorus. In: Craswell, E.T., Asher, C.J. and O'Sullivan, J.N. (ed) *ACIAR Proceedings No. 65: Mineral Nutrient Disorders of Root Crops in the Pacific*. Australian Centre for International Agricultural Research (ACIAR), Canberra, pp. 58–64.

Leonard, O.A., Anderson, W.S. and Gieger, M. (1949) Field studies on the mineral nutrition of the sweetpotato. *Proceedings of the American Society for Horticultural Science* 53, 387–392.

O'Sullivan, J.N., Asher, C.J. and Blamey, F.P.C. (1997) *Nutrient Disorders of Sweet Potato. ACIAR Monograph No. 48*. Australian Centre for International Agricultural Research (ACIAR), Canberra.

Plate 675. Youngest leaves papery white and middle leaves showing interveinal chlorosis. (Photo by Dr Prakash Kumar and Dr Manoj Kumar Sharma.)

Plate 676. Severely deficient papery white younger leaves showing marginal necrosis. (Photo by Dr Prakash Kumar and Dr Manoj Kumar Sharma.)

Plate 677. Leaf showing bright yellow interveinal tissues and a sharp network of green veins. (Photo by Dr Prakash Kumar and Dr Manoj Kumar Sharma.)

Plate 678. Papery white leaf with faded green veins and brown necrotic tissues appearing at the tip and in interveinal regions. (Photo by Dr Prakash Kumar and Dr Manoj Kumar Sharma.)

SWEET POTATO (*Ipomoea batatas* Linn.) IRON (Fe) DEFICIENCY

Symptoms

1. High iron concentration in sweet potato can affect the light reactions of photosynthesis, which may result in a declining net assimilation rate.
2. The visible symptoms of iron deficiency become obvious very early, even when crop growth is only mildly affected.
3. Interveinal chlorosis (yellowing of tissues between the veins) of the youngest leaves is the characteristic symptom of iron deficiency.
4. Iron is immobile within plants and is not rapidly mobilized from older to younger tissues (if iron supply to the plant becomes restricted). Therefore, the deficiency symptoms become evident first on younger tissues.
5. The interveinal tissues of young leaves become chlorotic with a sharp contrasting network of green veins, whereas older leaves remain green. The chlorosis becomes more pronounced in those leaves which are exposed to sunlight.
6. The chlorotic young leaves then turn white and the veins also lose their green colour (Plate 675).
7. Eventually, the leaves become papery white and show necrosis that usually starts from the tip and margins, spreading into interveinal tissues.

Developmental stages

Stage I: In mild deficiencies, interveinal tissues become yellow with contrasting green veins of youngest leaves.
Stage II: If deficiency advances, the chlorotic tissues turn bright yellow with the fading of green veins (Plate 677).
Stage III: In severe deficiency conditions, the entire leaf turns white and the veins are only slightly visible (Plate 675).
Stage IV: In the later stage, white leaves sometimes develop brown necrosis (Plates 676 and 678).

Likely to occur in

1. Sandy soils having low total iron content.
2. Alkaline and calcareous soils where solubility of iron is very low.
3. Peat and muck soils where organic matter ties up iron and reduces its availability in soil solution.
4. Soils where a high quantity of lime or phosphorus fertilizer is applied.
5. Acid soils with excessively high levels of soluble zinc, manganese, copper or nickel that can hinder the uptake of iron despite high iron availability.
6. Alkaline soils having pH above 7.5.

Integrated nutrient management

1. Analyse the soil before sowing to measure the 'available' iron in the soil.
2. Reclaim problematic alkaline soils.
3. Add organic manure well before sowing.
4. Apply a basal dose of soluble iron fertilizers such as $FeSO_4$ (commonly at 25 kg/ha) or Fe chelates (10 kg/ha). Use of organic chelates proves to be more promising as they maintain iron in soil solution.
5. In standing crops, apply $FeSO_4$ or Fe chelates (0.5% w/v solution) as a foliar spray. Foliar sprays must be repeated every 10–15 days.

Further reading

Adamski, J.M., Peters, J.A., Danieloski, R. and Bacarin, M.A. (2011) Excess iron-induced changes in the photosynthetic characteristics of sweet potato. *Journal of Plant Physiology* 168, 2056–2062.

Ames, T., Smit, N.E.J.M., Braun, A.R., O'Sullivan, J.N. and Skoglund, L.G. (1996) *Sweetpotato: Major Pests, Diseases, and Nutritional Disorders*. International Potato Center (CIP), Lima.

O'Sullivan, J.N., Asher, C.J. and Blamey, F.P.C. (1997) *Nutrient Disorders of Sweet Potato. ACIAR Monograph No. 48*. Australian Centre for International Agricultural Research (ACIAR), Canberra.

Spence, J.A. and Ahmad, N. (1967) Plant nutrient deficiencies and related tissue composition of the sweet potato. *Agronomy Journal* 59, 59–62.

Plate 679. Recently matured leaves showing interveinal chlorosis. (Photo by Dr Prakash Kumar and Dr Manoj Kumar Sharma.)

Plate 680. Young leaves are chlorotic, having pale mottled areas between minor veins. (Photo by Dr Prakash Kumar.)

Plate 681. Prominent yellow mottling and fine netting of major and minor veins. (Photo by Dr Prakash Kumar and Dr Manoj Kumar Sharma.)

Plate 682. Severely deficient leaves showing mottled interveinal chlorosis and brown necrosis. (Photo by Dr Prakash Kumar and Dr Manoj Kumar Sharma.)

SWEET POTATO (*Ipomoea batatas* Linn.) MANGANESE (Mn) DEFICIENCY

Symptoms

1. The manganese concentration in the blades, rather than that of the petioles or roots, is more closely associated with the growth of sweet potato.
2. Indistinct interveinal chlorosis, particularly on middle-aged leaves, is usually the first symptom of manganese deficiency in sweet potato.
3. Manganese is considered an immobile nutrient within plants. Under deficient conditions, it is not easily moved from older to younger leaves. Thus, the younger leaves display deficiency symptoms first.
4. The main veins with their associated tissues appear broad and green, while the minor veins appear relatively narrow and faded green.
5. Initially the recently matured leaves are the most affected, but as the symptoms progress both older and younger leaves are affected.
6. The interveinal chlorotic leaves may become limp or downward cupping may occur from the leaf margins.
7. On young recently matured leaves, the small areas between minor veins become pale and sunken, and eventually develop necrotic spots.
8. Severely affected leaves show necrosis that usually starts from the tip and margins and spreads into interveinal tissues (Plate 682).

Developmental stages

Stage I: In mild deficiency, indistinct interveinal chlorosis occurs particularly on recently matured leaves (Plate 679).
Stage II: In prolonged deficient conditions, the interveinal tissues turn yellow with a fine netting of minor veins (Plates 680 and 681).
Stage III: In severe deficiency, the pale areas between the minor veins become sunken and then turn into necrotic spots.
Stage IV: In the later stage, severely affected leaves become necrotic and shed prematurely.

Likely to occur in

1. Light-textured sandy soils that have been extensively leached by heavy rainfall or excessive irrigation.
2. Calcareous and alkaline soils where solubility of manganese is very low.
3. Waterlogged peaty soils where organic matter ties up manganese and reduces its availability in solution.
4. Soils derived from parent material that is inherently low in manganese.
5. Acid soils having pH below 5.0.
6. Alkaline soils having pH above 7.5.

Integrated nutrient management

1. Get the soil analysed before sowing to measure the amount of 'available' manganese in the soil.
2. Problematic alkaline soils should be reclaimed.
3. Add organic manures well before sowing.
4. Apply soluble salts of manganese such as manganese sulphate as basal.
5. If deficiency occurs in the standing crop, apply manganese sulphate (0.2 to 0.3% w/v solution) as a foliar spray. Repeated sprays may be required if symptoms reappear.

Further reading

Mishra, U.N. and Kelley, J.D. (1967) Manganese nutrition of sweet potatoes in relation to manganese content, deficiency symptoms and growth. *Agronomy Journal* 59, 578–581.

O'Sullivan, J.N., Asher, C.J. and Blamey, F.P.C. (1997) *Nutrient Disorders of Sweet Potato. ACIAR Monograph No. 48*. Australian Centre for International Agricultural Research (ACIAR), Canberra.

O'Sullivan, J.N., Asher, C.J., Blamey, F.P.C. and Edwards, D.G. (1993) Mineral nutrient disorders of root crops of the Pacific: preliminary observations on sweet potato (*Ipomoea batatas*). *Plant and Soil* 155–156, 263–267.

Pillai, N.G., Mohankumar, B., Kabeerathumma, S. and Nair, P.G. (1986) Deficiency symptoms of micronutrients in sweet potato (*Ipomoea batatas* L.). *Journal of Root Crops* 12, 91–95.

PART VI

Nutrient Deficiencies in Fodder Crops

Plate 683. Nitrogen deficient lucern plant. (Photo by Dr Prakash Kumar.)

Plate 684. Light green upper leaves contrasted with yellow lower leaves. (Photo by Dr Prakash Kumar.)

Plate 685. Nitrogen-deficient small and uniformly yellow trifoliate. (Photo by Dr Prakash Kumar.)

Plate 686. Severely deficient white chlorotic trifoliate showing necrosis. (Photo by Dr Prakash Kumar.)

LUCERNE or ALFALFA (*Medicago sativa* Linn.) NITROGEN (N) DEFICIENCY

Symptoms

1. Deficient plants become stunted and have sparse growth.
2. The stem becomes thin and elongated. The leaves become smaller and initially the entire plant appears chlorotic.
3. Since nitrogen is a mobile nutrient within plants, under short supply conditions it is quickly mobilized from lower to upper leaves. Thus the lower leaves display deficiency symptoms first.
4. The older leaves turn uniformly light yellow to dark yellow, while the younger leaves may remain light green (Plate 684).
5. The yellow older leaves then turn white and become necrotic. Eventually, the leaves die and drop off early.

Developmental stages

Stage I: In mild deficiency the entire plant becomes stunted and uniformly pale green (Plate 683).
Stage II: If the deficiency advances, the upper leaves appear pale green and the older leaves turn light yellow to dark yellow (Plate 684).
Stage III: In severe deficiency, the leaves turn dark yellow to whitish yellow (Plate 685).
Stage IV: In acute deficiency, the old leaves become white and necrotic (Plate 686).

Likely to occur in

1. Soils having low organic matter.
2. Light-textured sandy soils that have been leached by heavy rainfall or excessive irrigation.
3. Soils exhausted by intensive cropping.
4. Waterlogged conditions.
5. Acid soils having pH below 6.0.
6. Alkaline soils having pH above 8.0.

Integrated nutrient management

1. Get the soil analysed before sowing to measure the amount of 'available' nitrogen in the soil.
2. Apply analysis-based recommended quantity of nitrogen as basal by using:
  a. organic manures;
  b. bio-fertilizers;
  c. nitrogenous fertilizers.
3. Top-dress soluble nitrogenous fertilizers such as urea in split doses.
4. For quick recovery, apply urea (2% w/v solution) as a foliar spray in the standing crop. Foliar sprays must be repeated every 10–15 days.

Further reading

Khan, M.G., Silberbush, M. and Lips, S.H. (1994) Physiological studies on salinity and nitrogen interaction in alfalfa. I. Biomass production and root development. *Journal of Plant Nutrition* 17, 657–668.

Kim, T.H., Ourry, A., Boucaud, J. and Lemaire, G. (1991) Changes in source–sink relationship for nitrogen during regrowth of lucerne (*Medicago sativa* L.) following removal of shoots. *Australian Journal of Plant Physiology* 18, 593–602.

Kim, T.H., Ourry, A., Boucaud, J. and Lemaire, G. (1993) Partitioning of nitrogen derived from N_2 fixation and reserves in nodulated *Medicago sativa* L. during regrowth. *Journal of Experimental Botany* 44, 555–562.

Willett, I.R., Jakobsen, P. and Zarcinas, B.A. (1985) Nitrogen-induced boron deficiency in lucerne. *Plant and Soil* 86, 443–446.

Plate 687. Severely deficient plant showing symptoms progressing up the plant to younger leaves. (Photo by Dr Manoj Kumar Sharma.)

Plate 688. White necrotic spots developed along the edges of leaves. (Photo by Dr Manoj Kumar Sharma.)

Plate 689. Marginal yellowing and scattered white necrotic spots appearing along the margins. (Photo by Dr Prakash Kumar.)

Plate 690. Close-up of a leaflet showing marginal yellowing and irregular white necrotic spots. (Photo by Dr Prakash Kumar.)

LUCERNE or ALFALFA (*Medicago sativa* Linn.) POTASSIUM (K) DEFICIENCY

Symptoms

1. Visible symptoms of potassium deficiency do not appear immediately. The growth rate is reduced initially and plants appear stunted (the phenomenon of hidden hunger is very common in the case of potassium deficiency).
2. The diamine putrescine is found to be a sensitive indicator of potassium requirements in lucerne.
3. Potassium is highly mobile in plants and is rapidly translocated from older to younger leaves. Therefore, deficiency symptoms appear first and become more pronounced on lower, older leaves and then work up the plant to upper leaves, if deficiency persists (Plate 687).
4. The potassium-deficient lucerne plant develops small white or yellow necrotic spots along the edges of older leaves (Plate 688).
5. The tissues between the necrotic spots then turn yellow and die.
6. In some cases, white or yellow necrotic spots and marginal chlorosis can occur simultaneously (Plates 689 and 690).
7. Later, the entire margins become dry or scorched and the white necrotic spots spread inwards.

Developmental stages

Stage I: In mild deficiency, plants become stunted and have dark green leaves.
Stage II: When deficiency becomes severe, necrotic white or yellow spots develop along the leaf edges (Plate 688).
Stage III: If deficiency persists, the tissues between necrotic spots turn yellow and die (Plate 687).
Stage IV: In the later stage, the affected leaves become dry or scorched from the margins and shed prematurely.

Likely to occur in

1. Soils formed from parent material low in potassium.
2. Light-textured soils where potassium has been leached by heavy rainfall or excessive irrigation.
3. Soils low in organic matter.
4. Soils with wide Na:K, Mg:K or Ca:K ratio.
5. Acid soils having pH below 6.0.

Integrated nutrient management

1. Analyse the soil before planting to measure the amount of plant-available potassium.
2. Problematic acid/alkaline/saline soils should be reclaimed.
3. Add organic manures well before planting.
4. Apply potassium chloride, potassium sulphate or potassium nitrate to the soil at or before planting as per soil testing recommendations.
5. In standing crops, apply soluble potassium salts with irrigation water.

Further reading

James, D.W., Dhumal, S.S., Rumbaugh, M.D. and Tindal, T.A. (1995) Inheritance of potassium–sodium nutritional traits in alfalfa. *Agronomy Journal* 87, 681–686.

Mengel, K. and Kirkby, E.A. (2001) *Principles of Plant Nutrition*. Kluwer Academic Publishers, Dordrecht, the Netherlands.

Smith, G.S., Lauren, D.R., Cornforth, I.S. and Agnew, M.P. (1982) Evaluation of putrescine as a biochemical indicator of the potassium requirements of lucerne. *New Phytologist* 91, 419–428.

Stevens, G., Motavalli, P., Scharf, P., Nathan, M. and Dunn, D. (2002) *Crop Nutrient Deficiencies & Toxicities. IPM 1016, Plant Protection Programs*. MU Extension, University of Missouri–Columbia, Columbia, Missouri.

Plate 691. Pale green younger leaves contrasted with dark green older leaves. (Photo by Dr Prakash Kumar and Dr Manoj Kumar Sharma.)

Plate 692. Leaf lamina and veins are evenly chlorotic in young leaves. (Photo by Dr Prakash Kumar and Dr Manoj Kumar Sharma.)

Plate 693. Younger leaves turn yellow. (Photo by Dr Prakash Kumar and Dr Manoj Kumar Sharma.)

Plate 694. Youngest leaves appear sharply chlorotic in severe conditions. (Photo by Dr Prakash Kumar and Dr Manoj Kumar Sharma.)

LUCERNE or ALFALFA (*Medicago sativa* Linn.) SULPHUR (S) DEFICIENCY

Symptoms

1. Sulphur deficiency affects nitrogen fixation in lucerne and decreases chlorophyll concentration markedly.
2. The adequate supply of sulphur in lucerne increases protein synthesis by accelerating protein metabolism.
3. Sulphur-deficient plants become stunted and slender.
4. Sulphur is immobile in plants and is not readily mobilized from older to younger growth in the plant when its supply is reduced. Hence, the deficiency symptoms tend to occur first and become more severe on younger leaves.
5. The uniform chlorosis appears on younger leaves while older leaves remain green (Plate 691).
6. The entire leaf lamina and veins become uniformly chlorotic.
7. In prolonged deficient conditions, the chlorosis spreads rapidly down the plant to older leaves.
8. In severe deficiency, all the leaves of the plant turn yellow.

Developmental stages

Stage I: In mild deficiency conditions, the plant turns light green contrasted with darker older leaves (Plate 692).
Stage II: If deficiency persists, the younger leaves turn yellow (Plates 693 and 694).
Stage III: In severe deficiency conditions, the entire plant turns yellow.

Likely to occur in

1. Soils having low organic matter.
2. Light-textured sandy soils that have been leached by heavy rainfall or excessive irrigation.
3. Soils exhausted by intensive cropping.
4. Soils derived from parent material that is inherently low in sulphur.
5. Acid soils having pH below 6.0.
6. Soils of higher elevated areas in the field.

Integrated nutrient management

1. Get the soil analysed before sowing to measure the amount of 'available' sulphate in the soil.
2. Apply analysis-based recommended quantity of sulphur by mixing into the surface soil, well before sowing, either:
  a. elemental sulphur; or
  b. gypsum.
3. In deficient standing crops apply ammonium sulphate, magnesium sulphate or potassium sulphate with irrigation water.

Further reading

Aulakh, M.S., Dev, G. and Arora, B.R. (1976) Effect of sulphur fertilization on the nitrogen–sulphur relationships in alfalfa (*Medicago sativa* L. Pers.). *Plant and Soil* 45, 75–80.

DeBoer, D.L. and Duke, S.H. (1982) Effects of sulphur nutrition on nitrogen and carbon metabolism in lucerne (*Medicago sativa* L.). *Physiologia Plantarum* 54, 343–350.

McNaught, K.J. and Chrisstoffels, P.J.E. (1961) Effect of sulphur deficiency on sulphur and nitrogen levels in pastures and lucerne. *New Zealand Journal of Agricultural Research* 4, 177–196.

Nelson, W.L. and Barber, S.A. (1964) Nutrient deficiencies in legumes for grain and forage. In: Sprague, H.B. (ed.) *Hunger Signs in Crops – A Symposium*, 3rd edn. David McKay Co. Inc., New York, pp. 143–170.

Plate 695. Iron deficient lucern crop. (Photo by Dr Prakash Kumar.)

LUCERNE or ALFALFA (*Medicago sativa* Linn.) IRON (Fe) DEFICIENCY

Plate 696. Plant showing varying degree of chlorosis ranging from pale yellow chlorotic leaves to almost white leaves and green older leaves. (Photo by Dr Prakash Kumar and Dr Manoj Kumar Sharma.)

Plate 697. Leaflets showing yellow interveinal tissues and green veins. (Photo by Dr Prakash Kumar.)

Plate 698. Severely deficient papery white younger leaves (Photo by Dr Prakash Kumar.)

Symptoms

1. Iron is an immobile nutrient in plants and it is not rapidly transferred from older to younger tissues when the plants' supply of iron becomes restricted. Therefore, the visible deficiency symptoms are observed primarily in the younger leaves.
2. Initially, the younger leaves develop a pale green to pale yellow interveinal chlorosis (yellowing of tissues between the veins while veins remain prominently green) whereas the older leaves stay green and healthy.
3. The chlorosis extends to the full length of the leaves.
4. With the advancement of symptoms, the paleness intensifies into interveinal tissues with the fading of veins (Plate 697).
5. Severely affected leaves then turn almost white and the veins also disappear (Plate 698).
6. Later, the white chlorotic leaves develop brown necrotic spots in interveinal regions.

Developmental stages

Stage I: In early deficiency, the pale green interveinal chlorosis develops in younger leaves whereas older leaves remain dark green.
Stage II: If deficiency persists, the interveinal tissues turn yellow with faded green veins (Plates 696 and 697).
Stage III: In cases of severe deficiency, the entire leaf becomes almost white and veins also become invisible (Plates 695 and 698).
Stage IV: In the later stage, leaves are bleached and develop brown necrotic lesions.

Likely to occur in

1. Sandy soils having low total iron content.
2. Alkaline and calcareous soils where solubility of iron is very low.
3. Peat and muck soils where organic matter ties up iron and reduces its availability in soil solution.
4. Acid soils with excessively high levels of soluble zinc, manganese, copper or nickel that can hinder the uptake of iron despite high iron availability.
5. Alkaline soils having pH above 7.5.

Integrated nutrient management

1. Analyse the soil before sowing to measure the 'available' iron in the soil.
2. Reclaim problematic alkaline soils.
3. Add organic manure well before sowing.
4. Apply a basal dose of soluble iron fertilizers such as $FeSO_4$ (commonly at 25 kg/ha) or Fe chelates (10 kg/ha). Use of organic chelates proves to be more promising as they maintain iron in soil solution.
5. In standing crops, apply $FeSO_4$ or Fe chelates (0.5% w/v solution) as a foliar spray. Foliar sprays must be repeated every 10–15 days.

Further reading

Brown, J.C. (1961) Iron chlorosis in plants. In: Norman, A.G. (ed.) *Advances in Agronomy*. Academic Press, London, pp. 329–366.
Fehr, W.R. (1984) Current practices for correcting iron deficiency in plants with emphasis on genetics. *Journal of Plant Nutrition* 7, 347–354.
Glauser, R. and Jenny, H. (1960) Two-phase studies on availability of iron in calcareous soils. I. Experiments with alfalfa plants. *Agrochimica* 4, 263–278.
Katyal, J.C. and Vlek, P.L.G. (1984) Micronutrient problems in tropical Asia. *Fertilizer Research* 7, 69–94.

Index